CW00727558

RECENT ADVANCES IN ANIMAL NUTRITION — 1993

Cover design
From Poultry Paintings by C.S.Th. van Gink, published by Dutch Branch of the World's Poultry Science Association (WPSA), Beekbergen, the Netherlands

RED JUNGLEFOWL (Gallus gallus)
These fowl originated in southeast Asia where they live up to an altitude of 6000 feet. It appears that, long before the beginning of our era, Red Junglefowl were domesticated by the old civilisations in India and China (1400 BC) and are the basis of today's domesticated poultry breeds whatever their size, shape or colour.

Recent Advances in Animal Nutrition

1993

P.C. Garnsworthy, PhD

D.J.A. Cole, PhD

University of Nottingham School of Agriculture

Nottingham University Press
Sutton Bonington Campus,
Loughborough, Leicestershire LE12 5RD

NOTTINGHAM

First published 1993

British Library Cataloguing in Publication Data
Recent Advances in Animal Nutrition – 1993 :
University of Nottingham Feed Manufacturers
Conference (27th, 1993, Nottingham)
I. Garnsworthy, Philip C.
II. Cole, D.J.A.
636.08

ISBN 1-897676-00-X

Typeset by The Midlands Book Typesetting Company, Loughborough, Leicestershire
Printed and bound by Quorn Selective Repro Ltd, Loughborough, Leicestershire

PREFACE

This book contains the proceedings of the 27th University of Nottingham Feed Manufacturers Conference. As in previous conferences, a variety of topical issues were addressed by international speakers to provide the Animal Feed Industry with the up-to-date information required in today's economic climate.

In order to formulate diets for optimum performance of farm animals it is essential that feeds can be accurately and reliably described. Three chapters are devoted to the characterisation of forages, carbohydrates and nitrogen for ruminants. Each critically examines current systems of evaluation and discusses opportunities for future developments in this area. A fourth ruminant chapter discusses the amino acid nutrition of dairy cows. This topic has frequently been addressed for non-ruminants but this chapter presents evidence indicating that amino acid balance is important for high producing dairy cows.

With increasing awareness of animal welfare issues, those involved in animal production must take account of the public perception of their field. A chapter on social concerns and responses in animal agriculture presents a review of research into peoples attitudes towards animals in general.

The section on poultry nutrition presents new ideas on the nutritional management of broilers and the laying pullet and also examines the way in which nutrition can influence the growth of bones in poultry.

Major changes to legislation affecting feed manufacturers are to be introduced in 1993 and these are described in one chapter. They concern the marketing of feedstuffs, additives in feeds and specific aspects of the COSHH regulations. The Expert Group on Animal Feedingstuffs presented its report in 1992. This report covered spongiform encephalopathies, Salmonella, medicated feedingstuffs, feed composition and methods of analysis. The Chairman of the group has written a chapter on the implications for feed manufacturers of this important report. The third chapter on legislation discusses the importance of new regulations concerning maximum residue limits in animal products.

The section on pig nutrition covers three diverse but topical areas. The resurgence of interest in outdoor pig production justifies a close examination of their nutrient requirements under these conditions. New information on the action of probiotics in pigs comes from a study on the microbiology of the pig gut. New or improved technologies allow diets to be constructed for the newly weaned pig which will not stress the physiology of the immature gut and compromise health.

All chapters are written in a clear and informative manner and should be of considerable benefit to research workers, livestock advisers and anyone involved in the animal feed industry.

The Organisers of the Conference and the University of Nottingham are grateful to BP Nutrition for the support which they provided towards the running of this conference. They would also like to thank Butterworth-Heinemann for publication of previous Proceedings and wish Nottingham University Press success with the publication of this and future Proceedings.

P.C. Garnsworthy
D.J.A. Cole

CONTENTS

I

Ruminant Nutrition

1

CHARACTERISATION OF FORAGES: APPRAISAL OF CURRENT PRACTICE AND FUTURE OPPORTUNITIES

D.E. BEEVER
University of Reading, Department of Agriculture, Earley Gate, Reading, RG6 2AT, UK

Introduction

The diversity in forages consumed by ruminant livestock is large and hence the nutritive value of both grazed and conserved forages can be highly variable. Since the replacement of the Starch Equivalent and Digestible Crude Protein systems for the estimation of energy and protein value of ruminant feedstuffs, with systems based upon metabolisable energy (Agricultural Research Council, 1980; Agricultural and Food Research Council, 1990) and metabolisable protein (Agricultural Research Council, 1980; Agricultural and Food Research Council, 1984; Agricultural and Food Research Council, 1992) there has been considerable effort to provide suitable data to permit use of such systems in practice to formulate diets and to predict animal response. In part these data have been based upon *in vivo* estimates, albeit often obtained at modest levels of feed intake, of the digestibility of gross energy and subsequent determination of the metabolisable energy (ME) content, the extent of microbial protein synthesis in the rumen, and the contribution of undegraded dietary protein (UDP) to the total amount of protein absorbed from the small intestine. Such measurements have subsequently been followed by the development of *in vitro* and laboratory techniques to permit prediction of *in vivo* processes, with adjustments for level of feeding effects as considered appropriate. New analytical techniques have been established and emerging procedures such as near infra-red reflectance spectroscopy (NIRS) have been adapted to meet the routine demands of feedstuff and forage analysis.

From a resumé of the factors which influence the nutritive value of forages, this paper will examine those methods employed to predict the voluntary intake of forages, the overall efficiency of energy utilisation and the supply of absorbed protein in the animal and consider why current predictions of animal production often fail to reconcile with observed performance. The paper will conclude that current feeding systems for ruminants, whilst considerably improved compared with those available 20–30 years ago, still remain inadequate in relation to the demands of today's industry and that further refinement of the systems and the associated analytical techniques is likely to bring only marginal reward. Systems which consider individual nutrients and the important interrelationships between nutrients represent the only suitable alternative, and urgent support from both Industry and Government is required to fully develop such systems, as well as the associated chemical and physical analyses that will be required to permit routine use of such systems in practice.

Variation in forage composition

The chemical and physical characteristics of forages will be influenced by forage species, prevailing environmental conditions and management factors including fertiliser application and stage of growth when the crop is grazed or harvested. Forage nutrients can be broadly classified into cell contents and cell walls (van Soest and Wine, 1967), and in young vegetative material, the cell contents may constitute as much as 70% of the total dry matter. Consisting principally of free and water-soluble sugars (including fructosans), true- and non-protein nitrogen, and modest amounts of lipid and minerals, this fraction is considered to be extensively digested within the alimentary tract, with virtually all of this digestion occurring within the reticulorumen. Cell walls consist principally of hemicellulose and cellulose (holocellulose) and lignin. Holocellulose can be effectively degraded within the rumen, and provided the associated lignin content is low, the overall digestibility of holocellulose may exceed 80% in young vegetative forage. As forages mature, the relative proportions of cell contents decrease whilst cell walls increase (Gill, Beever and Osbourn, 1989a), the overall effect being a progressive decline in 'protein' concentration, associated with substantially increased concentrations of holocellulose, and, in particular, lignin. Lignin is not extensively digested in the rumen and it is this increased concentration of lignin which causes overall forage digestibility to decline as the spatial distribution of lignin within the crop reduces the accessibility of the other nutrients, and in particular holocellulose to ruminal processes of digestion. Starch levels are insignificant in most forages and as such quantification of this fraction is not appropriate. One exception is maize silage where the content of starch may be substantial. Furthermore, it should be recognised that the digestion characteristics of maize starch are unlikely to be similar to those of other starch sources such as the cereal grains (Beever, Coehlo da Silva and Armstromg, 1970).

The crude protein fraction of forages consists largely of three different proteins, as well as varying amounts of non protein moieties including peptides, free amino acids, amines and nitrates. Peptides and amino acids will be digested in the rumen and become available for the synthesis of microbial protein, but it is unlikely that the other non protein nitrogen containing moieties make any positive contribution to the nitrogen requirements of rumen microbes. Fraction 1 protein (ribulose-1,5-biphosphate carboxylase) is the major forage protein (Mangan, 1982) owing to its important function as an enzyme involved in photosynthesis. It is considered to be largely available for digestion in the rumen and its relatively rapid rate of digestion following ingestion can have a significant negative effect upon energy and protein synchronisation within the rumen as observed with high nitrogen containing forages by Beever, Dhanoa, Losada, Evans, Cammell and France (1986). Fraction 2 protein and chloroplast membrane protein comprise the other major proteins, and whilst their digestion characteristics within the rumen have not been fully elucidated, Mangen (1982) did suggest that their relatively lower rates of availability may substantially influence the pattern of nitrogen availability to the rumen microbes, or possibly contribute to a rumen by pass fraction.

Current forage characterisation

Against this variation in the relative proportions of the individual nutrients which comprise the principal carbohydrate and 'protein' fractions of forages, current forage analysis is largely restricted to methodology such as that proposed by van Soest and Wine (1967) which used neutral-detergent solutions to estimate the relative proportions of cell contents and cell walls, with further differentiation of the cell-wall fraction using acid-detergent solution. Equally there is no systematic attempt to determine the nutritionally specific nitrogen fractions, and for many years chemical analysis has been restricted to determination of crude protein content (Total nitrogen × 6.25) and in the case of silages, an estimation of ammonia nitrogen concentration. The apparent digestibility of forage dry matter and organic matter has been considered to be an important attribute of nutritive value and consequently much time and effort has been spent on providing such estimates. The development of the two stage *in vitro* technique by Tilley and Terry (1963) with the consecutive simulation of ruminal and pepsin digestion represented a major advance in this respect, and this technique gained universal acceptance in the evaluation of both temperate and tropical forages. Alternatives such as the use of neutral cellulase have been proposed, but in many laboratories, where access to rumen fluid is not guaranteed, dry matter and organic matter digestibility are often predicted from equations relating digestibility to forage parameters such as modified acid-detergent fibre. On the basis of such predictions, it now seems acceptable to predict the energy value of the feed, in particular the ME content. Whilst the need for such estimates is recognised, provided the ME system remains the preferred method for meeting the energy requirements of ruminants, concern does exist over the variety of methods used to derive the ME estimates. Against a database of *in vivo* estimates of forage ME content, obtained largely with sheep fed at maintenance levels of feeding, it is not surprising that the ME estimates for high producing dairy cows fed at 35+ g DM/kg body weight may be questionable. ME concentrations for grass silage diets exceeding 12.0 MJ/kg DM are often quoted, yet studies conducted over several years with dairy cows at Institute for Grassland and Environmental Research, Hurley failed to identify any grass silages which had ME concentrations even approaching 12.0 MJ/kg DM. In part this overestimation may be related to difficulties experienced in accurately determining the gross energy content of grass silages, where accounting for the calorific value of the volatile components maybe problematic, or in the determination of the dry matter content of the ensiled forage. Equally, substantial errors in determination of ME content will arise if the factor for conversion of digestible energy to ME is assumed to be constant at 0.82. The studies reported by Beever, Cammell, Sutton, Spooner, Haines and Harland (1989) and Sutton, Cammell, Beever, Haines, Spooner and Harland (1992) demonstrate that even with modest yielding cows, the ME/DE ratio may approach 0.88 on grass silage:concentrate based rations.

One recent development with respect to the evaluation of forage protein has been the distinguishing of rumen degradable protein (RDP) from ruminally undegradable protein (UDP) as first advanced by Agricultural Research Council (1980) and Agricultural and Food Research Council (1984) and the recently proposed concept of effective ruminally degradable protein (ERDP) as advanced by the Interdepartmental Working Party (IDWP) report (Agricultural and Food Research Council, 1992). Associated with these proposals appropriate analytical methods have been developed. These largely involve the use of surgically modified animals in conjunction with the *in situ* artificial fibre bag technique in order to provide estimates of the rate and extent of feed nitrogen degradation. Suitable alternatives not involving the use of fistulated animals have been suggested but as yet such techniques have not been adequately validated, even against the *in situ* technique. Of possibly greater concern however is the fact that there appears to be little evidence indicating that the *in situ* technique truly replicates *in vivo* measurements of feed protein degradation and the extent to which undegraded feed protein may contribute directly to small intestinal protein supply.

Most other analyses conducted on forages relate specifically to silage diets, where estimates of pH, ammonia concentration, and possibly some measure of fermentation acid contents are often provided. Apart from extreme situations, it is unlikely that silage pH is of significant value in predicting animal performance, but there is now considerable interest in discounting the fermentation end products from any estimation of the total energy available for the processes of fermentation and synthesis occurring in the rumen. Lipids are not routinely analyzed, but this situation is likely to change given the requirement to discount ruminally non productive energy. Currently there is little regard for the mineral content of forages with no attempt to distinguish the soluble from the insoluble components. Recognition that some elements such as phosphorus can play an important role in the regulation of microbial biomass synthesis suggests that further attention to the individual mineral elements may be appropriate.

Prediction of feed intake

The quantity of food consumed voluntarily by a ruminant is influenced by the rate at which previously ingested meals are removed from the rumen by the competing processes of degradation and passage. Rate of degradation is substantially affected by the nature of the forage offered with respect to the relative proportions of the various fibre fractions, but can be significantly reduced when large quantities of readily fermentable carbohydrate are fed as shown by Beever, Sutton, Thomson, Napper and Gale (1988b) when concentrates based on contrasting energy sources were fed to dairy cows receiving grass silage. The rate at which undegraded feed particles leave the rumen is controlled by the size of the particles in relation to the dimensions of the reticulo-omasal orifice and studies by Moseley and Jones (1984) and others have demonstrated that the nature of the forage, in this case perennial ryegrass and white clover, can have a major influence on the rate of disappearance of small particles within the rumen. Equally, this should stimulate the rate of digestion. With ensiled forages the end products of the fermentation process in the silo can also affect the voluntary consumption of the forage, with the relative concentrations of metabolites such as butyric acid, ammonia and possibly some of the amines (e.g. γ-amino butyric acid, Buchanan-Smith, 1982) being particularly important. Furthermore, feed consumption will be regulated by the rate at which absorbed nutrients are removed from circulation by the processes of metabolism.

Against this apparent complexity relating to the physical and metabolic control of feed intake, it is of considerable concern that most of the intake prediction equations fail to recognise the composition of the forage as well as the nature of any concentrate fed, relying largely upon the liveweight of the animal and its current level of productivity. This was particularly the case with the equations proposed by Ministry of Agriculture, Fisheries and Food (1975) and Agricultural

Research Council (1980), whilst two equations proposed by Vadiveloo and Holmes (1979) recognised concentrate intake level but did not include any description of concentrate composition. The equation of Bines, Napper and Johnson (1977) attempted to predict total dry matter intake solely on the basis of animal liveweight and the rate of liveweight change. The equations proposed by Lewis (1981) represented a substantial change in so much that they considered the intake of silage separately in a mixed feeding situation by first attempting to predict the intake of the silage as a sole feed with some recognition of silage composition and then predicting actual silage intake on the basis of concentrate allowance. Subsequently, Neal, Thomas and Cobby (1984) examined the behaviour of these equations against an independent data set and found considerable variation in the prediction of dry matter intake compared with observed values. Mean square prediction errors (MSPE) in a comparison which included cows and heifers and using weekly liveweights ranged from 2.1 to 10.8 $(kg\ DM)^2$, and Neal *et al.* (1984) concluded that the lack of accuracy in the equations was cause for concern, whilst indicating that future research effort in this area was required. By making assumptions regarding the ME contents of feeds for lactating dairy cows and the efficiency of ME utilisation for lactation they estimated that an MSPE of 3.0 $(kg\ DM)^2$ equated to an intake error of 1.7 kg DM which could lead to an error of ± 4.0 kg milk output at zero body weight change.

Subsequently, Rook, Dhanoa and Gill (1990a; 1990b) and Rook and Gill (1990) re-examined the mathematical procedures used to develop empirical equations for the prediction of feed intake and by the use of techniques such as ridge regression analysis to remove collinearity between variables proposed a number of alternative equations which took into account an increased number of feed attributes, including the concentrations of butyric acid and ammonia in silages. The subsequent publication *Voluntary Intake of Cattle* (Agricultural and Food Research Council, 1991) concluded that such approaches should continue and suggested that the use of near infra-red spectroscopy to improve the chemical characterisation of feedstuffs, along with estimates of the rate of digestibility, as opposed to digestibility *per se* should be encouraged. Given such changes were implemented, it is likely that further improvements in the prediction of feed intake by ruminants could be achieved, but it is accepted by many that mechanistic approaches, which aim to describe the underlying biological principles will ultimately replace the less than adequate empirical approach where relationships are casual not causal.

Prediction of energy utilisation

The extent of energy retention in growing cattle is largely a function of total ME intake, the estimated ME requirement of the animal for maintenance and the estimated efficiency with which the ME available above maintenance is used for production. A similar situation prevails with lactating ruminants provided the energy contribution from tissue mobilisation in early lactation, or the requirements for tissue repletion in later lactation are taken into account. It is beyond the scope of this paper to examine the estimates of maintenance ME requirements which have been derived or to adequately challenge the concept of a fixed maintenance energy requirement, as discussed by Cammell, Thomson, Beever, Haines and Dhanoa (1986) and Gill, Cammell, Haines, France and Dhanoa (1989b). Concern must be expressed however over the way in which estimates of efficiency of ME utilisation for maintenance (k_m), growth (k_f) and lactation (k_l) are derived, seemingly with scant regard for the nature of the diet being fed, the composition of absorbed nutrients arising from that diet or variations which may occur in the composition of the final animal product. Thus it is not surprising that in a comparison of estimates of energy retention by comparative slaughter, calorimetry and prediction equations proposed by Agricultural Research Council (1980), in growing steers fed four silage based diets (Thomas, Gibbs, Beever and Thurnham, 1988; Beever, Cammell, Thomas, Spooner, Haines and Gale 1988a), the discrepancies were often large and not consistent between diets. Equally, in a study with dairy cows, Thomas, Chamberlain, Martin and Robertson (1987) found that as the level of digestible energy intake was increased by additional amounts of concentrates in the diet, the output of milk energy exhibited a curvilinear response which fell short of the linear output of milk energy predicted from Agricultural Research Council (1980). Subsequent examination of the data indicated that the effect could be explained by a change in the composition of the milk produced with a substantial reduction in milk fat content, as would predicted on the basis of the studies reported by Sutton, Hart, Broster, Elliot and Schuller (1985) with increased concentrate feeding, reducing the energy concentration of milk. Equally when Cammell, Beever, Sutton, Spooner and Haines (1992) examined the consequence of increased concentrate feeding on the energy balance of dairy cows (by calorimetric studies) receiving *ad libitum* grass silage, the results did not accord with those which would have been predicted from Agricultural Research Council (1980) or the recent IDWP publication (Agricultural and Food Research Council, 1991) relating to energy utilisation by ruminants. As expected milk energy output increased in response to the additional concentrates at all stages of lactation examined and the size of this response was reduced at the second increment of concentrate feeding. What was most surprising however was the observation that the first increment of concentrate caused a substantial increase in the extent of body mobilisation in early lactation or a reduced body energy gain in later lactation compared with the cows receiving the lowest level of concentrates. It would be inappropriate within this paper to speculate excessively about the cause of this response; suffice to conclude that is likely to be regulated through endocrinological changes occurring within the cow, more so than changes in the pattern of digestion and the profile of resulting nutrients absorbed from the alimentary tract. In this respect, it is appropriate to add that current energy systems do not adequately represent the important nutritional factors which may influence the way in which absorbed nutrients are partitioned for use, in the dairy cow between mammary and body metabolism, and in all species between protein and fat accretion in the body.

The importance of the animals endocrine system on the partition of nutrients between milk constituent and body tissue synthesis is demonstrated by the study of Sutton, Hart, Morant, Schuller and Simmonds (1988) in which two diets based on increasing amounts of concentrates were offered in two or six equal meals/day at similar levels of DM intake to lactating cows. Increased concentrate intake led to increased milk yields, but overall fat contents and yield were significantly reduced. However the magnitude of these depressions were considerably reduced when the diets were fed more frequently. Examination of plasma metabolite and hormone concentrations revealed high pulsatile changes in insulin in response to twice daily feeding, whilst more frequent feeding led to considerable attenuation of these peaks. Insulin is known to stimulate synthetic pathways, particularly lipogenic activity, in peripheral tissues, and it is this repartitioning of nutrients towards body tissues which caused the dramatic reductions in fat content and yield when the concentrate level was increased on the twice daily feeding regime.

Given the apparent unacceptability of the situation, it is difficult to recommend how current systems can be substantially improved through the use of more comprehensive feed characterisation. One major concern relates to the accuracy of estimation of the gross energy content of the feed as this will affect the ultimate accuracy of prediction of energy utilisation in terms of estimating total ME intake and diet metabolizability. Estimation of the gross energy content of silage is particularly difficult, and determination of the energy contribution of the volatile components does not lend itself to routine laboratory analysis. However this fraction may constitute up to 15% of the total gross energy content of the silage and given its renewed importance with respect to estimation of fermentable contents of feeds as will be discussed later, serious consideration should be given to the development of techniques which could provide rapid and accurate assessments of volatile components suitable for use in routine feed characterisation.

Prediction of absorbed protein supply

Current systems relating to the utilisation of protein by ruminants are largely concerned with prediction of the quantity of protein absorbed from the small intestine, whilst the principles relating to the utilisation of absorbed protein appear to be accepted with minimal debate. It may be reasonable to conclude that the factors governing the ultimate supply of amino acids are of prime importance but there is growing belief that the efficiency of utilisation of absorbed amino acids is often much lower than the values proposed in the recent IDWP publication (Agricultural and Food Research Council, 1992). In a study involving growing cattle fed grass silage with or without fish meal supplementation, Beever, Gill, Dawson and Buttery (1990) were able to calculate the efficiency of utilisation of absorbed amino acids, using whole body protein retention data obtained earlier by Gill, Beever, Buttery, England, Gibb and Baker (1987) for similar cattle fed identical diets. On an incremental basis relating to amino acid supply and protein retention they obtained an estimate of the efficiency of utilisation of amino acids of 0.53, whilst if likely costs of body protein maintenance costs were taken into account they estimated a value of 0.51 for both the unsupplemented and the fish meal supplemented diet. These estimates are surprising in two respects. They are clearly much lower than the 'ideal' value quoted in current protein systems and this could have a major effect upon the calculated requirement of absorbed amino acids to meet a desired level of body protein retention. Furthermore, the finding that the estimates of efficiency were low even on the control diet where protein supply was clearly insufficient to meet the potential of the animals to deposit protein suggests that much greater attention should be given to the composition of the absorbed amino acid fraction arising from forage diets.

In relation to the suitability of current procedures to predict the supply of protein to the small intestine of forage fed ruminants, a detailed discussion will not be presented here as this forms the basis of the chapter by Cottrill (1993) in this book. Suffice to say, considerable interest exists in the routine determination of the ruminally degradable and undegradable protein fractions of feeds. The use of the artificial fibre bag method still appears to be the recognised technology as to date no acceptable alternative has emerged, despite efforts to develop a system which relies neither on the use of surgically modified animals, nor on the provision of rumen liquor for an *in vitro* system.

Whilst the concepts behind the proposed improvements in the characterisation of the protein fraction of forages and other feedstuffs are acceptable, extensive use and faith placed upon the artificial fibre bag method appears to have two major weaknesses. As indicated earlier, there is little reported evidence to suggest that use of the estimates of RDP and UDP derived from the method to calculate ERDP and ultimately absorbed protein supply gives actual estimates of protein supply which agree with *in vivo* observations, particularly with high intakes of energy and protein as occurs in the high producing dairy cow. Furthermore, there is considerable concern over the repeatability of the technique which in essence is a bio-assay, and inter laboratory comparisons lack the repeatability that would be expected from more standard laboratory procedures. The recent protein publication (Agricultural and Food Research Council, 1992) drew attention to this, and even when attempts were made to introduce standardised procedures, considerable variation was still evident in the results obtained. The consequence of this is illustrated in Table 1.1 in which it was estimated by the IDWP that the use of the lowest and highest estimates of the ruminally degradable protein content of barley and soya obtained from a standardised 'laboratory ring test' could lead to a 100% error in the prediction of absorbed protein supply. Clearly such error is unacceptable, a point fully recognised by the IDWP in its recommendation that technique developments were still required.

Table 1.1 PREDICTION OF TRULY ABSORBED DUP FROM A BARLEY:SOYA CONCENTRATE (79:21) USING HIGH AND LOW ESTIMATES OF EFFECTIVE DEGRADABILITY

	Effective digestibility	
	High estimates	*Low estimates*
Barley/Soya	0.89/0.69	0.60/0.56
Effective degradability	0.79	0.58
Truly absorbed DUP (g/d) from 8 kg concentrate	211	444

DUP = Digestible undegradable dietary protein

Over and above this concern however, it is necessary to question if the concepts relating to the ruminal utilisation of protein and in particular the control of microbial protein synthesis are adequately encompassed within the new protein system. There are many examples from *in vivo* experimentation to indicate that the efficiency of microbial protein synthesis is both variable and often markedly less than the values used in the IDWP recommendations (McAllan, Siddons and Beever, 1987). An overestimation of the value of ruminally degradable protein to support microbial protein synthesis may be one possible reason for the lower values for microbial protein synthesis often seen on grass silage based diets, and this aspect was addressed by the study of Rooke, Lee and Armstrong (1987), in which the importance of an adequate supply of good quality protein for the microbes was established. Equally the failure to recognise the need to synchronise carbohydrate and protein availability in the rumen, with particular emphasis upon the rate at which key substrates become available for microbial biomass synthesis is likely to have a major effect upon the resultant level of microbial synthesis and the ultimate supply of microbial protein to the small intestine. Methodology which includes assessment of the rate of protein degradation as well as extent is important, but equally, some attempt should be made to assess the rate of availability of the energy/carbohydrate components of the feed. The relevance of this is apparent from the data of Jacobs and McAllan (1992) who examined the influence of rape seed meal supplementation of grass silages made by formic acid addition or by enzyme treatment at the time of harvesting. Differences in the chemical composition of the two silages, which were made from the same grass crop, were generally small, with changes in the fibre fractions being in line with expectations. However, response to the protein supplement was markedly different between the two silages (Table 1.2). Addition of rape seed meal to the formic acid silage led to a modest increase in the flow of amino acid-N to the small intestine (+12%) with no marked change in the extent of microbial protein synthesis. In contrast on the enzyme treated silage the response in duodenal amino-N flow was much greater (+37%) accompanied by a substantial increase in the net synthesis of microbial-N (+37%). It is clear from the feed characterisation data presented for the two silages by Jacobs and McAllan (1992) that the magnitude of this response could not have been predicted; a response suspected to be associated with an increased availability of carbohydrate following enzyme treatment of the grass which possibly led to an improved synchronisation of energy and protein within the rumen.

Table 1.2 EFFECT OF PROTEIN SUPPLEMENTATION UPON NITROGEN KINETICS IN THE RUMEN OF GROWING CATTLE RECEIVING FORMIC ACID OR ENZYME TREATED SILAGE

	Formic acid			*Enzyme*		
Silage composition (g/kg DM)						
Total nitrogen		24.4			24.1	
Lactic acid		101			134	
Water soluble carbohydrate		8			23	
Neutral detergent fibre		537			492	
Acid detergent fibre		333			303	
Rapeseed meal (g/kg total DM)	0	60	120	0	60	120
Duodenal non-ammonia nitrogen (g/d)	99	96	106	90	109	121
Duodenal amino-nitrogen (g/d)	58	63	65	54	68	74
Microbial nitrogen (g/d)	64	68	67	59	75	81

Fermentable metabolisable energy

One of the major weaknesses of the current systems used in the UK for meeting the energy and protein requirements of ruminants is that they are largely independent of each other. This aspect was discussed in detail by MacRae, Buttery and Beever (1988) whilst acknowledging that in the protein system, some consideration is given to the supply of energy in the rumen in relation to the availability of nitrogen for microbial protein synthesis, albeit energy supply was simply expressed as a function of total ME intake. Thomas (1982) and others have drawn attention to the fact that on fermented feeds, such as grass silage, as much as 15% of the dietary ME may be present as volatile components which are of limited, or indeed no value to the microbial synthetic processes. In the recent IDWP publication the concept of fermentable ME (FME) supply was advanced as an appropriate index of energy availability in the rumen, defined as the total ME of the diet less the energy content of the volatile components and the lipid fraction of the feed. This development is supported by the study of Gill, Siddons, Beever and Rowe (1986) which considered the metabolic fate of lactic acid in silage fed sheep. Lactate concentration in the silage amounted to 139 g/kg DM, whilst measured ruminal flux of lactate was 1.83 moles/day, of which 1.37 moles was attributable to dietary lactate intake. It was estimated that 54%, 31% and 5% of total lactate flux was converted, within the rumen, to acetate, propionate and butyrate respectively, with the remaining lactate (10%) presumed to be absorbed directly. From these results, Gill, Siddons, Beever and Rowe (1986) were able to conclude that almost 30% of the fermentable organic matter of the silage was metabolised in the rumen via lactic acid with no associated release of energy to support microbial growth.

A major concern exists however over the approach used to estimate FME contents. If both the volatile and lipid components are measured and suitable energy values ascribed to them, then it is likely that the values obtained will be acceptable for assessment of nutritive value. Assumptions regarding the content of volatile and lipid components, in those situations where measured data are not available, is a potentially dangerous approach which could lead to substantial errors. Equally prediction from the proposed Agricultural Development and Advisory Service equation in which the content of volatiles is inversely related to the DM content of the crop and the extent of fermentation, is likely to add further confusion to the estimates of FME. This was demonstrated in a recent study by Seyoum and Chamberlain (personal communication) in which measured FME contents of over 70 grass silages were related to predictions based on Agricultural Development and Advisory Service (1991) and Agricultural and Food Research Council (1992) proposals by regression analysis, but only 39.1 and 43.4% of the variation respectively was accounted for. The consequences of possible errors in the estimation of FME are examined further in Table 1.3. From a consideration of the measurements of silage DM, gross energy, lipid and volatile contents,

Table 1.3 POSSIBLE CONSEQUENCES OF ACCUMULATIVE ERRORS IN THE ESTIMATION OF FERMENTABLE ENERGY CONTENTS (ALL UNITS MJ/kg SILAGE DM)

Component		*Estimates*	
	High	*Actual*	*Low*
Gross energy[a]	19.73	19.25	18.76
Digestible energy[b]	14.56	13.86	13.17
Metabolisable energy (ME)[c]	12.52	11.37	10.54
Oil[b]	0.99	1.05	1.10
Volatiles[b]	1.56	1.65	1.73
Fermentable ME (FME)	9.97	8.67	7.71
Fermentability (FME/ME)	0.80	0.76	0.73

[a] Assumes 2.5% analytical error; [b] Assumes 5% analytical error;
[c] Assumes ME/DE range from 0.80 to 0.86

and predictions of energy digestibility and metabolizability, an FME content of 8.67 MJ/kg DM was estimated, giving an FME/ME ratio (fermentability?) of 0.76. In contrast, by assuming 2.5% analytical errors in the estimates of silage DM and gross energy content, 5% errors in the estimates of digestibility, and the contents of volatiles and lipids, and applying the range of estimated values for ME/DE ratio which occurs in the literature (0.80–0.86), and then compounding the errors as opposed to allowing them to cancel out, which is every nutritional advisers hope, the range in estimated FME contents was from 7.71 to 9.97 MJ/kg DM. This magnitude of discrepancy is not permissable, but the data do indicate the relative ease with which errors can be compounded simply because current feeding systems are largely modifications/add-ons to previous systems.

Conclusions

This paper has attempted to review current procedures being used to evaluate feeds, and in particular, forages, for ruminants, with specific reference to the prediction of feed intake, energy utilisation and the small intestinal absorption of protein. It is recognised that considerable progress has been made over the last 20 years in this respect, and the need to provide an improved characterisation of the forage component of the ration is unquestionable. In this regard some further attention is required to aspects such as the methodology associated with determination of silage DM and gross energy contents, along with a more rigorous assessment of the techniques used to estimate the RDP and UDP contents of forages. Equally, the determination of FME contents will need to be further refined if we are to be able to place full reliance upon this entity as a valuable component in the prediction of microbial biomass synthesis in the rumen. It may be necessary to accept that FME represents a mixed and variable substrate and further refinement, possibly to include discounting both RDP and UDP, may be required.

It is encouraging that more attention is being directed towards determination of the rate of degradation of nutrients within the rumen, particularly with the established need to improve energy and protein synchronisation for microbial metabolism. This is already occurring with respect to the ruminally degradable protein fractions and should be extended to the energy yielding components, and in particular the individual carbohydrate moieties, especially starch. The development of alternative methodologies which do not include the routine use of animals would be helpful in this respect. Recent development of the pressure transducer technique (PTT) by Theodorou, Williams, Brooks, Dhanoa and Gill (1993) following the earlier work of Menke, Raab, Salewski, Steingass, Fritz and Schneider (1979) appears to be a possible alternative, with measurement of the rate of production of fermentation gases being used to predict the rate of organic matter digestion in a simulated rumen situation. Such methodology is of value in attempting to provide some relative ranking between different forages, but the absolute nature of the results may be more doubtful. As suggested by Beever (1993), the relative partition of ruminally available hexose between fermentation and direct incorporation into microbial biomass may vary widely depending upon the availability of other essential nutrients for microbial growth. Furthermore, the pattern of fermentation can vary with respect to the yield of individual VFAs and both of these will undoubtedly affect the yield of fermentation gas per unit of hexose digested. In a recent publication, Chamberlain (1993) concluded that the PTT method appeared to give reliable estimates concerning the total extent but not rate of digestion, and as such may only be considered as an alternative to the method proposed by Tilley and Terry (1963).

As to the longer term, it is difficult to establish what further refinements to present systems of feed characterisation are possible. There is a growing opinion that current systems need major reappraisal but despite this view most initiatives have been confined to adjusting current systems, in many instances compounding errors that already existed in the existing approach. Consequently, the major reappraisal which is urgently required has been delayed by this 'add-on' mentality. Presently we are obliged to accept the weaknesses of the current systems, although this will undoubtedly compromise our ability to predict animal performance, and to provide appropriate nutritional evaluations of novel feedstuffs, including alternative forage sources such as fermented and urea-treated whole crop wheat silage. Until systems which truly represent the dynamic nature of forage digestion and utilisation by ruminants are developed, future advances in forage characterisation are likely to be limited to 'fine tuning' of the present systems which as this paper has demonstrated, will never meet the demanding requirements of the ruminant animal feeding industry.

References

Agricultural Development and Advisory Service (1991) Fermentable metabolisable energy content of grass silages. Technical Bulletin 91/5, ADAS Feed Evaluation Unit: Stratford-upon-Avon

Agricultural Research Council (1980) The Nutrient requirements of Farm Livestock, No. 2, Ruminants. Farnham Royal: Commonwealth Agricultural Bureaux

Agricultural and Food Research Council (1984) The Nutrient requirements of Farm Livestock, No. 2, Ruminants (Suppl. No. 1) Commonwealth Agricultural Bureaux: Farnham Royal

Agricultural and Food Research Council (1990) Technical Committee on Responses to Nutrients, Report No. 5, Nutrient Requirements of Ruminant Animals, *Nutrition Abstracts and Reviews, Series B; Livestock Feeds and Feeding*, **60**, 729–804

Agricultural and Food Research Council (1991) Technical Committee on Responses to Nutrients, Report No. 8, Voluntary Intake of Cattle, *Nutrition Abstracts and Reviews, Series B; Livestock Feeds and Feeding*, **61**, 815–823

Agricultural and Food Research Council (1992) Technical Committee on Responses to Nutrients, Report No. 9, Nutrient Requirements of Ruminant Animals; Protein, *Nutrition Abstracts and Reviews, Series B; Livestock Feeds and Feeding*, **62**, 787–835

Beever, D.E. (1993) Rumen function. In *Quantitative Aspects of Ruminant Digestion and Metabolism*, pp. 187–215. Edited by J.M. Forbes and J. France. Commonwealth Agricultural Bureaux: Wallingford

Beever, D.E., Cammell, S.B., Sutton, J.D., Spooner, M.C., Haines, M.J. and Harland, J.I. (1989) In *Energy Metabolism of Farm Animals*, pp. 33–36. Edited by Y. van der Honing and W.H. Close. EAAP Publication No. 43. Wageningen: Pudoc

Beever, D.E., Cammell, S.B., Thomas, C., Spooner, M.C., Haines, M.J. and

Gale, D.L. (1988a) The effect of date of cut and barley substitution on gain and on the efficiency of utilisation of grass silage by growing cattle. 2, Nutrient supply and energy partition. *British Journal of Nutrition*, **60**, 307–319

Beever, D.E., Coehlo da Silva, J.F. and Armstrong, D.G. (1970) The effect of processing maize on its digestion in sheep. *Proceedings of Nutrition Society*, **29**, 43A

Beever, D.E., Dhanoa, M.S., Losada, H.R., Evans, R.T., Cammell, S.B. and France, J. (1986) The effect of forage species and stage of harvest on the processes of digestion occurring in the rumen of cattle. *British Journal of Nutrition*, **56**, 439–454

Beever, D.E., Sutton, J.D., Thomson, D.J., Napper, D.J. and Gale, D.L. (1988b) Comparison of molassed and unmolassed sugar beet pulp and barley as energy supplements on nutrient digestion and supply in silage fed cows. *Animal Production*, **46**, 490

Beever, D.E., Gill, M., Dawson, J.M. and Buttery, P.J. (1990) The effect of fishmeal on the digestion of grass silage by growing cattle. *British Journal of Nutrition*, **63**, 489–502

Bines, J.A., Napper, D.J. and Johnson, V.W. (1977) Long term effects of level of intake and diet composition on the performance of lactating dairy cows. 2. Voluntary intake and ration digestibility in heifers. *Proceedings of Nutrition Society*, **36**, 146A

Buchanan-Smith, J.G. (1982) Voluntary intake in ruminants affected by silage extracts and amines in particular. In *Forage Protein in Ruminant Animal Production*, pp. 180–182. Edited by D.J. Thomson, D.E. Beever and R.G. Gunn. BSAP Occasional Publication No. 6, Thames Ditton: BSAP

Cammell, S.B., Thomson, D.J., Beever, D.E., Haines, M.J. and Dhanoa, M.S. (1986) The efficiency of energy utilisation in growing cattle consuming fresh perennial ryegrass or white clover. *British Journal of Nutrition*, **55**, 669–680

Cammell, S.B., Beever, D.E., Sutton, J.D., Spooner, M.C. and Haines, M.J. (1992) Body composition and performance of autumn-calving Holstein-Friesian dairy cows during lactation: Energy partition. *Animal Production*, **54**, 475A

Chamberlain, A.T. (1993) The use of the gas pressure transducer technique for the assessment of the ruminal degradation of the dry matter of protein supplements. *Animal Production*, (in press)

Cottrill, B.R. (1993) Characterisation of Nitrogen in Ruminant Feeds. In *Recent Advances in Animal Nutrition – 1993* pp. 39–53. Edited by P.C. Garnsworthy and D.J.A. Cole. Nottingham: Nottingham University Press

Gill, E.M., Siddons, R.C., Beever, D.E. and Rowe, J.B. (1986) Metabolism of lactic acid isomers in the rumen of silage-fed sheep. *British Journal of Nutrition*, **55**, 399–407

Gill, M., Beever, D.E., Buttery, P.J., England, P., Gibb, M.J. and Baker, R.D. (1987) The effect of oestradiol-17β implantation on the response in voluntary intake, liveweight gain and carcase composition, to fishmeal supplementation of silage offered to growing calves. *Journal of Agricultural Science, Cambridge*, **108**, 9–16

Gill, M., Beever, D.E. and Osbourn, D.F. (1989a) The feeding value of grass and grass products. In *Grass — Its Production and Utilisation*, 2nd edition. Edited by W. Holmes. Oxford: Blackwell Scientific

Gill, M., Cammell, S.B., Haines, M.J., France, J. and Dhanoa, M.S. (1989b) Energy balance in cattle offered a forage diet at submaintenance levels. In

Energy Metabolism of Farm Animals, pp. 300–303. Edited by Y. van der Honing and W.H. Close. EAAP, Publication No. 43. Wageningen: Pudoc

Jacobs, J.L. and McAllan, A.B. (1992) Protein supplementation of formic acid and enzyme-treated silages. 2. Nitrogen and amino acid digestion. *Grass and Forage Science*, **47**, 114–120

Lewis, M. (1981) Equations for predicting silage intake by beef and dairy cattle. *Proceedings of VIth Silage Conference, Edinburgh, 1981*, pp. 35–36

MacRae, J.C., Buttery, P.J. and Beever, D.E. (1988) Nutrient interactions in the Dairy Cow. In *Nutrition and Lactation in the Dairy Cow*, pp. 55–75. Edited by P.C. Garnsworthy. London: Butterworths

Mangen, J.L. (1982) The nitrogenous constituents of fresh forages. In *Forage Protein in Animal Production*, pp. 25–40. Edited by D.J. Thomson, D.E. Beever and R.G. Gunn. BSAP Occasional Publication No. 6, Thames Ditton: BSAP

McAllan, A.B., Siddons, R.C. and Beever, D.E. (1987) The efficiency of conversion of degraded nitrogen to microbial nitrogen in the rumen of sheep and cattle. In *Feed Evaluation and Protein Requirement Systems for Ruminants*, pp. 111–128. Edited by R. Jarrige and G. Alderman. Luxembourg: CEC

Menke, K.H., Raab, L., Salewski, A., Steingass, H., Fritz, D. and Schneider, W. (1979) The estimation of the digestibility and metabolizable energy content of ruminant feedingstuffs from the gas production when they are incubated with rumen liquid *in vitro*. *Journal of Agricultural Science, Cambridge*, **93**, 217–222

Ministry of Agriculture, Fisheries and Food (MAFF) (1975) Energy allowances and Feeding Systems for Ruminants. *Technical Bulletin, No. 33*, London: HMSO

Moseley, G. and Jones, J.R. (1984) The physical digestion of perennial ryegrass (*Lolium perenne*) and white clover (*Trifolium repens*) in the foregut of sheep. *British Journal of Nutrition*, **52**, 381–390

Neal, H.D. St.C., Thomas, C. and Cobby, J.M. (1984) Comparison of equations for predicting voluntary intake by dairy cows. *Journal of Agricultural Science, Cambridge,* **103**, 1–10

Rook, A.J., Dhanoa, M.S. and Gill, M. (1990a) Prediction of the voluntary intake of grass silages by beef cattle. 2. Principal component and ridge regression analyses. *Animal Production*, **50**, 439–454

Rook, A.J., Dhanoa, M.S. and Gill, M. (1990b) Prediction of the voluntary intake of grass silages by beef cattle. 3. Precision of alternative prediction models. *Animal Production,* **50**, 455–466

Rook, A.J. and Gill, M. (1990) Prediction of the voluntary intake of grass silages by beef cattle. 1. Linear regression analysis. *Animal Production*, **50**, 425–438

Rooke, J.A., Lee, N.H. and Armstrong, D.G. (1987) The effects of intraruminal infusions of urea, casein, glucose syrup and a mixture of casein and glucose syrup on nitrogen digestion in the rumen of cattle receiving grass-silage diets. *British Journal of Nutrition*, **57**, 89–98

Sutton, J.D., Hart, I.C., Broster, W.H., Elliott, R.J. and Schuller, E. (1985) Feeding frequency for lactating cows: effect on rumen fermentation and blood metabolites and hormones. *British Journal of Nutrition*, **56**, 181–192

Sutton, J.D., Hart, I.C., Morant, S.V., Schuller, E. and Simmonds, A.D. (1988) Feeding frequency for lactating cows: diurnal patterns of hormones and metabolites in peripheral blood in relation to milk fat concentration. *British Journal of Nutrition*, **60**, 265–274

Sutton, J.D., Cammell, S.B., Beever, D.E. Haines, M.J., Spooner, M.C. and Harland, J.I. (1992) The effect of energy and protein sources on energy and nitrogen balances in Friesian cows in early lactation. In *Energy Metabolism of Farm Animals*, pp. 288–291. Edited by C. Wenk and M. Boessinger. EAAP Publication No. 58. Zurich: ETH-Zentrum

Theodorou, M.K., Williams, B.A., Brooks, A., Dhanoa, M.S. and Gill, E.M. (1993) In *Animal Production in Developing Countries*. Edited by M. Gill, E. Owen, G.E. Pollot and T.L.J. Lawrence. BSAP Occasional Publication No. 6, (in press)

Thomas, C., Gibbs, B.G., Beever, D.E. and Thurnham, B.R. (1988) The effect of date of cut and barley substitution on gain and on the efficiency of utilisation of grass silage by growing cattle. 1. Gains in liveweight and its components. *British Journal of Nutrition,* **60**, 297–306

Thomas, P.C. (1982) Utilization of Conserved Forages. In *Forage Protein in Ruminant Animal Production*, pp. 67–77. Edited by D.J. Thomson, D.E. Beever and R.G. Gunn. BSAP Occasional Publication No. 6. Thames Ditton: BSAP

Thomas, P.C., Chamberlain, D.G. Martin, P.A. and Robertson, S. (1987) Dietary energy intake and milk yield and composition in dairy cows. In *Energy Metabolism of Farm Animals*, pp. 188–191. Edited by P.W. Moe, H.F. Tyrrell and P.J. Reynolds. EAAP Publication No. 32, New Jersey: Rowman Littlefield

Tilley, J.M.A. and Terry, R.A. (1963) A two-stage technique for the *in vitro* digestion of forage crops. *Journal of the British Grassland Society*, **18**, 104–111

Vadiveloo, J. and Holmes, W. (1979) The prediction of the voluntary feed intake of dairy cows. *Journal of agricultural Science, Cambridge,* **93**, 553–562

Van Soest, P.J. and Wine, R.H. (1967) Use of detergents in the analysis of fibrous feeds. IV Determination of plant cell wall constituents. *Journal of the Association of Official Analytical Chemists,* **50**, 50–55

2

CHARACTERIZATION OF CARBOHYDRATES IN CONCENTRATES FOR DAIRY COWS

H. DE VISSER
Research Institute for Livestock Feeding and Nutrition (IVVO-DLO) Lelystad, The Netherlands

Introduction

Recently, successful breeding programmes have increased the production potential of our dairy cows considerably and modern breeding techniques give cause for further optimism (van der Schans, Westerlaken, van der Wit, Eyestone and Boer, 1991). However, in order to exploit the higher genetic potential, the feeding of animals has to improve as well. Further improvements by increasing the quality of the roughages may be possible using careful breeding and harvesting programmes, converting from hay-making to ensiling, earlier cutting of grass, strategic use of nitrogen fertilizers and new ensiling methods to increase the energy value of the roughage (Salette, 1982; McDonald, 1983; Thomas and Thomas, 1985; Frame, Harkess and Talbot, 1989). However, in spite of all the progress made, the potential intake of energy and protein from forage will not be sufficient meet the increased requirements of high producing dairy cows so the addition of concentrates to basal diets is unavoidable. Large differences are found between countries in the amount and type of compound feeds used (De Visser and Steg, 1988). Differences are determined by availability, price, level of production, type of basal diet, etc. When a choice of feeds is available adequate yardsticks are required to evaluate quality. 'Quality' can be defined in various ways. Traditionally, the principle factors estimated are the energy and protein concentration of a feedstuff. However, milk production and composition are not just related to energy and nitrogen balance, but to the nutrient supply, e.g. production of acetic acid, glucose, amino acids, etc (MacRae, Buttery and Beever, 1988), so increasing emphasis is given to the nutrient availability of feedstuffs. Information on the availability of these nutrients to the mammary gland is essential to meet the genetic potential for milk production on the one hand and to be able to adapt milk composition to the needs of dairy industry and consumers on the other. Moreover, dairying should be performed under economically profitable conditions whilst causing minimal environmental pollution.

To provide the mammary gland with the nutrients required, information of the composition, fermentation, digestion and utilization of carbohydrates in concentrates is essential. However, effects of compound feed carbohydrates on milk performance can only be discussed in relation to the basal diet fed and therefore will be discussed in this context.

Identification of carbohydrates

Carbohydrates are the most important type of organic components on Earth. Most roughages and concentrates, with the exception of oil seeds and products of animal origin, contain more than 60% carbohydrates (Centraal Veevoederbureau, 1991; DLG, 1991). They therefore make a large contribution to the energy value of feedstuffs.

The structure and functional properties of various types of carbohydrates are very different. Their occurrence in nature, especially in plants, is strongly related to functional properties (Aman and Graham, 1990). Several roles can be identified — primary energy carrier, intermediary metabolism, energy storage, energy transport and structural elements. The large variations are related to their chemical properties and reactivities. Most of the carbohydrates occurring in feedstuffs are molecules with 3–6 carbon-atoms. Well known 5 carbon-atoms (pentose) are arabinose and xylose. Glucose, mannose and galactose are 6 carbon-atoms (hexose). Only a few carbohydrates are present in nature as monomers. Due to their water solubility and reactivity, these monomeric molecules are used as intermediary and primary energy carriers in plants. Monomers are very often combined to form disaccharides, such as maltose, lactose and saccharose, which are important feed carbohydrates (oligosaccharides).

A further combination of these singular molecules into long chains of saccharides provide the so-called non-structural polysaccharides. These polysaccharides form a very important group of carbohydrates in feeding comprising the starches. They are the basic form of energy storage in plants (Aman and Graham, 1990). Due to their long chains they are not water soluble.

Structural polysaccharides are complex carbohydrates, which can be divided into pectins, celluloses and hemicelluloses (Meier and Reid, 1982; van Soest, 1982). These carbohydrates are the structural elements in plants. Hemicellulose and cellulose are usually referred to as 'structural carbohydrates, cell wall constituents or Non Starch Polysaccharides (NSP)'. These polymers also include non-carbohydrate structures, such as phenolic acids and lignins, making the structural carbohydrates more difficult to digest as a result of a special type of bonding (van Soest, 1982).

Chemical analysis of carbohydrates

For many years carbohydrates have been analyzed using empirical methods which do not divide carbohydrates into individual chemical structures. At first the carbohydrates were divided in the crude fibre fraction (CF) and Nitrogen Free Extract (NFE). Crude fibre contains part of the structural polysaccharides; cellulose, hemicellulose and the lignin fraction (Ely and Moore, 1955). It is determined by subjecting the feedstuffs to an acid and alkaline extraction. The remainder of the carbohydrates are calculated as NFE:

NFE = DM – ash – protein – fat – crude fibre

This analytical model is still frequently used for the routine estimation of energy values of feedstuffs (Barber, Offer and Givens, 1989; Centraal Veevoederbureau, 1991; DLG, 1991). Considerable information is now available concerning the

'Weende' proximate analysis of feedstuffs. Variation in organic matter-digestibility has been predicted with satisfactory precision from crude fibre content (Givens, Everington and Adamson, 1990a,b,c; Centraal Veevoederbureau, 1991). Carbohydrates are, however, present in various forms and compositions.

In order to determine nutrient availability, methods must be available to identify different types (fractionalize) and to be able to characterize the composition of each fraction into original material (analyze).

Goering and van Soest (1970) developed a series of fractional analyses to define the cell wall constituents of roughages, by using insolubility in three types of increasingly aggressive solvents. Cell wall constituents, were fractionated into Neutral Detergent Fibre (NDF), Acid Detergent Fibre (ADF) and Acid Detergent Lignin (ADL). Hemicellulose was calculated as NDF — ADF. Cellulose was calculated as ADF — ADL. Although developed for roughages, compound feed ingredients are frequently analyzed using the same methods.

Non-structural carbohydrates may be characterized in various ways mainly as sugars (using the alcohol soluble method; total sugars by the Luff-Schoorl method) and starch (using polarimetric or enzymatic methods).

More recently, advanced methods have been developed to analyze the fractions in more detail, involving physical as well as chemical properties. Use of gas chromatography (GLC), liquid chromatography (HPLC) in combination with analysis of methylation and mass spectrometry make it possible to quantify the composition and amount of individual monosaccharides (sugars) and the structure and branching of the carbohydrate-chains. (Boon, 1989). These advanced methods are not yet available for routine analysis of feedstuffs, but are encouraging signs for the modelling of carbohydrate digestion (Cone, Tas and Wolters, 1992). They become especially important when feedstuffs are processed by cell-wall degrading enzymes (Merry and Braithwaite, 1987; Chamberlain and Robertson, 1989; van Vuuren, Bergsma, Krol-Kramer and van Beers, 1989; Spoelstra, 1991). However, in practical ruminant nutrition, the positive effects on animal performance due to enzymatic processing of feedstuffs have been limited so far.

Degradability of carbohydrates

Chemical characterization of feed carbohydrates into various fractions alone is insufficient to provide quantitative information about the nutrients that will become available for milk production. Information is required concerning rate and extent of digestion in the various compartments of the digestive tract (rumen or intestines) in order to predict the amount and type of nutrients made available for maintenance and/or milk production. Therefore, information on rumen degradation characteristics of organic components is essential. For many years, nitrogen components of feedstuffs have been divided into rumen fermentable and rumen resistant fractions. Several protein evaluation systems are already using this type of information regarding the ruminal behaviour of nitrogen components (Agricultural Research Council, 1984; National Research Council, 1988; van Straalen and Tamminga, 1991). However, for the prediction of microbial protein synthesis in the rumen, additional information is needed concerning the rate and extent of carbohydrate degradation in the rumen. Methods have been developed to rank feedstuffs according to their degradation or resistance to degradation in the rumen. The *in situ* nylon bag technique (Mehrez and Ørskov,

1977; van Vuuren *et al.*, 1989; Tamminga, van Vuuren, van der Koelen, Ketelaar and van der Togt, 1990) is frequently used for characterizing rumen degradation of organic components. The method divides the various OM fractions (N, starch, NDF) into water washable (S), rumen undegradable (U) and potentially rumen degradable (D) fractions. The rate of degradation (k_d, %/h) is mathematically estimated assuming a certain rumen passage rate. Calculations can then provide an estimate for the effective rumen degradable and undegradable fractions (Verite, Michalet-Doreau, Chapoutot, Peyraud and Poncet, 1987; Tamminga, Robinson, Vogt and Boer, 1989).

Information concerning the carbohydrate degradability (starch, NDF) in concentrate ingredients and roughages is required in order to quantify microbial protein and other nutrients (energy sources) which become available after rumen fermentation or digestion in the intestines of the host animal. In ruminants, structural-polysaccharides and rumen degradable starch are fermented in the rumen and, through the action of the microflora, become available to the host animal as volatile fatty acids (VFA). The amount of VFA produced depends on the substrate availability of both the structural polysaccharides and rumen degradable starch. Starch resistant to rumen degradation becomes, if digested, available to the host animal as glucose after hydrolysis and absorption in the small intestine (Nocek and Tamminga, 1991).

Differences in nutrient supply between roughages and concentrate ingredients are related to the chemical composition as well as to physical factors, such as particle size. Murphy, Baldwin and Koong (1982) showed differences in the proportion of volatiles produced from varying carbohydrate sources (sugars, starch, cellulose, hemicellulose) as well as between roughage and concentrate ingredients. However, in dairy cows predictions of total VFA and VFA composition remain inaccurate, due to the relatively limited supply of data from high-yielding (and therefore high intake) dairy cows (Neal, Dijkstra and Gill, 1992).

Starch degradability

Large differences occur between feedstuffs in the percentage of starch that is resistant to rumen degradation. The Dutch Feeding Table (Centraal Veevoederbureau, 1991) gives information concerning the percentage of starch resistant to rumen degradation in compound feedstuffs, based on synthetic nylon bag incubations and validated with *in vivo* data (Nocek and Tamminga, 1991). Information regarding the rate of starch degradation is not included. However, this factor could be of importance, when optimizing microbial protein synthesis (Chamberlain, Thomas, Wilson, Newbold and MacDonald, 1985; Rooke, Lee and Armstrong, 1987; Tamminga *et al.*, 1990), because nitrogen and energy sources must be available at the same time for optimal microbial protein synthesis in the rumen. Table 2.1 gives an example of starch degradations assuming a rumen passage rate of 6%/h. The assumption of a single passage rate irrespective of feed type may be criticised, but *in vivo* measurements for each individual ingredient do not exist. Another problem in tabulating starch degradability of feedstuffs involves the influence of treatment during the production process and, in particular, treatment during the production of compound feeds, such as grinding and exposure to high temperatures.

Table 2.1 RESULTS FROM NYLON BAG INCUBATIONS CONCERNING STARCH DEGRADABILITY AND RESISTANCE TO RUMEN DEGRADABILITY IN COMPOUND FEEDSTUFFS

Feedstuff	*Starch (g/kg DM)*	*Fraction*[a] *S* (%)	*D* (%)	*Rate of degradation,* K_d (% of D/h)	*Effective rumen resistant starch*[b] (%)
Milo	652	32	67	3.6	41
Maize	676	27	73	4.0	42
Maize gluten feed	403	70	30	7.8	13
Maize gluten meal	205	23	77	28.6	12
Barley	561	62	38	24.2	7
Wheat	654	68	32	17.5	8
Wheat bran	134	81	19	20.8	4
Tapioca	726	75	25	16.8	6

[a] S = soluble fraction, D = potentially degradable fraction
[b] Assuming rumen outflow rate of 6%/h
After Nocek and Tamminga (1991)

Another method used to rank starch degradability in feedstuffs is an *in vitro* method described by Cone (1991). This method describes the disappearance of starch after incubation with rumen fluid *in vitro*. In a comparative experiment using both methods (nylon bag and *in vitro*) ranking of feedstuffs was similar but differences were observed in the extent of degradation (Table 2.2).

Table 2.2 PERCENTAGE OF RUMEN RESISTANT STARCH IN FEEDSTUFFS, USING NYLON BAG OR *IN VITRO* TECHNIQUE

Feedstuff	*Resistance to rumen degradation* *Nylon bag*	In vitro
Barley	13	26
Maize	45	34
Tapioca	9	28
Concentrate mixture	22	26

After Tamminga *et al.* (1989)

There is clearly no point in optimizing compound feeds for rumen degradable starch without first obtaining information about the degradability of starch contained in roughages (maize silage) or home-grown ‘concentrate’ feedstuffs, such as corn cob mix (CCM) and husk cob silage (CHS). In forage maize, the amount of starch and its degradability are clearly related to stage of maturity (Table 2.3) although interactions due to variety, climate and growing season may be expected.

The supply of starch (resistant to rumen breakdown) at the small intestine is related to the rate of degradation, the rate of passage and the quantity fed.

Table 2.3 RELATIONSHIP BETWEEN STARCH RESISTANT TO RUMEN BREAKDOWN AND MATURITY IN MAIZE SILAGE

Maize variety	*Dry Matter (g/kg)*	*Starch (g/kg DM)*	*Proportion of resistant starch (%)*
Anjou	198	154	6
	237	270	17
	315	340	39
Scana	228	156	7
	275	233	17
	331	292	25

De Visser and Klop, unpublished data (1992)

Owens, Zinn and Kim (1986), Nocek and Tamminga (1991) and Harmon (1992) questioned the idea that intestinal capacity for starch digestion was unlimited. Dietary influences on small intestinal carbohydrases are apparently minimal and do not significantly increase the hydrolic capacity. Also, the residence time and surface area exposure may limit starch digestion in the small intestine. As a result, part of the rumen resistant starch may enter the large intestine and become fermented into volatiles or escape digestion completely and be excreted in the faeces. The mechanism of absorption and utilization of glucose from the small intestine remains unclear because it only partially enters the portal drained viscera (PDV) of dairy cows as glucose (Huntington, personal communication). Huntington and Reynolds (1986) and Kreikemeier, Harmon, Brandt, Avery and Johnson (1991) found a lower net flux of glucose entering the PDV when infusing starch rather than glucose, into the intestine of steers. This was probably the result of the limited hydrolic capacity of the intestine to digest starch. The role of glucose in proximity of the intestinal wall may also be important in this matter, but information is lacking.

Structural polysaccharide (NDF) degradability

The need for information concerning the degradation of structural polysaccharides (NDF) is increasing in importance since this can be used for the prediction of microbial protein synthesis (Chamberlain *et al.*, 1985; Rook *et al.*, 1987; van Straalen and Tamminga, 1991) and the supply of nutrients to the host animal (Murphy *et al.*, 1982; Dijkstra, Neal, France and Beever, 1992). Large differences occur between feedstuffs as is shown in Table 2.4.

However, information concerning ruminal degradation of NDF in concentrate ingredients and roughages remains limited. Large differences occur in the undegradable fraction between feedstuffs (roughage and concentrate ingredients), which influences the amount of substrate available for VFA production in the rumen.

Table 2.4 RESULTS OF NDF DEGRADATION IN COMPOUND FEEDSTUFFS AND ROUGHAGES

Feedstuff	*NDF* (*g/kg DM*)	*Fraction*[a] *D* (%)	*U* (%)	*Rate of degradation Kd* (%/h)
Sugar beet pulp	468	94	6	5.0
Maize bran	563	93	7	1.6
Barley	220	73	27	14.5
Maize gluten meal	349	86	14	6.5
Rice bran	197	27	73	4.8
Grass silage	446	89	11	5.9
Grass silage	641	76	24	3.9
Maize silage	441	66	34	1.8

[a] D = potentially degradable, U = undegradable;
Adapted from De Visser *et al.* (1992); Tamminga *et al.* (1990); Bosch (1991)

Feed intake in relation to carbohydrate composition

Feed intake is a constraining factor in high-level dairy performance (Bines, 1976). Milk yield is considered an important factor in controlling feed intake, due to its role in defining the physiological status of the animal (Vadiveloo and Holmes, 1979; Neal, Thomas and Cobby, 1984; Bosch, 1991) and the extraction of nutrients from the metabolic pool by the excretion of milk (MacRae *et al.*, 1988).

Especially in early lactation, animals are unable to increase DM intake to meet their energy requirements. Dairy cows at peak production may consume up to four times their maintenance requirements, but may have a production level of five times their maintenance requirements. Factors associated with the procurement and processing of feeds assume considerable importance in the lactating cow; these include nutritional and physical characteristics of feed as well as gastrointestinal and metabolic factors (Baile and Della-Fera, 1988). Carbohydrates are strongly related to these feed characteristics, digestion and metabolism.

The physical capacity of the rumen to digest roughages is very important in this matter. Conrad, Pratt and Hibbs (1964) found a negative relationship between the digestibility of roughage, which is strongly related to the cell wall fraction, and total DM intake. Jarrige, Demarquilly, Dulphy, Hoden, Robelin, Beranger, Geay, Journet, Malterre, Micol and Petit (1986), Mertens (1987) and Bosch (1991) used 'rumen fill' as a factor restricting feed intake. NDF content, particle size, degradability and the passage rate of undegraded particles are the main factors influencing rumen fill. Hoover (1986) reported a linear relationship between DM intake and dietary NDF concentration; Bosch (1991) found similar tendencies when increasing the NDF content of grass silage by using mature grass.

Low quality roughages, with fibre of low digestibility or a high undegradable fraction of the fibre (U) are more sensitive to a decrease in ruminal pH and show a more pronounced decrease in DM intake than roughages with higher quality fibre (Malestein and van't Klooster, 1986; De Visser, van der Togt and Tamminga, 1990).

Increasing the proportion of concentrates in dairy diets reduces the cellulolytic activity of the microbial population in the rumen (Porter, Balch, Coates, Fuller, Latham and Sharpe, 1972; Taylor and Aston, 1976; Russell, Sharp and Baldwin, 1979; Hiltner and Dehority, 1983; Hoover, 1986) and causes a shift towards amylolytic activity. This results from an increase in the amount of easily fermentable carbohydrate substrate (sugars, rumen degradable starch) causing a rapid fermentation in the rumen, increasing the total concentration of volatile fatty acids and thus reducing the pH of the rumen fluid. Therefore, optimal conditions for ruminal digestion aimed at maximization of total DM intake, have to provide a balance between a high level of cellulolytic activity, attempting to digest as much NDF (roughage) as possible (Russell and Sniffen, 1984), and a high rate of digestion of rapidly degradable concentrate carbohydrates (sugars, rumen degradable starch), which may induce rumen acidosis (Giesecke, Bartelmus and Stangassinger, 1976; Counotte, 1981).

Castle and Watson (1975), Meys (1985), Thomas, Aston, Daley and Bass (1986) and De Visser *et al.*, (1990) found reduced total DM intake when feeding concentrates high in rapidly degradable starch (barley, tapioca, wheat) compared with concentrates based on cell wall constituents (beet pulp, maize bran, rice bran, dried roughages). Effects were more pronounced when high concentrate levels were fed in combination with high amounts of rumen degradable starch. De Visser and de Groot (1980) found a dramatic reduction in intake and an increase in the number of animals suffering from rumen acidosis after feeding large quantities of concentrates high in rumen degradable starch immediately after parturition.

De Visser (1984) fed two levels of concentrates, with a starch content between 150 and 450 g/kg DM, and showed that the total amount of feed negatively influenced conditions for cellulolytic activity (increase in $H+$ concentration four hours after feeding). However, these effects were more pronounced when feeding concentrates high in rumen degradable starch (Figure 2.1).

Robinson, Tamminga and van Vuuren (1986) confirmed these findings, when feeding increasing total amounts of totally mixed rations, which were high or low in starch content. The greatest effects of rumen degradable starch content were present at the highest levels of intake. In particular, the diurnal pattern of ruminal pH was negatively influenced and showed a significantly longer period of values below 6.0. Since degradation was reduced NDF accumulated in the rumen.

As shown previously, the use of concentrates high in rumen degradable starch negatively influences rumen fermentation conditions, which may reduce NDF degradation, especially from roughage. These negative effects can be reduced by decreasing the substrate available for rumen fermentation by increasing the amount of rumen-resistant starch, or by changes in the feeding system.

Several researchers (Waldo, 1973; Malestein and van't Klooster, 1986; Casper and Schingoethe, 1989; Herrera-Saldena and Huber, 1989; McCarthy, Klusmeyer, Vicini, Clark and Nelson, 1989; De Visser, van der Togt, Huisert and Tamminga, 1993) showed that an increase in rumen-resistant starch content of concentrates, as compared to rumen degradable starch, positively influenced rumen fermentation conditions, depicted by a higher ruminal pH, reduced total VFA concentrations and an increase in the acetic:propionic acid ratio (Ørskov, 1975).

Feeding concentrates and roughage as totally mixed rations reduces the negative influence of high amounts of rumen degradable starch in compound feeds (Rohr and Schlunsen, 1986), because the animals are forced to eat slowly degradable as well as more rapidly degradable feedstuffs at the same time. Computerized systems

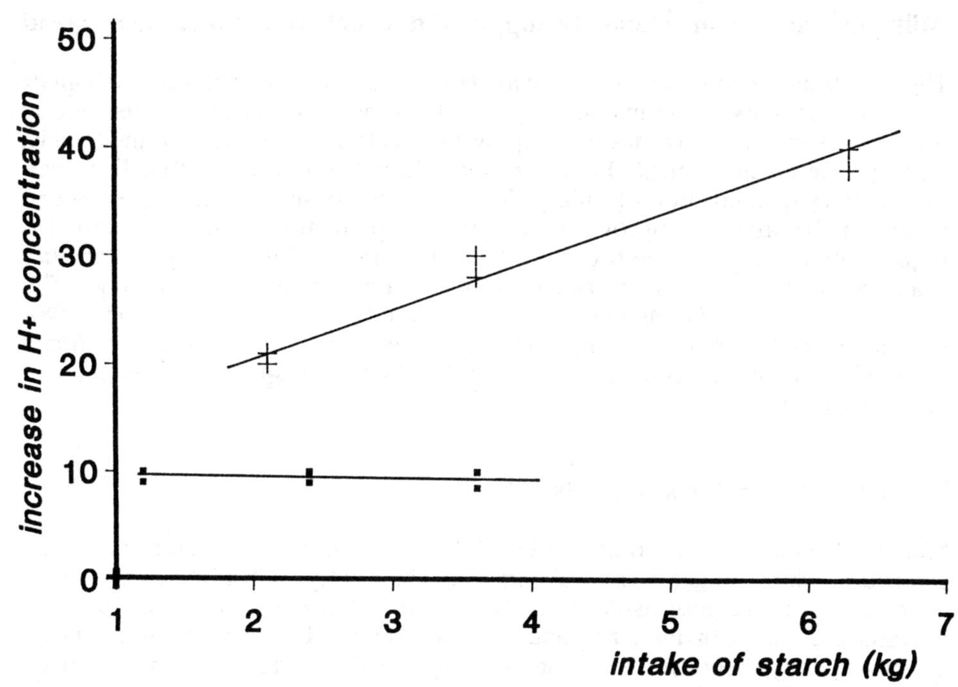

Figure 2.1 Increase in rumen fluid H+ concentration of dairy cows fed forage (8 kg DM/day) plus concentrate mixtures containing 150, 300 or 450 g rumen degradable starch/kg DM (8 kg DM/day = ■; 14 kg DM/day = +. Markers indicate range of 12 observations for each starch concentration)

in which concentrates can be fed in several small portions throughout the day also reduce the negative influence on rumen fermentation, by limiting the amount of rapidly fermentable sugars and rumen degradable starch ingested at each meal.

High moisture diets, based on moist ensiled roughages or moist by-products reduce total DM intake (Waldo, 1978; Steen and Gordon, 1980; Lahr, Otterby, Johnson, Linn and Lundquist, 1983; Rohr and Thomas, 1984; Gordon and Peoples, 1986; De Visser and Tamminga, 1987; Peoples and Gordon, 1989; De Visser and Hindle, 1990; De Visser and Hindle, 1992). The reduced intake is partly the result of an increase in the total bulk which must be eaten by the animals, due to the lower DM content of moist diets. In early lactation, the animals are further hindered by a limited intake capacity (Jarrige *et al.*, 1986).

The effects may also be related to the higher amounts of end products of fermentation (VFA, lactic acid) produced in the silo, replacing rapidly degradable carbohydrates (sugars, rumen degradable starch) and the increase in degradability of the nitrogen components. Nitrogen components and rumen degradable carbohydrates are out of balance in this situation, which in turn reduces the capacity of the rumen to digest slowly degradable NDF, due to a lack of energy (van Straalen and Tamminga, 1991; Hvelplund and Madsen, 1985).

Addition of small amounts of rumen digestible starch and/or sugars may improve microbial activity under these circumstance, which may partly reduce the negative effects on roughage DM intake (Tamminga, 1981).

Milk performance in relation to supply of nutrients to the mammary gland

High levels of milk production from dairy cows are impossible without an adequate supply of nutrients to the mammary gland. Currently, energy and protein intake are used as indicators of nutrient supply (Agricultural Research Council, 1984; National Research Council, 1988; Centraal Veevoederbureau, 1991). However, the accuracy of prediction of milk yield or milk composition using these systems is limited. In order to improve the accuracy of prediction, more information is required on the supply of nutrients to the mammary gland in terms of metabolites available for biosynthesis of each milk constituent (Oldham and Sutton, 1979; Larson and Smith, 1974; Mepham, 1982, 1988; MacRae *et al.*, 1988; Vernon, 1988). It is necessary to consider the roles of major precursors for milk production and their influence on milk constituents and milk volume in response to manipulation of dietary inputs.

Precursors for milk components

Most lipids in milk fat are triglycerides. A large quantity is obtained from plasma glycerides, which originate from dietary long chain fatty acids (fats) or mobilized energy reserves (adipose tissue) in early lactation. The remainder of the milk fat is synthesized in the mammary gland from acetate and ß-hydroxybutyrate, which are provided by carbohydrate fermentation (rumen degradable starch and NDF) in the rumen (ketogenic precursors).

Milk lactose is synthesized from glucose and UDP-galactose, both derived from glucose either directly or from gluconeogenesis from amino acids or propionic acid. Glucose is provided by digestion of starch and its absorption in the small intestine or from propionic acid provided by carbohydrate fermentation (rumen degradable starch and NDF) in the rumen (glucogenic precursor).

Milk proteins are mainly synthesized in the mammary gland from the free amino acids pool (DePeters and Cant, 1992). The remainder is supplied by pre-formed blood proteins (DePeters and Cant, 1992). Amino acids are provided by the microbial protein synthesized in the rumen and from rumen undegradable protein (aminogenic precursor) absorbed in the small intestine.

The ketogenic, glucogenic and aminogenic precursors absorbed from the gastrointestinal tract are not only used for milk production. After liver metabolism, they are partitioned for maintenance requirements, the mammary gland, reproduction, muscle tissue and adipose tissue. Several researchers have created models relating the availability of precursors from the gastro-intestinal tract with their metabolism and partitioning for different tissues (Baldwin, France, Beever, Gill and Thornley, 1987a; Baldwin, France and Gill, 1987b; Baldwin, Thornley and Beever, 1987c; Danfaer, 1991). These models are complex and have unfortunately not been evaluated for their accuracy in predicting milk production due to limited *in vivo* data from high yielding dairy cows, as well as their limitations in predicting nutrient supply (Neal *et al.*, 1992). However, in the future more emphasis must be given to the validation of these models and, where possible, improvements should be made because they are very useful in screening and understanding the relationships between feeding, nutrient supply and metabolism in dairy cows.

The ketogenic, glucogenic and aminogenic nutrient supply available for milk production can be altered by changing the type of carbohydrate, the rate and extent of degradation and balancing the rate and extent of degradation in the rumen between carbohydrates and protein.

Milk fat content

Feeding large quantities of concentrates containing rumen-degradable starch, instead of rumen-degradable NDF, increases amylolytic activity in the rumen, inducing a higher concentration of propionic acid in rumen fluid (Thomas *et al.*, 1986; Tremere, Merrill and Loosli, 1968; Lees, Garnsworthy and Oldham, 1982; Sutton, Bines, Napper, Willis and Schuler, 1984; Thomas, Aston, Daley and Hughes, 1984; De Visser *et al.*, 1990; Valk and Hobbelink, 1992). Cows fed these rumen-degradable starch diets showed decreased levels in milk fat content and milk fat yield, compared to NDF-rich diets. Casse, Huntington and Rulquin (1993a,b) found a similar decrease in milk fat content in dairy cows, when propionic acid was infused into the blood of portal drained viscera (PDV). Total net flux of propionate was increased in PDV, as was total liver net flux (splanchnic) of glucose, initiating an increase in glucogenic precursors available to the mammary gland.

Westerhuis and De Visser (1974) and van Beukelen (1983), feeding flaked maize containing very rapidly rumen degradable starch, found dramatic decreases in acetic:propionic acid ratios in the rumen which resulted in extremely low milk fat contents (<20 g/kg milk) and yields.

Feeding large quantities of concentrates containing rumen-resistant starch, instead of rumen degradable starch, has a positive influence on rumen fermentation because the starch escapes rumen fermentation, reducing the supply of rumen degradable organic matter. Malestein and van't Klooster (1986), Waldo (1973), Sutton (1985), De Visser *et al.*, (1990), De Visser *et al.*, (1993), Casper and Schingoethe (1989), Herrera-Saldena, Gomez-Alarcon, Torabi and Huber (1990) and McCarthy *et al.*, (1989) fed diets, which differed in the rumen undegradable/ rumen degradable starch content. The milk fat content was lowest on rumen undegradable starch diets, compared to rumen degradable starch diets. Measurements of the rumen fermentation characteristics showed higher pH, reduced concentrations of total VFA and higher acetic:propionic acid ratios in rumen fluid. All of these are indications of reduced amylolytic activity in the rumen of cows fed rumen undegradable starch. However, milk fat content on these diets indicated an increase in glucogenic precursors to the mammary gland. Calculations with data obtained from Huntington (personal communication), measuring net flux in PDV and total splanchnic tissue, showed an increase in the net flux of glucose over total splanchnic tissue when feeding rumen undegradable starch instead of rumen degradable starch, initiating an increase in supply of glucogenic precursors to the mammary gland. However, the amount of glucose appearing in the peripheral blood was approximately 25% of the expected amount of glucose, using average degradability values (Centraal Veevoederbureau, 1991). The variation in rumen degradability within products can not explain these differences. Probably a larger amount of rumen undegradable starch escapes digestion in the small intestine (Harmon, 1992), or glucose is used by intestinal tissue as an energy source, or microbes attached to the gut wall use part of the glucose. Further research is required in this area. However, Nocek and Tamminga (1991) reviewed some feeding experimental data, which contradict these results. Stage of lactation, basal diets and protein supplementation are the main factors involved.

Manipulation of milk fat output from dairy cows can only be initiated by changing the rumen degradable starch:NDF ratio, or the rumen undegradable: rumen degradable starch ratio, when large amounts of concentrates are fed. Changes in rumen fermentation or the balance between rumen fermentation and intestinal digestion are limited when small amounts of concentrates are fed. Under such circumstances the substrate pools will not be changed much, nor will be the supply of nutrients to the mammary gland.

Milk protein content

Although milk protein is synthesized from aminogenic precursors (amino acids), carbohydrates are related to milk protein output. Firstly, carbohydrates are involved in microbial protein synthesis in the rumen and secondly, a limited supply of propionic acid or glucose during negative energy balance stimulates the use of amino acids as glucose precursors (gluconeogenesis), which may reduce the availability of amino acids for milk protein synthesis. Milk protein output appears difficult to manipulate. However, during the lactation cycle several feeding conditions reduce the milk protein output. Protein content of milk rapidly decreases after parturition and lowest values are often measured at the peak of lactation. Broster and Thomas (1981) showed a negative relationship between energy intake and milk protein output. Under these circumstances, a lower substrate supply of nitrogen and carbohydrates reduces total microbial protein synthesis in the rumen and the availability of rumen undegradable protein, both reducing the supply of aminogenic precursors to the mammary gland.

Rumen degradable starch or rumen digestible NDF in concentrates have similar effects on milk protein content and yield when equal amounts of rumen degradable organic matter is supplied to the rumen microbes (De Visser *et al.*, 1990; Thomas *et al.*, 1986; Sutton *et al.*, 1984). Comparisons were made between barley starch and beet pulp. However, De Visser *et al.* (1990), found lower milk protein content and yield when a rumen degradable starch (barley) and NDF (maize bran) were compared. Differences in degradation characteristics between beet pulp and maize bran were found. In this matter the solubility of protein and carbohydrates (S), as well as the rate of degradation of the ruminal digestible fraction (k_d) have to be taken into account. Chamberlain, Thomas and Quig (1986) and Rooke *et al.*, (1987) showed an increase in microbial protein synthesis in the rumen when the rate of degradation of the protein and the carbohydrates were in balance. Under these more balanced conditions the aminogenic supply to the mammary gland may be improved.

Feeding more rumen undegradable starch instead of rumen degradable starch decreased the supply of energy available to the rumen microbes, which may result in lower microbial protein synthesis. However, milk protein content and yield were not affected (De Visser *et al.*, 1990; Waldo, 1973). The higher supply of glucose to the small intestine, when feeding rumen undegradable starch, may have reduced the utilization of amino acids by the intestinal wall, which would compensate for the reduction in microbial protein synthesis in the rumen. Huntington (personal communication) found tendencies towards higher net fluxes in PDV of α-amino-N, when infusing starch into the small intestine of steers. Although not confirmed by experiments with high yielding dairy cows in early lactation, the extra glucose available at the wall of the small intestine may increase the supply of amino acids to the mammary gland at low energy intakes.

Feeding moist diets (roughage and/or by-products) often decreases milk protein yield and content (De Visser and Tamminga, 1987; De Visser and Hindle, 1990; De Visser and Hindle, 1992) due to a reduced energy supply available for microbial protein synthesis (carbohydrates versus fermentation end products) and an increase in the extent of degradation of protein from ensiled products. However, results from Gordon and Peoples (1986), Peoples and Gordon (1989), Steen and Gordon (1980) and Zimmer and Wilkins (1984), comparing wilted grass silage and moist grass silage did not show lower milk protein yields and contents, although these experiments were performed with concentrate supplements based on barley, which may have offset the loss of rapidly fermentable carbohydrates in roughages.

Partitioning of aminogenic precursors is related to the stage of lactation. Meys (1985) found higher milk protein content and yield when increasing the energy supply in the rumen of cows fed fresh herbage (*Lolium perenne*) in the second half of lactation, by the addition of carbohydrates (Table 2.5). Valk, van Vuuren and Langelaar (1992) repeated this experiment in early lactation and found an increase in milk yield (lactose production) and milk protein yield, but milk protein content was not affected.

Table 2.5 MILK PERFORMANCE ON HERBAGE DIETS BALANCED FOR NITROGEN:CARBOHYDRATE COMPOSITION BY ADDING CARBOHYDRATES (STARCH AND NDF)

	Meys (1985)		*Valk, van Vuuren and Langelaar (1992)*	
Period of lactation	*Late*		*Early*	
Diet	*Grass*	*Grass/maize*	*Grass*	*Grass/maize*
DM intake (kg/day)	12.9	14.4	18.6	18.7
Milk yield (kg/day)	22.0	22.8	26.9	29.7
Milk protein (g/kg)	32.9	34.2	31.4	31.1
Milk protein yield (g/day)	683	779	845	923

Both experiments showed an increase in microbial protein synthesis. The animals in early lactation were able to channel the extra protein supplied to the mammary gland towards milk lactose and milk protein yield, while the animals in the later stage of lactation were unable to stimulate milk yield.

Conclusion

Carbohydrates are very important ingredients in dairy diets. Nutritionists use empirical methods to divide carbohydrates into various fractions (sugars, starch and cell wall constituents). Modern methods can quantify the chemical composition as well as physical structures and bonds. However these methods are not yet available for practical use. Carbohydrates vary in complexity, due to chemical composition and rumen and intestinal digestion. Carbohydrates may be ranked according to these characteristics, which in turn may alter rumen fermentation, intestinal digestion and the supply of nutrients to the mammary gland.

Due to these differences in supply of nutrients to the mammary gland of the cow, milk lactose, milk fat and milk protein production may be affected by altering dietary carbohydrates.

References

Agricultural Research Council (1984) *The Nutrient Requirements of Ruminant Livestock, Supplement 1.* Commonwealth Agricultural Bureau

Aman, P. and Graham, H. (1990) Chemical evaluation of polysaccharides in animal feeds. In *Feedstuff Evaluation*, pp. 161–178. Edited by J. Wiseman and D.J.A. Cole. London: Butterworths

Baile, C.A. and Della-Fera, M.A. (1988) Physiology of control of food intake regulation of energy balance in dairy cows. In *Nutrition and Lactation in the Dairy Cow*, pp. 251–261. Edited by P.C. Garnsworthy. London: Butterworths

Baldwin, R.L., France, J., Beever, D.E., Gill, M. and Thornley, J.M. (1987a) Metabolism of the lactating cow. III. Properties of mechanistic models suitable for evaluation of energetic relationships and factors involved in the partition of nutrients. *Journal of Dairy Research*, **54**, 133–145

Baldwin, R.L., France, J. and Gill, M. (1987b) Metabolism of the lactating cow: I. Animal elements of a mechanistic model. *Journal of Dairy Research*, **54**, 77–105

Baldwin, R.L., Thornley, J.H.M. and Beever, D.E. (1987c) Metabolism of the lactating cow. II Digestive elements of a mechanistic model. *Journal of Dairy Research*, **54**, 107–131

Barber, G.D., Offer, N.W. and Givens, D.I. (1989) Predicting the nutritive value of silage. In *Recent Advances in Animal Nutrition — 1989*, pp. 141–158. Edited by W. Haresign and D.J.A. Cole. London: Butterworths

Bines, J.A. (1976) Regulation of food intake in dairy cows in relation to milk production. *Livestock Production Science*, **3**, 115–128

Boon, J.J. (1989) An introduction to pyrolysis mass spectrometry of lignocellulosic material: case studies on barley, straw, corn stem and Agropyron. In *Physico-chemical characterisation of plant residues for industrial and feed use*, pp. 25–49. Edited by A. Chesson and E.R. Ørskov. Essex: Elsevier Applied Science

Bosch, M.W. (1991) *Influence of stage of maturity of grass silages on digestion processes in dairy cows.* Thesis of the Agricultural University, Wageningen, 150 pp

Broster, W.H. and Thomas, C. (1981) The influence of the level and pattern of concentrate input on milk output. In *Recent Advances in Animal Nutrition*, pp. 49–69. Edited by W. Haresign. London: Butterworths

Casper, D.P. and Schingoethe, D.J. (1989) Lactational response of dairy cows to diets varying in ruminal solubilities of carbohydrates and crude protein. *Journal of Dairy Science*, **72**, 928–941

Casse, E.A., Huntington, G.B. and Rulquin, H. (1993a) Effect of mesenteric infusion of propionate on splanchnic metabolism in first-lactation Holstein heifers. 1. Net uptake or output of fatty acids, ketones and oxygen, *Journal of Dairy Science*, in press

Casse, E.A., Huntington, G.B. and Rulquin, H. (1993b) Effect of mesenteric vein infusion of propionate on splanchnic metabolism in first-lactation Holstein heifers. 2. Net uptake or output of glucose, l-lactate, nitrogenous compounds and hormones, *Journal of Dairy Science*, in press

Castle, M.E. and Watson, J.N. (1975) Silage and milk production. A comparison between barley and dried grass as supplements to silage of high digestibility. *Journal of the British Grassland Society*, **30**, 217–222

Chamberlain, D.G. and Robertson, S. (1989) In *Silage for Milk Production. Occasional Symposium no 23*, pp. 187–195. Edited by C.S. Mayne. British Grassland Society

Chamberlain, D.G., Thomas, P.C. and Quig, J. (1986) Utilization of silage nitrogen in sheep and cows: amino acid composition of duodenal digesta and rumen microbes. *Grass and Forage Science*, **41**, 31–38

Chamberlain, D.G., Thomas, P.C., Wilson, W., Newbold, C.J. and MacDonald, J.C. (1985) The effects of carbohydrate supplements on ruminal concentrations of ammonia in animals given diets of grass silage. *Journal of Agricultural Science, Cambridge*, **104**, 331–340

Centraal Veevoederbureau (1991) De Veevoedertabel, Centraal Veevoederbureau, Lelystad, The Netherlands

Cone, J.W. (1991) Degradation of starch in feed concentrates by enzymes, rumen fluid and rumen enzymes. *Journal the Science of Food and Agriculture*, **54**, 23–34

Cone, J.W., Tas, A.C. and Wolters, M.G.E. (1992) Pyrolysis mass spectrometry (PyMS) and degradability of starch granules. *Stärke*, **44**, 55–58

Conrad H.R., Pratt, A.D. and Hibbs, J.W. (1964) Regulation of feed intake in dairy cows. 2. The yield and composition of milk. *Agricultural Systems*, **1**, 261–279

Counotte, G.H.M. (1981) *Regulation of Lactate Metabolism in the Rumen*. Thesis of the State University, Utrecht, The Netherlands

Danfaer, A. (1990) A dynamic model of nutrient digestion and metabolism in lactating dairy cows. *Beretning fra Statens Husdyrbruksforsog*, pp. 1–480. Foulum

DePeters, E.J. and Cant, J.P. (1992) Nutritional factors affecting the nitrogen composition of milk: A review. *Journal of Dairy Science*, **75**, 2043–2070

De Visser, H. and de Groot, A.M. (1980) The influence of the starch and sugar content of concentrates on feed intake, rumen fermentation, production and composition of milk. In *Proceedings of the 4th International Conference on Production Disease in Farm Animals*, pp. 41–48. Edited by D. Giesecke, G. Dirksen and M. Stangassinger Munich: Tierarzliche Fakultat

De Visser, H. (1984) Krachtvoer voor hoogproduktief melkvee in rantsoenen met snijmais. *Bedrijfsontwikkeling*, **15**, 383–388

De Visser, H. and Tamminga, S. (1987) Influence of wet versus dry by-product ingredients and addition of branched chain fatty acids and valerate to dairy cows. 1. Feed intake, milk production and milk composition. *Netherlands Journal of Agricultural Science*, **35**, 163–175

De Visser, H. and Steg, A. (1988) Utilization of by-products for dairy cow feeds. In *Nutrition and Lactation in the Dairy Cow* pp. 378–394. Edited by P.C. Garnsworthy. London: Butterworths

De Visser, H. and Hindle, V.A. (1990) Dried beet pulp, pressed beet pulp and maize silage as substitutes for concentrates in dairy cow rations. 1. Feeding value, feed intake, milk production and milk composition. *Netherlands Journal of Agricultural Science*, **38**, 77–88

De Visser, H., van der Togt, P.L. and Tamminga, S. (1990) Structural and non-structural carbohydrates in concentrate supplements of silage-based dairy cow rations. 1. Feed intake and milk production. *Netherlands Journal of Agricultural Science*, **38**, 487–498

De Visser, H. and Hindle, V.A. (1992) Autumn-cut grass silage as roughage component in dairy cow rations. 1. Feed intake, digestibility and milk performance. *Netherlands Journal of Agricultural Science*, **40**, 147–158

De Visser, H, van der Togt, P.L., Huisert, H. and Tamminga, S. (1993) Structural and non-structural carbohydrates in concentrate supplements of silage based dairy cow rations. 2. Rumen degradation, fermentation and kinetics. *Netherlands Journal of Agricultural Science*, **40**, in press

Dijkstra, J., Neal, H.D.St.C., France, J. and Beever, D.E. (1992) Simulation of nutrient digestion, absorption and outflow in the rumen: Model description. *Journal of Nutrition*, **122**, 2239–2256

DLG-Futterwert Tabellen fur Wiederkauer (1991) DLG-verlag, Frankfurt am Main

Ely, R.E. and Moore, L.A. (1955) Hollocellulose and summative analysis of forages. *Journal of Animal Science*, **14**, 718–724

Frame, J., Harkess, R.D. and Talbot, M. (1989) The effect of cutting frequency and fertilizer nitrogen rate on herbage productivity from perennial rye grass. *Research and Development in Agriculture*, **62**, 99–105

Giesecke, D., Bartelmus, G. and Stangassinger, M. (1976) Untersuchungen zur Genese und Biochemie der Pansen Acidose. *Zentralblatt für Veterinar Medizin A*, **23**, 353–363

Givens, D.I., Everington, J.M. and Adamson, A.H. (1989a) The nutritive value of spring-grown herbage produced on farms throughout England and Wales over four years. I. The effect of stage of maturity and other factors on chemical composition, apparent digestibility and energy values *in vivo*. *Animal Feed Science and Technology*, **27**, 157–172

Givens, D.I., Everington, J.M. and Adamson, A.H. (1989b) The nutritive value of spring-grown herbage produced on farms throughout England and Wales over four years. II. The prediction of apparent digestibility *in vivo* from various laboratory measurements. *Animal Feed Science and Technology*, **27**, 173–184

Givens, D.I., Everington, J.M. and Adamson, A.H. (1989c) The nutritive value of spring-grown herbage produced on farms throughout England and Wales over four years. III. The prediction of energy values from various laboratory elements.various laboratory measurements. *Animal Feed Science and Technology*, **27**, 185–196

Gordon, F.J. and Peoples, A.C. (1986) The utilization of wilted and unwilted silages by lactating cows and the influence of changes in the protein and energy concentration of the supplement offered. *Animal Production*, **43**, 355–366

Goering, H.K. and van Soest, P.J. (1970) *Forage Fiber Analysis*. Agricultural Handbook No 379, ARC, USDA, Washington, 1–12

Harmon, D.L. (1992) Dietary influences on carbohydrases and small intestine over four years. I. The effect of stage of maturity and other factors on starch hydrolysis capacity in ruminants. *Journal of Nutrition*, **122**, 203–210

Herrera-Saldena, R. and Huber, J.T. (1989) Influence of varying protein and starch degradabilities on performance of lactating cows. *Journal of Dairy Science*, **72**, 1477–1483

Herrera-Saldena, R., Gomez-Alarcon, R., Torabi, M. and Huber, J.T. (1990)

Influence of synchronizing protein and starch degradation in the rumen on nutrient utilization and microbial protein synthesis. *Journal of Dairy Science*, **73**, 142–148

Hiltner, P. and Dehority, B.A. (1983) Effect of soluble carbohydrates on digestion of cellulose by pure cultures of rumen bacteria. *Applied and Environmental Microbiology*, **46**, 642–648

Hoover, W.H. (1986) Chemical factors involved in ruminal fiber digestion, *Journal of Dairy Science*, **69**, 2755–2766

Huntington, G.B. and Reynolds, P.J. (1986) Net absorption of glucose, L-lactate, volatile fatty acids, and nitrogenous compounds by bovine given abomasal infusions of starch or glucose. *Journal of Dairy Science*, **69**, 2428–2436

Hvelplund, T. and Madsen, J. (1985) Amino acid passage to the small intestine in dairy cows compared with estimates of microbial protein and undegraded dietary protein from analysis on the feed. *Acta Agriculture of Scandinavia Supplement*, **25**, 21–36

Jarrige, R., Demarquilly, C., Dulphy, J.P., Hoden, A., Robelin, J., Beranger, C., Geay, Y., Jounet, M., Malterre, C., Micol, D. and Petit, M. (1986) The INRA 'fill unit' system for predicting the voluntary intake of forage-based diets in ruminants: a review. *Journal of Animal Science*, **63**, 1737–1758

Kreikemeier, K.K., Harmon, D.L., Brandt, R.T., Avery, T.B. and Johnson, D.E. (1991) Small intestinal starch digestion in steers: effect of various levels of abomasal glucose, corn starch and corn dextrin infusion on small intestinal disappearance and net glucose absorption. *Journal of Animal Science*, **69**, 328–335

Lahr, D.A., Otterby, D.E., Johnson, D.G., Linn, J.G. and Lundquist, G. (1983) Effects of moisture content of complete diets on feed intake and milk production by cows. *Journal of Dairy Science*, **66**, 1891–1900

Larson, B.L. and Smith, R.S. (1974) *The Biosynthesis of Milk, Lactation, a Comprehensive Treatise, Part 1.* pp. 1–143. London and New York: Academic Press

Lees, J.A., Garnsworthy, P.C. and Oldham, J.D. (1982) The response of dairy cows in early lactation to supplements of protein given with rations designed to promote different patterns of rumen fermentation. *British Society of Animal Production, Occasional publication*, **6**, 157–159

MacRae, J.C, Buttery, P.J. and Beever, D.E. (1988) Nutrient interactions in the dairy cow. In *Nutrition and Lactation in the Dairy Cow*, pp. 55–75., Edited by P.C. Garnsworthy. London: Butterworths

Malestein, A. and van't Klooster, A.Th. (1986) Influence of ingredient composition of concentrates on rumen fermentation rate *in vitro* and *in vivo* and on roughage intake of dairy cows. *Journal of Animal Physiology and Animal Nutrition*, 551–13

McCarthy, R.D., Klusmeyer, T.H., Vicini, J.L., Clark, J.H. and Nelson, D.R. (1989) Effect of source of protein and carbohydrate on ruminal fermentation and passage of nutrients to the small intestine of lactating cows. *Journal of Dairy Science*, **72**, 2002–2016

McDonald, P. (1983) Hannah Research Institute Annual Report, p. 59–67

Mehrez, A.Z. and Ørskov, E.R. (1977) A study of the artificial fibre bag technique for determining the digestibility of feeds in the rumen. *Journal of Agricultural Science, Cambridge*, **88**, 645–650

Meier, H. and Reid, J.S.G. (1982) Reserve polysaccharides other than starch in higher plants. In *Plant Carbohydrates 1. Intracellular Carbohydrates*, pp. 418–471 Edited by F.A. Loewus and W. Tanner. Berlin: Springer Verlag

Mepham, T.B. (1982) Amino acid utilization by lactating mammary gland. *Journal of Dairy Science*, **65**, 289–298

Mepham, T.B. (1988) Nutrient uptake by the mammary gland. In *Nutrition and Lactation in the Dairy Cow*, pp. 15–31. Edited by P.C. Garnsworthy. London: Butterworths

Merry, R.J. and Braithwaite, G.D. (1987) The effect of enzymes and inoculants on chemical and microbiological composition of grass and legumes. In *Proceedings of the 8th Silage Conference, Institute of Grassland and Animal Production, Hurley*, pp. 27–28

Mertens, D.R. (1987) Predicting intake and digestibility using mathematical models of ruminal function. *Journal of Animal Science*, **64**, 1548–1558

Meys, J.A.C. (1985) Comparison of starchy and fibrous concentrates for grazing dairy cows. *British Grassland Society, Occasional Symposium no 19*, pp. 129–137. Hurley

Murphy, M.R., Baldwin, R.L. and Koong, L.J. (1982) Estimation of stoichiometric parameters for rumen fermentation of roughage and concentrate diets. *Journal of Animal Science*, **55**, 411–421

National Research Council (1988) *Nutrient Requirements of Dairy Cattle*. Washington DC: National Academy Press

Neal, H.D.St.C., Thomas, C. and Cobby, J.M. (1984) Comparison of equations predicting voluntary intake by dairy cows. *Journal of Agricultural Science, Cambridge*, **103**, 1–10

Neal, H.D.St.C., Dijkstra, J. and Gill, M. (1992) Simulation of nutrient digestion, absorption and outflow in the rumen: Evaluation. *Journal of Nutrition*, **122**, 2257–2272

Nocek, J.E. and Tamminga, S. (1991) Site of digestion of starch in the gastrointestinal tract of dairy cows and its effect on milk yield and composition. *Journal of Dairy Science*, **74**, 3598–3629

Oldham, J.D. and Sutton, J.D. (1979) Milk composition and the high yielding dairy cow. In *Feeding Strategy for the High Yielding Dairy Cow*, p. 114–147. Edited by W.H. Broster and H. Swan. London: Granada

Ørskov, E.R. (1975) Manipulation of rumen fermentation for maximum food utilisation. *World Review of Nutrition and Dietics*, **22**, 152–182

Owens, F.N., Zinn, R.A. and Kim, Y.K. (1986) Limits to starch digestion in the ruminant small intestine. *Journal of Animal Science*, **63**, 1634–1648

Peoples, A.C. and Gordon, F.J. (1989) The influence of wilting and season of silage harvest and the fat and protein concentration of the supplement on milk production and food utilization by lactating cattle. *Animal Production*, **48**, 305–317

Porter, J.W.G., Balch, C.C., Coates, M.E., Fuller, R., Latham, M.J. and Sharpe, M. (1972) The influence of the gut flora on the digestion, absorption and metabolism of nutrient in animals *Biennial Reviews*, pp. 13–36, National Institute for Research and Dairying, Reading

Robinson, P.H., Tamminga, S. and van Vuuren, A.M. (1986) Influence of declining level of feed intake and varying the proportion of starch in the concentration rumen fermentation in dairy cows. *Livestock Production Science*, **15**, 173–189

Rohr, K. and Thomas, C. (1984) In *Eurowilt*. Edited by E. Zimmer and R.J. Wilkins Landbauforschung Völkenrode, Sonderheft **69**, 1–70

Rohr, K. and Schlunsen, D. (1986) The bearing of feeding methods on digestion and performance of dairy cow's. In *New Developments and Future Perspectives in Research on Rumen Function*, pp. 227–242. Edited by Neiman-Sorensen. Brussels: ECSC-EEG-EAEC

Rooke, J.A., Lee, N.H. and Armstrong, D.G. (1987) *Journal of Agricultural Science, Cambridge*, **107**, 263–272

Russell, J.B., Sharpe, W.M. and Baldwin, R.L. (1979) The effect of pH on maximal bacterial growth rate and its possible role as a determinant of bacterial competition in the rumen. *Journal of Animal Science*, **48**, 251–255

Russell, J.B. and Sniffen, C.J. (1984) Effect of carbon-4 and carbon-5 volatile fatty acids on growth of mixed rumen bacteria *in vitro*. *Journal of Dairy Science*, **67**, 987–994

Salette, J. (1982) The role of fertilizers in improving herbage quality and optimization of its utilization. In *Proceedings of the 12th IPI-Congress, Goslar*, 117–144

Spoelstra, S.F. (1991) Chemical and biological additives in forage conservation. In *Forage Conservation Towards 2000*. Edited by G. Pahlow and H. Honig, Landbauforschung Völkenrode, Sonderheft **123**, 48–70

Steen, R.W.J. and Gordon, F.J. (1980) The effect of type of silage and level of concentrate supplementation offered during early lactation on total lactation performance of January/February calving cows. *Animal Production*, **30**, 341–354

Sutton, J.D. (1985) Digestion and absorption of energy substrates in the lactating cow. *Journal of Dairy Science*, **68**, 3376–3393

Sutton, J.D., Bines, J.A., Napper, D.J., Willis, J.M. and Schuller, E. (1984) Ways of improving the efficiency of milk production and of altering milk composition by manipulation of concentrate feeding. In *Annual Report 1983*, pp. 74–75, National Institute for Research in Dairying, Reading

Tamminga, S. (1981) *Nitrogen and Amino Acid Metabolism in Dairy Cows*. Thesis of the Agricultural University, Wageningen, The Netherlands

Tamminga, S, Robinson, P.H., Vogt, M. and Boer, H. (1989) Rumen ingesta kinetics of cell wall components in dairy cows. *Animal Feed Science and Technology*, **25**, 89–98

Tamminga, S, van Vuuren, A.M., van der Koelen, C.J., Ketelaar, R.S. and van der Togt, P.L. (1990) Ruminal behaviour of structural carbohydrates, non-structural carbohydrates and crude protein from concentrate ingredients in dairy cows. *Netherlands Journal of Agricultural Science*, **38**, 513–526

Taylor, J.C. and Aston, K. (1976) Milk production from diets of silage and dried forage. *Animal Production*, **23**, 197–209

Thomas, C., Aston, K., Daley, S.R. and Hughes, P.M. (1984) The effect of composition of concentrate on the voluntary intake and milk output. *Animal Production*, **38**, 519

Thomas, C. and Thomas, P.C. (1985) Factors affecting the nutritive value of grass silage In *Recent Advances in Animal Nutrition — 1985*, pp. 223–256. Edited by W. Haresign and D.J.A. Cole. London: Butterworths

Thomas, C., Aston, K., Daley, S.R. and Bass, J. (1986) Milk production from silage. 4. The effect of the composition of the supplement. *Animal Production*, **42**, 315–325

Tremere, A.W., Merrill, W.G. and Loosli, J.K. (1968) Adaptation to high concentrate feeding as related to acidose and digestive disturbances in dairy heifers. *Journal of Dairy Science*, **51**, 1065–1072

Vadiveloo, J. and Holmes, W. (1979) The prediction of voluntary feed intake of dairy cows. *Journal of Agricultural Science, Cambridge*, **93**, 553–562

Valk, H. and Hobbelink, M.E.J. (1992) Bijvoedering aan grazende melkkoeien. De invloed van grasaanbod en soort bijvoer (maiskolvensilage, pulp en mais) op de gras-, bijvoeropname en melkproduktie van weidende melkkoeien. *IVVO Rapport 238*

Valk, H., van Vuuren, A.M. and Langelaar, S.J. (1992) Bijvoeren verhoogt voederopname en melkproduktie en verlaagt de stikstofuitscheiding in de urine. *Mededelingen IVVO-DLO no. 8*

Van Beukelen, P. (1983) *Studies on Milk Fat Depression in High Producing Dairy Cows*. Thesis of the State University, Utrecht, The Netherlands, 268 pp.

Van der Schans, A., Westerlaken, L.A.J., van der Wilt, A.A.C., Eyestone, W.H. and Boer, H.A. (1991) Ultra-sound guided transvaginal collection of oocytes in the cow. In *Proceedings Annual Conference of the International Embryo Transfer Society*, Bournemouth, England, *Theriogenology*, **35**, 288–296

Van Soest, P. (1982) *Nutritional Ecology of the Ruminant*. 374 pp Corvallis, OR, USA, O & E Books

Van Straalen, W.M. and Tamminga, S. (1991) Protein degradation of ruminant diets. In *Feedstuff Evaluation*, p. 55–72. Edited by J. Wiseman and D.J.A. Cole. London: Butterworths

Van Vuuren, A.M., Krol-Kramer, F., van der Lee, R.A. and van Beers, J.A.C. (1989) Effects of addition of cell wall degrading enzymes on the chemical composition and the *in sacco* degradation of grass silage. *Grass and Forage Science*, **44**, 223–230

Verite, R., Michalet-Doreau, B., Chapoutot, P., Peyraud, J.L. and Poncet, C. (1987) Revision du systeme des proteine digestables dans l'intestin (PDI). *Bulletin Technique CRVZ*, **70**, 19–34

Vernon, R.G. (1988) The partition of nutrients during the lactation cycle. In *Nutrition and Lactation in the Dairy Cow*, pp. 32–52. Edited by P.C. Garnsworthy. London: Butterworths

Waldo, D.R. (1973) Extent and partition of cereal grain starch digestion in ruminants. *Journal of Animal Science*, **37**, 1062–1074

Waldo, D.R. (1978) The use of direct acidification in silage production, In *Fermentation of Silage a Review*, pp. 120–180. Edited by M.E. McCullough. Iowa, USA: National Feed Ingredients Association

Westerhuis, J.H. and de Visser, H. (1974) De invloed van de krachtvoersamenstelling op fermentatie stoornissen in de pens bij melkkoeien na de partus en tijdens de verdere lactatieperiode. *Intern Rapport IVVO no 74* 34 pp

Zimmer, E. and Wilkins, R.J. (1984) Efficiency of silage systems: a comparison between unwilted and wilted silages. In *Eurowilt*. Edited by E. Zimmer and R.J. Wilkins. Landbauforschung Völkenrode, Sonderheft **69**, 71–72

3

CHARACTERISATION OF NITROGEN IN RUMINANT FEEDS

B.R. COTTRILL
ADAS, Starcross, Devon, U.K.

Introduction

During the last decade or so a number of new protein evaluation systems have been published, aimed at improving our estimation of requirements of ruminant livestock for dietary nitrogen (N) and introducing new ways of describing dietary N to permit diet formulation to meet these requirements (Agricultural Research Council, 1980; 1984; Institut National de la Recherche Agronomique, 1978; 1988; Madsen, 1985; National Research Council, 1985; Rohr, 1987; CSIRO, 1990; Fox, Sniffen, O'Conner, Russell and van Soest, 1992; Agricultural and Food Research Council, 1992). A common feature of all of them is that they differentiate between dietary N that is degraded in the rumen, and that which escapes degradation and passes to the small intestine for digestion by the animal. It follows, therefore, that to successfully implement any of these systems requires an ability to measure, or predict, the degree to which dietary N is degraded in the rumen, and the digestibility of the undegraded nitrogen (UDN).

The total 'protein' in feedingstuffs (N × 6.25) generally consists of true proteins and non-protein nitrogen (NPN). A number of different types of protein have been identified in ruminant feeds (van Straalen and Tamminga, 1990), and are generally classified according their solubility in different solutions, or their function in the plant. NPN may account for up to 30% of total nitrogen in forages (Tamminga, 1986). The proportions of the different proteins and NPN fractions vary considerably between feeds, and this will be a major factor in influencing the degree and rate at which they are degraded. There are, however, insufficient data available to allow requirements to be expressed, or feeds characterised on this basis, and consequently all the protein systems currently proposed describe feeds in terms of total N (or crude protein, CP; N × 6.25).

In vivo measurement of nitrogen degradability

A number of *in vivo* methods have been developed that aim to measure N degradation. Numerous studies, using animals fitted with duodenal cannulae (either re-entrant or T-piece) have been published. They are, however, plagued with problems of methodology (MacRae, 1975). Undegraded N flow at the duodenum is estimated as the difference between total N and microbial N (with or without a correction for endogenous N). A number of different markers have been developed, designed to estimate microbial N production; all have their limitations and have given conflicting results (Siddons, Beever and Nolan, 1982). It has been argued that measurements of degradability obtained *in vivo* most closely reflect N degradability under normal feeding conditions. However, level of feeding significantly affects the degree to which dietary N is degraded in the rumen (Tamminga, 1979), and thus the results obtained from these studies are applicable only to the conditions under which the measurements were made. Furthermore, *in vivo* studies are expensive and time consuming, and not appropriate for routine feed evaluation.

In situ measurement of nitrogen degradability

The *in situ* method (Mehrez and Ørskov, 1977) is now widely used as a means of measuring N degradability. This approach involves feed samples, sealed within nylon or polyester fibre (Dacron) bags, being suspended in the rumen of cattle or sheep for varying periods of time (typically between 0 and 48 hours). The degree to which dietary N is degraded is calculated as the difference between the amount of nitrogen contained within the bag prior to incubation in the rumen and that remaining after incubation. This technique gives characteristic disappearance curves for different feedstuffs; combining the rate of degradation with an appropriate outflow rate from the rumen provides estimates of effective degradability (Ørskov and McDonald, 1979). Despite its widespread use, the technique is not without limitations:

1. Feed particles in the bag are not subject to particle size reduction through chewing and rumination, and it is therefore likely that *in situ* degradability measurements underestimate degradability rates.
2. Undegraded feed particles within the bags may be contaminated with microbial N (Mathers and Aitchison, 1981; Kennedy, Hazlewood and Milligan, 1984). Failure to correct for this will lead to further underestimation of degradability.
3. Implicit in the calculation of degradability is the assumption that N leaving the bag at 0 hour incubation is soluble and immediately degradable, although this may not always be so (Nugent and Mangen, 1978; Chen, Sniffen and Russell, 1987; Waters, 1991). Extensive loss of this material will lead to overestimation of degradability.
4. Differences in the prediction of degradability between laboratories, for the same feeds, can be considerable. A number of methodology reviews have been published (Weakley, Stern and Satter, 1983; Lindberg, 1985; Nocek, 1988; Michalet-Doreau and Cerneau, 1991) which attempt to identify the sensitivity of different components of the technique — including bag size, bag pore size, bag:sample size ratio, sample characteristics and type of animal used — and the contribution they might make to differences between laboratories.

To identify the scale of these differences, the Inter-Departmental Protein Working Party (IDWP) organised two ring tests amongst laboratories in the UK (Oldham, 1986, Agricultural and Food Research Council, 1992). The results for hay and soya bean meal were disappointing, and showed marked between-laboratory variation in the estimates of degradability (Oldham, 1986; Agricultural and Food Research

Council, 1992). Methodology differed markedly between laboratories. It was not possible in this study to identify the main causes of this variation, although there were no obvious differences between values for sheep and cattle nor basal diets used. However, significant differences in the estimates of the immediately soluble N were reported. A subsequent ring test was undertaken in which an attempt was made to standardise on bag size, bag pore size, sample size:bag size ratio, proportion of concentrates in the DM fed and the incubation and washing procedures for the bags, according to the standard procedure recommended by the Working Party (Agricultural and Food Research Council, 1992). As in the first ring test, there were substantial differences in the zero-time estimate, although (with a smaller number of laboratories participating) reproducibility was improved.

More recently, a European ring test has been undertaken (Madsen and Hvelplund, 1992), involving 23 laboratories and a standardised procedure to measure the degradability of five concentrate feeds. Again, wide differences in predicted degradability were reported. Using an outflow rate of 8%/h, calculated effective degradability (and standard deviations) for soya bean meal, cottonseed meal, coconut meal, fish meal and barley were 0.63 (0.11), 0.51 (0.09), 0.47 (0.08), 0.23 (0.06) and 0.72 (0.08), respectively. While the mean values were as might have been expected from other published data, the variability between laboratories for individual feeds was again large. The effect of this variation in predicted metabolisable protein (MP) supply from an 80:20 mix of barley and soya bean meal is given (Table 3.1) using the lowest and highest degradability values obtained.

Table 3.1 CALCULATED METABOLISABLE PROTEIN SUPPLY (g/kg DM) FROM A BARLEY/SOYA MIX (80:20) USING EXTREMES OF SOLUBILITY/DEGRADABILITY MEASUREMENTS: EUROPEAN *IN SITU* RING TEST (MADSEN AND HVELPLUND, 1992)

Barley/soya degradability	*Low estimate* 0.58/0.34	*High estimate* 0.85/0.79
Effective Rumen Degradable Protein (g/kg DM)	93	162
Digestible Undegraded Protein (g/kg DM)	96	71
Metabolisable Protein (g/kg DM)	155	174

Outflow rate: 8%/hour

Part of the reason for this variability could be accounted for by differences in N determinations between laboratories. For the sample of soya bean meal, for example, the mean crude protein content was 504 g/kg, but a range of 454 to 561 g/kg (s.d. = 21) was reported. There were also significant differences between laboratories in the estimation of water soluble N. These are clearly problems of procedure, rather than fundamental problems of the dacron bag technique itself.

Undoubtedly, the *in situ* procedure is potentially an extremely valuable tool for measuring protein degradability. However, since it requires the use of rumen fistulated animals, it is likely to be available only where extensive research facilities are available. For this reason also it should not be considered as an appropriate means for routine feed evaluation. While the within laboratory variation is frequently acceptably low, the results of these ring tests suggest that some caution should be exercised when comparing degradability data produced by different laboratories. They also highlight the need for better standardisation of the technique, a need that is given added urgency since, with the limitations associated with the *in vivo* technique, this method is likely to be the one by which other methods are assessed.

Prediction of degradability requires not only the characteristics described above, but an estimate of rumen retention time, or the rate of outflow of the feedingstuffs from the rumen (Ørskov and McDonald, 1979). While not entirely a characteristic of the feed itself, this estimation will influence the prediction of the proportions of rumen degradable and undegradable N, and hence the potential value of a feed to meet N requirements. Estimates of outflow rate have generally been based on studies that have examined the behaviour of external markers, which may or may not mimic the flow of feed particles (Tamminga, van der Koelen and van Vuuren, 1979). It is likely that outflow rates will be influenced by a number of factors, and in particular the level of feeding, particle size and feed digestibility. Agricultural Research Council (1980) proposed the adoption of three different outflow rates, namely 2, 5 and 8%/h for cattle and sheep at low planes of nutrition, up to twice maintenance (or 15 kg milk) and for dairy cows giving >15 kg/day, respectively. The Metabolisable Protein (MP) System (Agricultural and Food Research Council, 1992) proposes the incorporation of a variable outflow rate, based on level of feeding. Using the example diet given in Agricultural and Food Research Council (1992), and for an outflow rate of 8%/h, effective rumen degradable protein (ERDP) and digestible undegraded protein (DUP) supplies are 1964 and 533 g/d, respectively. Assuming an adequate supply of fermentable metabolisable energy (FME), this would be expected to yield 1781 g MP/day. With a lower outflow rate, e.g. 5%/h, ERDP supply is increased (2080 g/d) but DUP supply reduced (436 g/d). Assuming no change in FME supply, MP supply is 1684 g/day at the lower outflow rate. The difference (97 g MP) is equivalent to the requirement of 2.1 kg milk containing 32 g protein/kg. Despite the potentially significant effect of rumen retention time, or outflow rate, on estimates of degradability and MP supply, relatively little research effort appears to have been devoted to improving the accuracy of prediction of this aspect of digestion.

Alternative methods of predicting N degradability

The *in vitro* and *in situ* methods clearly have major limitations in terms of time, cost, speed of analysis and the reliance on surgically modified (rumen and/or duodenal cannulated) animals. If the protein systems recently published are to become workable means of formulating diets, then it is essential that other methods of predicting degradability are available to the industry. A number of criteria will need to be met in establishing viable alternative techniques. Webster (1992) has suggested that a pre-requisite for the successful adoption of a new feed rationing system is that it should be based on measurements of feed chemistry, physical form or biological degradation that can be adopted as routine by the feed compounder. To this may be added that any process designed to achieve this objective should:

- be reasonably accurate, and correlate well with results obtained from *in vivo* studies;

- be sufficiently precise and well defined, such that within and between laboratory errors are acceptably low;
- offer significant time and cost savings over current techniques, and should not involve the use of animals in the routine evaluation of feedstuffs.

PREDICTING FEED PROTEIN DEGRADABILITY FROM FEED CHEMISTRY

The results of studies published in the 1970s (Burroughs, Nelson and Mertens, 1975; Crooker, Sniffen, Hoover and Johnson, 1978) led to the view that there was a strong relationship between the solubility of protein in a feedingstuff and its degradability in the rumen. This stimulated extensive research effort to explore the relationship between different solvents, including distilled and hot water, borate-phosphate buffer, 70% ethanol, sodium chloride, McDougal's artificial saliva, Burrough's mineral mixtures and autoclaved rumen fluid (see Nocek, 1988) and degradability. Subsequent studies, with a wider range of feedstuffs, have suggested that these relationships may only hold true for short incubation times and a limited range of feedingstuffs. The main reason for this is likely to be that feeds contain several different N fractions (in different proportions) which vary considerably in their rates of degradation (see above). Estimates of solubility are most likely to be of value in determining the 'a' (Ørskov and McDonald, 1979), or quickly degradable N (QDN, Agricultural and Food Research Council, 1982) fractions of a feed, although even for this the technique has limitations; reference has already been made to the fact that some soluble proteins may not be rapidly degraded.

Chaudhry, Webster and Marsden (1992) have attempted to improve the potential of the solubility procedure by incorporating a description of protein structure (as determined by electrophoresis). For six different concentrate feedstuffs, they concluded that a relationship existed between the degradability, as measured *in situ* at 18 h incubation, and protein type and structure determined by solubility and electrophoresis. However, they concluded that the relationship is unlikely to be strong enough to allow this approach to replace other alternative methods of predicting degradability.

For concentrate feedingstuffs, Madsen and Hvelplund (1985) examined the relationship between solubility, in buffer solution, of a range of feeds and their degradability. Their studies indicated a curvilinear relationship, but the accuracy of prediction was not good ($r^2 = 0.64$, SD = 10.7). However, when the concentrates were grouped into feed categories, the inclusion of a term that described, for that feed type, the relationship between *in situ* degradability (8%/h outflow rate) and buffer solubility, prediction was significantly improved ($r^2 = 0.94$, SD = 5.2). Thus, for a given concentrate feed, and with a measure of buffer solubility, it is possible to predict dacron bag degradability with an error of less than 5.2 percentage units. Prediction of degradability of a known mixture of feeds was lower ($r^2 = 0.88$, SD = 6.4). Aufrere, Graviou, Demarquilly, Verite, Michalet-Doreau and Chapoutot (1991) also confirmed that solubility in buffer solution on its own was a poor predictor of degradability measured *in situ* (6%/h outflow rate). For 97 different concentrates, variance accounted for (r^2) was 0.75. Again, distinct differences in feed types were observed. When constant terms, specific to individual feed classes were applied, prediction was significantly improved ($r^2 = 0.97$). The equation derived for feed mixtures (n = 49) also showed a non-linear trend, and variance accounted for was low ($r^2 = 0.67$). The method used in this study overestimated the degradability of mixtures containing large proportions of oil seed meals and extruded seeds, confirming the need to include a description of feed type if this approach is taken.

While measurements of solubility, in isolation, may be poor predictors of degradability, the results of these studies indicate the potential of buffer solubility, together with other information about the feedingstuff, to predict N degradability of raw materials, and of mixtures of known composition. At the present time it would appear less applicable to compound feeds of unknown composition. These relationships would benefit from further verification with a wider range of feedingstuffs. For many compounders, however, who do not have access to extensive laboratory facilities, this approach offers a relatively quick, simple and cost-effective means of estimating N degradability of concentrate feeds (at a fixed outflow rate) with an apparently reasonable degree of accuracy.

Regrettably, the same does not appear to hold true for forages. Examples of published data (Table 3.2), based on *in situ* studies, for fresh and conserved forages illustrate the potential range of degradability for these feeds.

Table 3.2 DEGRADABILITY OF FRESH AND CONSERVED FORAGES

	Outflow rate (%/h)	*Degradability*
Fresh grass		
ADAS (1989)	8	0.41 — 0.82
Waters and Givens (1992)	5	0.59 — 0.79
Grass silage (clamp)		
Madsen and Hvelplund (1985)	8	0.62 — 0.81
ADAS (1989)	8	0.66 — 0.88
Grass silage (big bale)		
ADAS (1989)	8	0.58 — 0.83
Grass hays		
ADAS (1989)	8	0.46 — 0.56

A number of attempts have been made to relate degradability characteristics to conventional feed chemistry analyses. A decade ago, Filmer (1982) proposed that the coefficient of degradability (*Dg*) could be predicted, with an acceptable degree of accuracy, using the *CP* content:

$$Dg\ (\%,\ 90\%\ DM\ \text{basis}) = 100 - 148 \times 10^{-0.274\sqrt{CP}} \quad (1)$$

CP is % in partially dried material; r = –0.92; RSD = 0.063;

A similar relationship was proposed for grass hays. While these equations have the appeal of statistical accuracy, the biological significance of the terms are not altogether clear. For grass and grass/clover silages, Madsen and Hvelplund (1985) also noted a relationship between degradability and crude protein content:

$$Dg\ (\%) = 56.2 + (0.96\ CP) \quad (2)$$

CP is % in DM; $r^2 = 0.47$; SD = 4.3

Despite the lower variance accounted for, both equations highlight the positive relationship between degradability and crude protein content, and the apparent influence of the soluble component of silages (intercept term) on overall degradability.

As grass matures, total protein content tends to decline and the proportion of N bound to indigestible fibre is likely to increase. As a consequence, degradability tends to decline with increasing maturity. Webster (1992) proposed that effective degradability might be related to both crude protein and neutral detergent fibre (NDF) contents, together with outflow rate:

$$Dg\ (\%) = (0.9\text{–}2.4r)\left[\frac{(CP\text{–}0.059\ NDF)}{CP}\right] \tag{3}$$

r = outflow rate (%/hour); *CP* and *NDF* are % DM

The relationship with maturity has also been examined by Tamminga, Ketelaar and van Vuuren (1991) who propose a 'days after 1 April' factor:

$$Dg\ (\%) = 100\text{–}(17.9+0.33DM\text{–}0.078CP+0.084\ \textit{days after April}\ 1) \tag{4}$$

DM = g/kg; *CP* = g/kg DM; RSD = 3.222, $r^2 = 0.81$.

The effect of these different equations on the prediction of degradability for a single sample of grass silage (DM = 220 g/kg, CP and NDF = 165 and 400 g/kg DM, respectively, cut date 15 May and outflow rate = 8%/h) are illustrated below:

Equation 1 (Filmer, 1982)	0.89
Equation 2 (Madsen and Hvelplund, 1985)	0.72
Equation 3 (Webster, 1992)	0.66
Equation 4 (Tamminga *et al.*, 1991)	0.83

It is clear that for forages, which form a major part of the diet of ruminant livestock, improved methods of predicting N degradability are necessary.

THE PREDICTION OF DEGRADABILITY USING BACTERIA OR ENZYMES

Procedures involving the use of bacteria or enzymes offer potential advantages over other techniques, particularly in terms of cost and speed of operation. However, this potential has not been fully exploited (Roe, Chase and Sniffen, 1991). Part, at least, of the problem appears to due to the complicated reactions between proteases and substrate. Laycock, Hazlewood and Miller (1985), for example, examined the degradability for soya bean meal N, using *Streptococcus griseus* protease, *Butyrivibrio* strain-7 and *S. bovis*, and compared the results with those obtained *in situ*. The different methods gave very different estimates of degradability, while the differences were neither consistent in magnitude or order at different incubation times. It is generally accepted that non-rumen proteases may be of limited value or give misleading results (Mahadevan, Sauer and Erfle, 1987). Published data have suggested that the results obtained using enzymes will be significantly influenced by the methodology adopted. Nocek, Herbein and Polan (1983), for example, have shown that the enzyme:protein ratio can significantly influence the rate and degree to which proteins are de-aminated. Other factors, including end product inhibition (Roe, Chase and Sniffen, 1991), the presence of anti-trypsin factors in some feedstuffs, buffer composition and pH, will also influence the results obtained. Nevertheless, the results of a number of studies would suggest that enzymatic procedures do have the potential for predicting feed protein degradability. Using five different proteolytic enzymes, Poos-Floyd, Klopfenstein and Britton (1985) reported significant correlations with protein degradability and incubations *in situ* of 1 and 4 h. The use of a protease extracted from *S. griseus* forms the basis of feed characterisation for the French PDI system (Aufrere *et al.*, 1991).

IN VITRO TECHNIQUES USING RUMEN FLUID

Since the development of the Tilley and Terry (1963) procedure, attempts have been made to develop *in vitro* techniques to predict dietary N degradability. A disadvantage of this approach is the need for a supply of rumen fluid, and the need to maintain cattle or sheep fitted with rumen cannulae as rumen fluid donors. Thus this approach is unlikely to be appropriate to commercial laboratories for routine feed analyses. However, as a means of producing base-line data on which to evaluate other systems, it might offer significant advantages over other *in vivo* or *in situ* techniques. The protagonists of this approach can generally be divided into those operating batch or continuous culture systems.

Batch culture. The most commonly used example of this approach is the Tilley and Terry (1963) technique. A major disadvantage of this procedure is the length of time needed for analyses. Consequently, a number of changes have been proposed that reduce both incubation time and the number of stages involved to measure N degradation (either directly or from ammonia production). Broderick (1987) has refined the technique by adding hydrazine sulphate to inhibit uptake of amino acids and ammonia by bacteria. Raab, Cafantaris, Jilg and Menke (1983) proposed a modification to the system, based on the production of ammonia when increments of starch were added for microbial protein synthesis. Extrapolation to zero starch was claimed to indicate potential degradability, but this technique appears to be susceptible to carbohydrate source used in relation to potential N degradability (Nocek, 1988).

Continuous culture. This approach is claimed to more closely mirror *in vivo* conditions, and allows the ability to alter 'rumen' conditions to explore the effects of changes on degradability and microbial protein synthesis. Originally only semi-continuous systems were available, but these were followed by more elaborate continuous culture systems, e.g. RUSETIC (Czerkawski and Breckenridge, 1977). While this technique, in theory, has the potential for estimating protein degradability relatively quickly and cheaply under a number of different dietary regimes, there appear to be few published data.

A major limitation to the data produced from any of these techniques (with, perhaps, the exception of continuous culture measurements), is that predictions of protein degradability are for a single outflow rate. Since the MP system applies variable outflow rates, depending on the level of feeding, use of single values of degradability may inaccurately predict the supply of ERDP to the rumen micro-organisms. Although degradability data are often published for both concentrates and forages, these generally do not include the a, b and c values (Ørskov and McDonald, 1979) needed to calculate ERDP and DUP supply at different rumen outflow rates. Waters and Givens (1992) attempted to predict, from chemical analyses, the a, b and c values for the fresh herbages in their study, but relationships between any of the conventional analyses was poor. It is possible, however, to calculate the 'c' value of a feed, from a knowledge of its protein solubility and degradability, and the outflow rate at which degradability has been calculated/measured, using a transformation of the Ørskov and McDonald (1979) model:

$$c = (r(a - p))/(p - a - b)$$

where r = outflow rate/hour
a = water soluble N
b = potentially degradable but insoluble N
p = effective degradability at the stated outflow rate

This approach requires a number of assumptions to be made:

1. Water soluble N is equivalent to the intercept term in the Ørskov and McDonald (1979) model, i.e. it equates to QDN.
2. There is no lag phase.
3. The sum of a+b can be estimated. Waters and Givens (1992) reported the sum of a+b for both primary and regrowth grass as 92.6% (SED 2.55), data that is in line with others reported elsewhere (Beever, Dhanoa, Losada, Evans, Cammell and France, 1986; Siddons, Wilkinson, Paradine and de Faria, 1990). For grass silages (ADAS, 1989) the sum of the mean 'a + b' was 91%. This method of estimating a, b and c contents of feedingstuffs clearly needs to be treated with caution. However, in the absence of any alternative means of obtaining these values, it does at least give an indication of the relative degradability of different feedingstuffs for which limited data are published (see, for example, Hvelplund and Madsen, 1990; Aufrere *et al.*, 1991)

The prediction of degradation characteristics using near infra-red reflectance spectroscopy

Near infra-red reflectance spectrometry (NIRS) is now widely and routinely used to predict the chemical composition of concentrate feedingstuffs, and it has been shown to have potential for rapid analysis of forages and the prediction of dry matter digestibility (Norris, Barnes, Moore, and Shenk, 1976; Reeves, Blosser and Colenbrander, 1989; Barber, Givens, Kridis, Offer and Murray, 1990; Clark and Lamb, 1991; Givens, Barber, Moss and Adamson, 1991) in the dried or undried state. It has the advantage, over conventional laboratory procedures, of speed of analysis and has the potential for significantly reducing analytical costs. It is a powerful tool for rapidly identifying organic structures, changes in those structures over time and the effects of chemical and biological processes on individual feedstuffs. NIRS is now used routinely by the UK advisory services for forage evaluation, and in particular to predict the CP, modified acid detergent fibre (MADF) and water soluble carbohydrate contents of fresh herbage, and the CP, MADF, and NDF contents and organic matter digestibility of grass silages (Baker and Barnes, 1990).

In contrast to the effort devoted to predicting degradability characteristics of feedstuffs from chemical or biological procedures, little effort appears to have been devoted to achieving these objectives using NIRS. Givens, Baker and Zamime (1992) attempted to define, in spectral terms, rumen degradation of different treated and untreated cereal straws, and to identify biologically important wavelength regions. They examined the residues of ten different cereal straws after incubation *in situ* for periods of between 0 and 72 h. Using standard normal variate and detrended transformation, they identified clear and significant relationships between incubation time and certain spectral regions, indicating a potential to predict degradability, if only for a single outflow rate. Waters and Givens (1992) measured, *in situ*, the degradability of fresh herbage (primary growths, n = 8, and regrowths, n = 11), and related these data to NIRS spectral analysis. The results (Table 3.3) suggest that NIRS was able to predict degradability characteristics for both primary and regrowth material separately, and combined, with varying — but not altogether unacceptable — degrees of accuracy.

Table 3.3 PREDICTION OF PROTEIN DEGRADABILITY CHARACTERISTICS IN HERBAGES BY NIRS (WATERS AND GIVENS, 1992)

Degradation characteristic	*Primary growth (n = 8)*		*Regrowths (n = 11)*	
	r^2	*SEC*	r^2	*SEC*
a	0.662	0.052	0.750	0.039
b	0.531	0.051	0.951	0.017
a + b	0.653	0.048	0.470	0.016
c	0.421	0.022	0.937	0.006
Effective degradability[a]	0.908	0.024	0.882	0.024

[a] at outflow rate of 5%/hour; r^2 = coefficient of determination; SEC = standard error of calibration

Because of the relatively small sample set, no independent population was available for validating the relationships observed, and a larger population of herbages, with measured degradability characteristics, is required to develop these relationships further. From these reports, however, a number of tentative conclusions may be drawn:

1 It may be possible to predict of the ‘a’ and ‘b’ fractions with a reasonable degree of accuracy, and at least as well as other laboratory methods currently available.
2 It is likely that the feedstuffs will need to be identified in terms of feed class.

3 It may not be possible to predict, with an acceptable degree of accuracy, degradation rate (%/hour). However, the data of Givens *et al.* (1992) would suggest that it might be possible to predict, for the feedstuff being examined, N degradability at a given outflow rate. If this is so, then with measured 'a' and 'b' values the 'c' value can be calculated.

The use of NIRS to predict nitrogen degradability appears to have considerable potential, but the rate at which it is developed is likely to be determined by the availability of sufficient numbers of samples with which to generate, and validate, the NIRS calibrations.

The digestibility of rumen undegraded protein

As well as requiring estimates of the degree to which dietary protein is degraded in the rumen, the new protein systems also take account of the digestibility of undegraded dietary protein post-rumen, with either fixed or variable factors assumed (see Alderman, 1987). Digestibility can be measured *in vivo* using the mobile bag technique, in which nylon bags, containing feed residue obtained using the *in situ* technique, are inserted through duodenal cannulae and recovered either at the terminal ileum or in the faeces. Nitrogen remaining within the bag is assumed to be indigestible (Hvelplund, 1985). Using this approach, it has been possible to measure post-rumen digestibility of a number of different feeds. For eight grass silages, van Straalen and Huisman (1991) measured intestinal digestibility of between 0.73 and 0.88, (mean 0.82) but they concluded that low rumen degradation was compensated for by higher intestinal digestibility, such that differences in total tract digestibility were low. Similarly, Hvelplund and Weisbjerg (1991) concluded, from studies with eight concentrate and three forage feeds, that intestinal digestibility of undegraded dietary protein cannot be considered constant.

While this approach may be helpful in establishing baseline data, it has a number of limitations. Firstly, the material remaining in the nylon bags after passage through the alimentary tract is, as with the *in situ* technique, subject to microbial contamination, leading to potential under-estimation of digestibility (Jarosz, Weisbjerg, Hvelplund and Jensen, 1991). A second, and more basic drawback is that it again relies on surgically modified animals, making it inappropriate as a routine tool for feed analyses. Antoniewicz, van Vuuren and van der Koelen (1991) developed an enzymatic (pepsin/pancreatin) *in vitro* technique for predicting the digestibility of *in situ* residue, but the method requires further validation.

The Metabolisable Protein system (Agricultural and Food Research Council, 1992) assumes that undegraded dietary nitrogen consists of potentially digestible and indigestible nitrogen. Furthermore, it assumes that the N which is insoluble in acid detergent (acid detergent insoluble N, ADIN) is completely undegradable in the rumen and indigestible in the abomasum and ileum. Webster, Kitcherside, Keirby and Hall (1986), have proposed that the digestibility of UDN might be described as:

$$\text{Digestible undegraded N, DUN} = 0.9\ (\text{UDN} - \text{ADIN})$$

where UDN and ADIN are as g/kg dry matter, $r^2 = 0.73$.

While this relationship may hold true for the N naturally bound within cell walls of feeds, ADIN may also be 'created' when raw materials are exposed to heat and moisture during processing. Under these conditions, Maillard reactions may occur, resulting in an increase in ADIN content. Maize gluten and distillers' grains appear particularly susceptible to these reactions. Using a very laborious collection procedure, Chaudhry and Webster (1992) have confirmed that 'natural' ADIN is indigestible, but that 'added' ADIN from Maillard reactions may, in part, be degraded in the rumen and contribute to RDN supply. For distillers products and other raw materials containing significant amounts of added ADIN during processing, they propose that DUN can be estimated from:

$$DUN = 0.9\ (UDN - 0.5\ ADIN)$$

Conclusions

To operate any of the 'new' protein systems requires a knowledge of the degree to which dietary N is degraded in the rumen, and the digestibility of the undegraded N. For feed compounders, and those involved in diet formulation, the immediate situation is not altogether encouraging. For concentrate feeds, the use of enzymes or buffer solubility measurements in conjunction with a feed category factor may prove sufficiently accurate to provide estimates of degradability at a given rumen outflow rate. Curve stripping (Sniffen, O'Conner, van Soest, Fox and Russell, 1992) may improve the usefulness of these data in those systems that require an estimation of degradation rate. For forages, however, which constitute half or more of the dry matter consumed by many high producing livestock, there do not appear to be any simple, reliable and cost-effective means of predicting degradability at present. While current techniques, involving enzymes or bacterial cultures, may be improved to permit forage protein N degradability with an acceptable degree of accuracy, NIRS would appear to offer the best prospects, provided sufficient samples of known degradability are available with which to develop, and validate, predictive equations. In the long term, it is likely that requirements for protein will be expressed in terms of specific amino acids (Rulquin and Verite, 1993), rather than nitrogen; feed characterisation using NIRS would be an extension of studies currently in progress using this technology.

References

ADAS (1989) *Nutrition Chemistry Feed Evaluation Unit Technical Bulletin No 89/10, Tables of Rumen Degradability Values for Ruminant Feedstuffs*. Kidlington, Oxford:ADAS

Agricultural and Food Research Council (1992) Technical Committee on Responses to Nutrients, Report No.9, Nutritive requirements of Ruminant Animals: Protein, *Nutrition Abstracts and Reviews, Series B*, Livestock Feeds and Feeding **62**, 787–835

Agricultural Research Council (1980) *The Nutrient Requirements of Ruminant Livestock. No. 2 Ruminants*, Farnham Royal: CAB

Agricultural Research Council (1984) *The Nutrient Requirements of Ruminant*

Livestock, Supplement No.1, Report of the Protein Group of the ARC Working Party. Farnham Royal: CAB

Alderman, G. (1987) In *Feed Evaluation and Protein Requirement Systems for Ruminants*, pp. 283–297. Edited by R. Jarrige and G. Alderman, Luxembourg: CEC

Antoniewicz, Anna M., van Vuuren, A.M. and van der Koelen, C.J. (1991) In *Proceedings of the 6th International Symposium on Protein Metabolism and Nutrition, Herning, Denmark* (**Vol 2**), pp. 110–112. Edited by B.O. Eggum, S. Biosen, C. Borsting, A. Danfaer and T. Hvelplund. EAAP Publication No. 59

Aufrere, J., Graviou, D., Demarquilly, C.,Verite, R., Michalet-Doreau, B. and Chapoutot, P. (1991) *Animal Feed Science and Technology*, **33**, 97–116

Baker, C.W. and Barnes, R. (1990) In *Feedstuff Evaluation*, pp. 337–351. Edited by J. Wiseman and D.J.A. Cole. London: Butterworths

Barber, G.D., Givens, D.I., Kridis, M.S., Offer, N.W. and Murray, I. (1990) *Animal Feed Science and Technology*, **28**, 115–128

Beever, D.E., Dhanoa, M.S., Losada, H.R., Evans, R.T., Cammell, S.B. and France, J. (1986) *British Journal of Nutrition*, **56**, 439–454

Broderick, G.A. (1987) *British Journal of Nutrition*, **58**, 463–475

Burroughs, W., Nelson, D.K and Mertens (1975) *Journal of Dairy Science*, **58**, 611–620

Chaudhry, A.S and Webster, A.J.F. (1992) *Animal Feed Science and Technology* (in press)

Chaudhry, A.S., Webster, A.J.F. and Marsden, S. (1992) *Animal Production*, **54**, 504A

Chen, G., Sniffen, C.J. and Russell, J.B. (1987) *Journal of Dairy Science*, **70**, 1211–1219

Clark, D.H. and Lamb, R.C. (1991) *Journal of Dairy Science*, **74**, 2200–2205

Crooker, B.A., Sniffen, C.J., Hoover, W.H. and Johnson, L.L. (1978) *Journal of Dairy Science*, **61**, 437–447

CSIRO (1990) *Feeding Standards for Australian Livestock: Ruminants. Standing Committee on Agriculture, Ruminants Sub-Committee*. Edited by J.L. Corbet. Melbourne: CSIRO Publications

Czerkawski, J.W. and Breckenridge, G. (1977) *British Journal of Nutrition*, **38**, 371–379

Filmer, D.G. (1982) In *Forage Protein in Ruminant Animal Production, BSAP Occasional Publication No. 6*, pp. 129–138. Edited by D.J. Thomson, D.E. Beever and R.G Gunn. Edinburgh: British Society of Animal Production

Fox, D.G., Sniffen, C.J., O'Conner, J.D., Russell, J.B. and van Soest, P.J (1992) *Journal of Animal Science*, **70**, 3578–3596

Givens, D.I., Barber, G.D., Moss, Angela R. and Adamson, A.H. (1991) *Animal Feed Science and Technology*, **35**, 83–94

Givens, D.I., Baker, C.W. and Zamime, B. (1992) *Animal Feed Science and Technology*, **36**, 1–12

Hvelplund, T. (1985) *Acta Agriculturae Scandinavica*, **suppl 25**, 132–144

Hvelplund, T. and Madsen, J. (1990) *A Study of the Quantitative Nitrogen Metabolism in the Gastro-Intestinal Tract, and the Resultant New Protein Evaluation System for Ruminants. The AAT-PBV System*. 215 pp. Copenhagen, Institute of Animal Science, The Royal Veterinary and Agricultural University

Hvelplund, T. and Weisbjerg, M.R. (1991) In *Proceedings of the 6th International Symposium on Protein Metabolism and Nutrition, Herning, Denmark*, (**Vol 2**), pp. 86–88. Edited by B.O Eggum, S. Biosen, C. Borsting, A. Danfaer and T. Hvelplund. EAAP Publication No 59

Institut National de la Recherche Agronomique (1978) *Alimentation des Ruminants*. Paris: INRA

Institut National de la Recherche Agronomique (1988) *Alimentation des Bovins, Ovins et Caprins*. Edited by R. Jarrige. Paris: INRA

Kennedy, P.M., Hazlewood, G.P. and Milligan, L.P. (1984) In *Control of Digestion and Metabolism in Ruminants. Proceedings of the 6th International Symposium on Ruminant Physiology*, pp. 285–306. Edited by L.P. Milligan, W.L. Grovum and A. Dobson. New Jersey, USA: Prentice Hall

Jarosz, L., Weisbjerg, M.R., Hvelplund, T. and Jensen, B. (1991) In *Proceedings of the 6th International Symposium on Protein Metabolism and Nutrition, Herning, Denmark*, (**Vol 2**) pp. 113–115. Edited by B.O Eggum, S. Biosen, C. Borsting, A. Danfaer and T. Hvelplund. EAAP Publication No 59

Laycock, Kathryn A., Hazlewood, G.P and Miller, E.L (1985) *Proceedings of the Nutrition Society*, **44**, 54A

Lindberg, J.E. (1985) *Acta Agriculturae Scandinavia*, **suppl 25**, 64–97

Madsen, J. (1985) *Acta Agriculturae Scandinavia*, **suppl 25**, 9–20

Madsen, J. and Hvelplund, T. (1985) *Acta Agriculturae Scandinavia*, **suppl 25**, 103–124

Madsen, J. and Hvelplund, T. (1992) *Livestock Production Science* (in press)

MacRae, J.C. (1975) In *Digestion and Metabolism in the Ruminant*, pp. 261. Edited by I.W. McDonald and A.C.I. Warner. Armidale: University of New England Publishing Unit

Mahadevan, S., Sauer, F.D. and Erfle, J.D. (1987) *Canadian Journal of Animal Science*, **67**, 55–64

Mathers, J.C. and Aitchinson, E.M. (1981) *Journal of Agricultural Science, Cambridge*, **96**, 691–693

Mehrez, A.Z. and Ørskov, E.R. (1977) *Journal of Agricultural Science, Cambridge*, **88**, 645–653

Michalet-Doreau, B. and Cerneau, P. (1991) *Animal Feed Science and Technology*, **35**, 69–81

National Research Council (1985) Ruminant Nitrogen Usage. Washington: US National Academy of Science

Nocek, J.E. (1988) *Journal of Dairy Science*, **71**, 2051–2069

Nocek, J.E., Herbein, J.H. and Polan, C.E. (1983) *Journal of Dairy Science*, **66**, 1663–1667

Norris, K.H., Barnes, R.F., Moore, J.E. and Shenk, J.S. (1976) *Journal of Animal Science*, **43**, 889–898

Nugent, J.H.A. and Mangen, J.L. (1978) *Proceedings of the Nutrition Society*, **37**, 48A

Oldham, J.D. (1987) *Feed Evaluation and Protein Requirement Systems for Ruminants*, pp. 269–281. Edited by R. Jarrige and G. Alderman. Luxembourg: CEC

Ørskov, R. and McDonald, I. (1979) *Journal of Agricultural Science*, **92**, 499–503

Poos-Floyd, Mary, Klopfenstein, T. and Britton, R.A. (1985) *Journal of Dairy Science*, **68**, 829–839

Raab, L., Cafantaris, B., Jilg, T. and Menke, K.M. (1983) *British Journal of Nutrition*, **50**, 569–582

Reeves, J.B., Blosser, T.H. and Colenbrander, V.F. (1989) *Journal of Dairy Science*, **72**, 79–88

Roe, M.B., Chase, L.E. and Sniffen, C.J. (1991) *Journal of Dairy Science*, **74**, 1632–1640

Rohr, K. (1987) In *Feed Evaluation and Protein Requirement Systems for Ruminants*, pp. 3–10. Edited by R. Jarrige and G. Alderman. Luxembourg: CEC

Rulquin, H. and Verite, R. (1993) In *Recent Advances in Animal Nutrition — 1993*, pp. 55–77. Edited by P.C. Garnsworthy and D.J.A. Cole. Nottingham: Nottingham University Press

Siddons, R.C., Beever, D.E. and Nolan, J.V. (1982) *British Journal of Nutrition*, **48**, 377–389

Siddons, R.C., Wilkinson, J.M., Paradine, J.M. and de Faria, V. (1990) *Animal Production*, **50**, 586 (abstr)

Sniffen, C.J., O'Conner, J.D., van Soest, P.J., Fox, D.G. and Russell, J.B. (1992) *Journal of Animal Science*, **70**, 3562–3577

Tamminga, S. (1979) *Journal of Animal Science*, **49**, 1615–1622

Tamminga, S. (1986) *Archives of Animal Nutrition*, **36**, 169–176

Tamminga, S., van der Koelen, C.J. and van Vuuren, A.M. (1979) *Livestock Production Science*, **6**, 255–262

Tamminga, S., Ketelaar, R and van Vuuren, A.M. (1991) *Grass and Forage Science*, **46**, 427–435

Tilley, J.M.A and Terry, R.A (1963) *Journal of the British Grassland Society*, **18**, 104

Van Straalen, W.M and Tamminga, S. (1990) In *Feedstuff Evaluation*, pp. 52–72. Edited by J. Wiseman and D.J.A. Cole. London: Butterworths

Van Straalen, W.M. and Huisman, G. (1991) *Proceedings of the 6th International Symposium on Protein Metabolism and Nutrition, Herning, Denmark*, (**Vol 2**) pp. 83–85. Edited by B.O.Eggum, S. Biosen, C. Borsting, A. Danfaer and T. Hvelplund. EAAP Publication No 59

Waters, C.J. (1990) *PhD Thesis*, University of Bristol, UK. 330 pp

Waters, C.J and Givens, D.I. (1992) *Animal Feed Science and Technology*, **38**, 335–349

Weakley, D.C, Stern, M.D. and Satter, L.D. (1983) *Journal of Animal Science*, **56**, 493–507

Webster, A.J.F. (1992) In *Recent Advances in Animal Nutrition — 1992*, pp. 93–110. Edited by P.C. Garnsworthy, W. Haresign and D.J.A. Cole. Oxford: Butterworth Heinemann

Webster, A.J.F., Kitcherside, M.A., Keirby, J.R. and Hall, P.A. (1984) *Animal Production*, **38**, 548 (abstr)

4

AMINO ACID NUTRITION OF DAIRY COWS: PRODUCTIVE EFFECTS AND ANIMAL REQUIREMENTS

H. RULQUIN and R. VERITE
Institut National de la Recherche Agronomique, Station de Recherches sur la Vache Laitière, INRA 35590 Saint Gilles, France.

It is now commonly accepted that ruminants, as well as non-ruminants, should be supplied at tissue level with sufficient essential amino acids (Buttery and Foulds, 1985). The complexity of the biochemical transformations occurring in the rumen and the relative shortage of relevant nutritional tests have restricted the advances in ruminant amino acid nutrition.

From earlier studies on low yielding ruminants it was concluded that intestinal amino acid profile is almost constant and reasonably adequate for milk protein synthesis (Oldham and Tamminga, 1980; Tamminga and Oldham, 1980; Smith, 1984; Buttery and Foulds, 1985). However, recent studies on high yielding dairy cows suggest that intestinal amino acid profile varies more than previously assumed (Clark, Klusmeyer and Cameron, 1992). Moreover, it could differ to some degree from optimal balance, as indicated by responses to supplementation of lysine and/or methionine (reviewed by Rulquin and Champredon, 1987; Rulquin, 1992).

Current interest in increasing milk protein content and decreasing nitrogen wastage promotes research in this area. Therefore this chapter deals first with the extent and origins of variation in intestinal amino acid profile in ruminants, then the effects of intestinal amino acid balance on productive responses of dairy cows and finally, some further developments of the PDI system (Institut National de la Recherche Agronomique, 1989a) are proposed to estimate supply and requirement of individual amino acids (here lysine and methionine) for more efficient protein feeding of dairy cows.

Intestinal amino acid supply

Any progress in nutrition depends on the refinement of the knowledge of absorbed nutrient flow. During the last 20 years a lot of work has been done to quantify the intestinal non-amino nitrogen flow, assumed to be representative of total amino acid flow, but few studies were effectively concerned with measurements of individual amino acid flows. Consequently, due to limited data, background noise, originating from the technique of digestive flow measurement, could mask all other variation. Therefore, in order to avoid this confounding effect, no attempt will be made here to discuss absolute flow of amino acids but only relative amino acid profiles (expressed as % of amino acid measured).

VARIABILITY OF INTESTINAL AMINO ACID PROFILE

Microbial proteins make up the main part of intestinal amino acid. They could account for at least 35–66% in dairy cows (Clark *et al.*, 1992) and even more in sheep (up to 60–90%, Smith, 1975). As a result they have smoothing effects on the variation in the amino acid profile of the whole duodenal protein. Therefore it has been generally assumed that duodenal amino acid profile varies little; thus reflecting microbial amino acid composition (Oldham and Tamminga, 1980; Smith, 1984).

Extensive literature studies on the extent and origins of the variation in intestinal amino acid composition (105 data for sheep, Le Henaff, 1991; 133 data for cattle, Guinard, Rulquin and Vérité, unpublished) were carried out in our laboratory. The variation (C.V.) in individual essential amino acid concentration ranged from 6–11% for lysine, threonine, phenylalanine and branched chain amino acids, 18% for histidine and arginine (only for sheep) and 18–24% for methionine (Table 4.1). In other words, the concentration of each individual amino acid could vary two or threefold under different feeding conditions.

Table 4.1 VARIABILITY OF AMINO ACID COMPOSITION OF INTESTINAL CONTENTS IN RUMINANTS

Species			*Cattle*[a]					*Sheep*[b]	
	mean (%)	*CV* (%)	*minimum* (%)	*maximum* (%)		*mean* (%)	*CV* (%)	*minimum* (%)	*maximum* (%)
Lys	6.88	9.7	4.82	8.42	Lys	7.46	10.6	5.03	9.69
His	2.21	11.9	1.34	2.89	His	2.55	17.8	1.63	3.89
Arg	4.96	10.6	3.76	7.07	Arg	5.46	18.2	2.92	9.65
Thr	5.32	7.0	4.36	6.16	Thr	6.00	11.2	4.38	8.46
Met	1.97	17.6	1.27	2.99	Met	2.35	24.1	0.78	3.58
Val	6.01	10.8	4.03	7.33	Val }				
Ile	5.45	8.0	4.44	6.73	Ile }	21.48	6.0	18.59	24.49
Leu	8.87	10.5	6.77	11.90	Leu }				
Phe	5.12	7.1	4.13	6.06	Phe	5.75	9.0	4.44	6.89
Asp	10.94	6.6	8.9	12.23	Asp }				
Ser	5.12	8.8	4.01	7.19	Ser }				
Glu	14.52	8.5	11.93	18.08	Glu }	48.95	2.5	44.78	52.52
Gly	6.24	19.2	3.2	10.10	Gly }				
Ala	6.92	5.6	5.54	7.79	Ala }				
Tyr	4.47	11.9	2.98	5.87	Tyr }				
Pro	5.01	15.3	3.63	7.56	Pro	nd			

nd = not determined; CV = coefficient of variation. Concentrations are expressed in g/100 g of total determined amino acids. [a] Literature review on 133 diets (Guinard *et al.*, unpublished); [b] Literature review on 105 diets (Le Henaff, 1991)

Part of the variation could be an artefact of measurements resulting from technical differences between laboratories (site of sampling, hydrolysate and/or oxidation, chromatographic methods, etc.). Moreover rather large within-laboratory variability still exists. For example, Hvelplund and Madsen (1989) reported a coefficient of variation of 5 to 11% (n = 33) for different amino acids and a twofold difference in lysine/methionine ratio, for more extreme diets. Such variations may be due, most probably, to changes in the relative contribution of microbial and by-pass protein and also, to variability of the amino acid profile of each fraction.

INFLUENCE OF THE TYPE OF DIET

When feeding naturally low degradable protein rather close similarities in amino acid profiles between whole dietary and whole duodenal protein composition or between supplemental feed protein and corresponding extra duodenal protein composition can be expected. These relations were reported for corn gluten meal (Tamminga, 1973; Cecava, Merchen, Berger and Fahey, 1988; Stern, Rode, Prange, Stauffacher and Satter, 1983), fish meal (Ørskov, Fraser and MacDonald, 1971; Zerbini, Polan and Herbein, 1988; Tigemeyer, Merchen and Berger, 1989; Beever, Gill, Dawson and Buttery, 1990), blood meal or feather meal (Waltz, Stern and Illg, 1989), formaldehyde-treated protein (Leibholz and Hartmann, 1972; MacRae and Ulyatt, 1974; Faichney, 1974; Tamminga, 1973; Vérité, Poncet, Chabi and Pion, 1978; Varvikko, 1987) and heat or ethanol

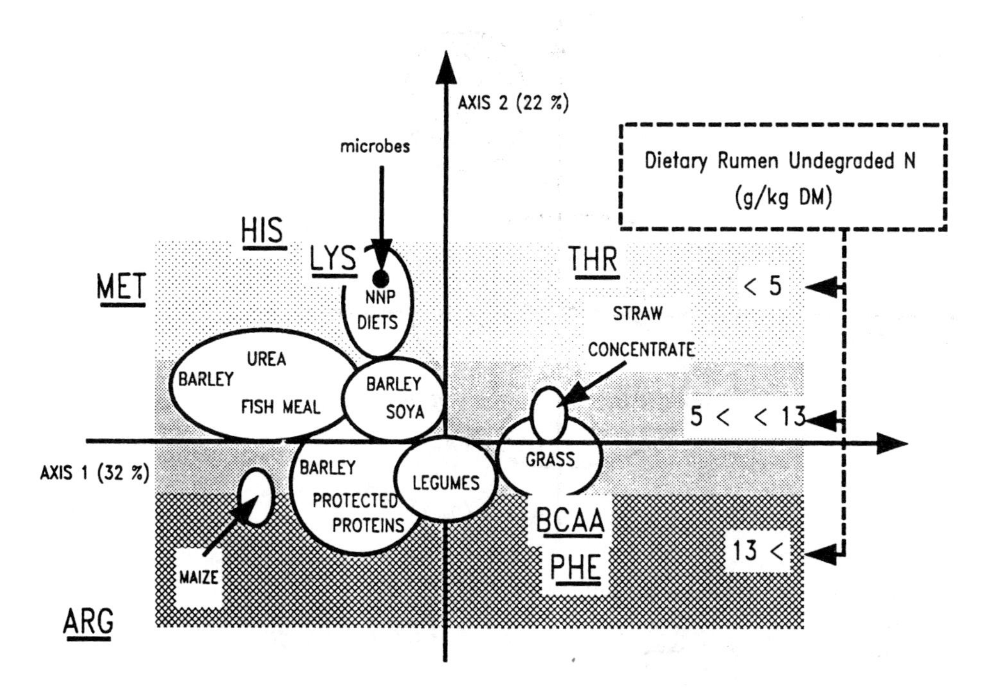

Figure 4.1 Sources of variation in intestinal amino acid profile (After a literature review of 105 data on sheep (Le Henaff, 1991))

treated protein (McMeniman and Amstrong, 1979; Pena, Tagari and Satter, 1986; Lynch, Berger, Merchen, Fahey and Baker, 1987; Cross, Vernay, Bayourthe and Moncoulon, 1992). For highly degradable proteins, the relationship would not be so close (Smith, 1980; Oldham, 1981).

However, from the survey of the two heterogeneous sets of duodenal amino acid composition data (see above), several important features may be identified using multivariate analysis (Figures 4.1 and 4.2). For a large range of diets, the different intestinal amino acid profiles aggregate significantly according to the nature of the diet from which they originate (Figure 4.1). In other words, the similarity between several intestinal amino acid profiles is better when they originate from similar diets than when they come from different types of diets. Thus intestinal proteins from high concentrate diets have a higher content of lysine, arginine, methionine, histidine and a lower content of branched-chain amino acids and phenylalanine than high roughage diets. Protected casein or soya bean meal lead to higher arginine and lower lysine and histidine duodenal content (Figure 4.1). When more mixed diets are fed to cows, similar trends can be observed: increasing the proportion of maize protein in the diet tends to increase leucine, proline and decrease lysine, threonine, valine, isoleucine and arginine intestinal content (Figure 4.2).

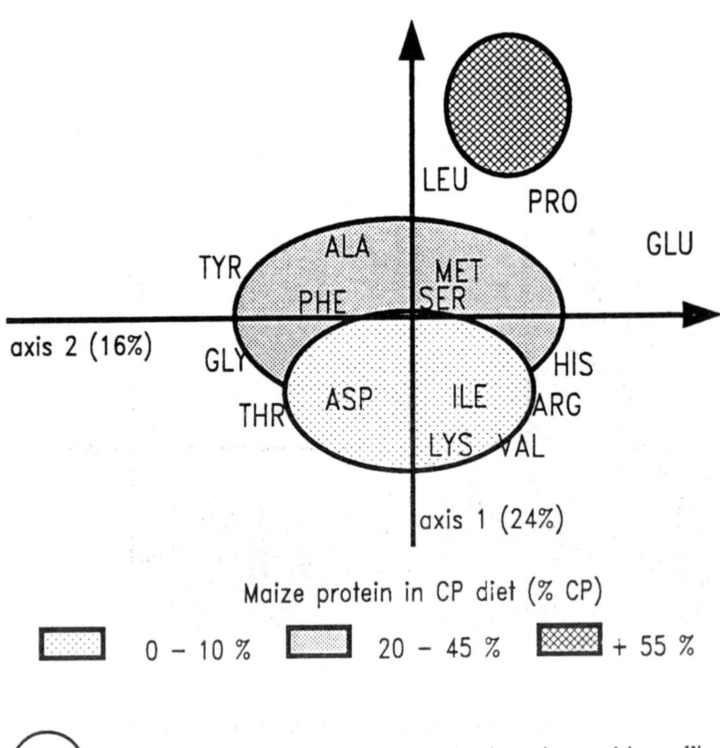

Figure 4.2 Effects of maize protein content of diet on intestinal amino acid profile of cattle (After a literature review of 133 data; Guinard *et al.*, unpublished)

EFFECT OF DIETARY AMINO ACID PROFILE

Gabel and Pope (1986), using their own data on young bulls, found significant relationships between the individual amino acid content in feeds and in the duodenum (r = 0.4–0.7), apart from methionine (r = 0.2). In apparent contrast, Hvelplund and Madsen (1989) concluded from their studies on dairy cows, that the prediction of the daily flow of individual amino acids to the duodenum was not improved by taking into account the amino acid composition of the diet. However, expressing their data in terms of relative amino acid composition (i.e. each amino acid as % of protein flow instead of amino acid flow in g/day), indicates that the relative amino acid content of duodenal protein is also fairly well correlated with the respective amino acid content of feed proteins for most individual essential amino acids (r = 0.49 to 0.82; Table 4.2), except for histidine and phenylalanine. Obviously, it can be expected that the similarity between dietary and intestinal amino acid profiles is related to the amount of the rumen undegradable protein used in the diet.

Table 4.2 RELATIONSHIPS BETWEEN INTESTINAL AND DIETARY AMINO ACID PROFILES

	Means		*Regression parameters*			*Significance*
	intestinal (%)	*dietary* (%)	*a*	*b*	*R*	
Arg	12.62	14.90	0.32	7.82	0.82	**
His	4.09	4.80	−0.10	4.56	0.14	
Ile	4.45	3.62	0.35	3.19	0.66	**
Leu	6.51	5.97	0.29	4.79	0.49	**
Lys	9.84	7.79	0.44	6.40	0.72	**
Met	1.34	1.26	0.36	0.88	0.75	**
Phe	3.19	2.90	0.20	2.62	0.16	
Thr	4.75	3.75	0.35	3.44	0.60	**
Val	5.70	5.03	0.30	4.19	0.64	**

After 31 on 33 data of Hvelplund and Madsen (1989). Equations are $y = ax + b$; y and x are respectively (intestinal and dietary) content (%) of N-amino acid in total respective N-amino acids: ** = significant at $P < 0.01$

EFFECT OF THE MICROBIAL/BY-PASS PROTEIN RATIO

The similarity between duodenal and microbial amino acid profile is expected to be closer when microbial protein flow is relatively high. This is borne out by the same data review. The composition of intestinal proteins from sheep fed only with non-protein nitrogen (NNP diets, Figure 4.1) is close to the composition of bacterial proteins. Moreover, as the proportion of undegraded feed protein in the diet increases, the duodenal amino acid composition differs from the microbial one (Figure 4.1). With more practical diets, such as those used for cattle, the tendency is the same: as the proportion of undegradable protein (maize protein, Figure 4.2) in the diet increases, intestinal protein composition differs from the microbial one mainly in lower lysine, histidine and methionine content and higher contents of leucine and proline.

Thus, the composition of absorbable amino acids is related to the amount and composition of undegraded dietary proteins. This could be of importance

Table 4.3 EFFECTS OF AMINO ACID COMPOSITION OF PROTEIN SUPPLEMENTS WITH MAIZE SILAGE BASED DIETS

Roughage[a] (%)	*Energy source*[a] (%)	*Protein meal*[a] (%)	*CP diet* (%)	*Milk* (kg/d)	*Proteins* (g/d)	*Protein content* (g/kg)	*Ref.*
Maize silage (76)	Cereals (13)	Prot. Groundnut (10)	17	24.8	803	32.4	1
Maize silage (76)	Cereals (11)	Prot. Soya/Rapeseed (6/6)	17	27.0	896	33.2	
Maize silage (60)	Maize (32)	Soya (5)	11	26.9	880	32.7	2
Maize silage (60)	Maize (33)	Maize Gluten (4)	11	26.6	790	29.9	
Maize silage (50)	Cereals (40)	Maize Gluten (7)	17	33.7	1016	29.8	3
Maize silage (50)	Cereals (42)	Fish (7)	17	38.9	1149	29.8	
Maize silage (74)	Cereals (11)	Prot. Soya/Rapeseed (8/2)	15	22.7	726	32.1	4
Maize silage (73)	Cereals (15)	Maize Gluten (7)	15	22.1	700	31.7	
Maize silage (68)	Cereals (24)	Maize Gluten (4)	14	24.4	770	31.8	5
Maize silage (68)	Cereals (15)	Blood (3)	14	24.5	764	31.5	
Maize silage (70)	Cereals (21)	Fish (6)	15	27.2	854	31.5	6
Maize silage (72)	Cereals (17)	Prot. Groundnut (8)	15	28.1	845	30.5	
Maize silage (83)		Soya (13)	17	24.6	780	33.0	7
Maize silage (83)		Soya/Fish (10/2)	17	25.6	830	32.5	
Grass/maize silage (26/19)	Barley (50)	Fish (3)	15	32.4	1010	31.4	8
Grass/maize silage (26/19)	Barley (48)	Soya (4)	15	32.6	1030	32.6	
Lucerne/maize silage (30/20)	Maize (40)	Soya/Fish (4/4)	17	41.7	1260	30.3	9
Lucerne/maize silage (30/20)	Maize (39)	Soya (9)	17	39.9	1180	29.4	
Lucerne/maize silage (15/40)	Maize (28)	Soya (15)	17	26.2	904	34.5	10
Lucerne/maize silage (15/40)	Maize (26)	Soya/Cotton (5/13)	17	27.0	818	30.3	

[a] figures in brackets show % dry matter in diet; Prot. = Formaldehyde protected; 1 Journet *et al.* (1983); 2 Klusmeyer *et al.* (1990); 3 Wohlt *et al.* (1991); 4 Rulquin *et al.* unpublished; 5 Rulquin *et al.* unpublished; 6 Rulquin *et al.* unpublished; 7 Spain, Alvarado, Polan, Miller and McGilliard (1990); 8 McCarthy *et al.* (1989); 9 Klusmeyer *et al.* (1991); 10 Mohamed *et al.* (1988)

in high yielding dairy cows because, owing to energy limitation of microbial protein synthesis, the proportion of these proteins increases as the production level increases. For this reason, relations between intestinal amino acid composition and production performances have been more thoroughly studied in dairy cows.

Productive effects of duodenal amino acid composition

DIETARY PROTEIN WITH DIFFERENT AMINO ACID PROFILES

In dairy cows, increasing the essential amino acid content of infused protein by changing from cotton meal to soya bean meal or casein and from gelatine to isolated soya protein or casein generally results in increased milk protein yield (Rogers, Clark and Drendel, 1984; Rulquin, 1986). Furthermore, the greater responses obtained with casein than with soya suggest that milk protein synthesis could be sensitive to essential amino acid composition. Similar conclusions could be drawn from 'classical' feeding trials comparing several protein sources differing in their amino acid profiles in diets with equal energy and 'metabolisable' protein supplies (Table 4.3 and 4.4). On maize silage diets supplemented with low degradable protein sources, milk protein yield was enhanced when the content of essential amino acids was changed from a low level (treated groundnut meal) to a high level (formaldehyde treated meal, fish meal) (Journet, Faverdin, Remond and Vérité, 1983; Rulquin *et al.*, unpublished). Protein sources with unbalanced amino acid composition such as corn gluten meal (high leucine and low lysine content), blood meal (low in methionine, isoleucine and high in histidine) would also be less efficient for milk protein synthesis than soya or fish meal (Klusmeyer, McCarthy, Clark, Nelson and Klusmeyer, 1990; Wohlt, Chmiel, Zajac, Backer, Blethen and Evans, 1991; Rulquin *et al.*, unpublished). On mixed silage diets (grass and maize, Table 4.3), supplementation with cottonseed meal (low lysine and high arginine) instead of soya bean meal led to lower protein yields whereas fish meal led to higher protein yields at least when the energy source in the concentrates was maize grain (Mohamed, Satter, Grummer and Ehle, 1988;

Table 4.4 EFFECTS OF AMINO ACID COMPOSITION OF PROTEIN SUPPLEMENTS WITH GRASS BASED DIETS

Roughage (%)	*Energy source* (%)	*Protein source* (%)	*CP diet* (%)	*Milk* (kg/d)	*Proteins* (g/d)	*Protein content* (g/kg)	*Ref.*
Kikuyu grass	Wheat	Protected Gelatine	14	14.3	441	30.9	1
Kikuyu grass	Wheat	Protected Casein	14	15.3	485	31.7	
Rye grass + Clover	Wheat	Protected Gelatine	22	20.6	628	30.5	1
Rye grass + Clover	Wheat	Protected Casein	22	22.5	704	31.3	
Grass sil. (47)	Barley (51)	Fish (2)	13	26.6	920	35.2	2
Grass sil. (50)	Barley (45)	Soya (5)	14	26.1	900	34.5	
Grass sil. (58)	Barley (30)	Soya (10)	17	26.2	790	30.2	3
Grass sil. (57)	Barley (32)	Soya/Fish (5/4)	18	26.2	810	30.9	
Grass sil. (54)	Barley (31)	Soya (11)	18	23.0	688	30.0	4
Grass sil. (54)	Barley (37)	Soya/Fish (3/3)	15	23.0	696	30.5	

[a] figures in brackets show % dry matter in diet; 1 Hamilton (1987); 2 Sloan *et al.* (1988); 3 Small *et al.* (1990); 4 Cody, Murphy and Morgan (1990)

Klusmeyer, Lynch, Clark and Nelson, 1991). Beneficial effects of high lysine and methionine protein sources would be associated mainly with maize diets (low lysine) according to Oldham (1981). However, the amino acid composition of protein supplements is also important for grass diets (Table 4.4). Even with pasture of high N content, supplementation with a protein source rich in essential amino acids, instead of gelatine, increased milk protein yield (Hamilton, 1987). Fish meal slightly increased milk protein yield over soya bean meal (Sloan, Rowlinson and Armstrong, 1988; Small and Gordon 1990).

Therefore it appears that, for different types of diet, optimization of amino acid composition of protein supply could be effective in increasing milk protein content and yield. Identification of the main limiting amino acids and determination of the related requirements only be achieved by postruminal supplies of synthetic amino acids.

POSTRUMINAL SUPPLEMENTATION WITH SYNTHETIC AMINO ACIDS

Predictions of limiting amino acids in ruminants are often derived from theoretical calculations based either on comparison of amino acid profiles of animal products and digesta (reviewed by Smith, 1984; Buttery and Foulds, 1985; Oldham, 1987), on variations in blood amino acid content, or on mammary uptake of blood amino acids compared to their secretion levels (reviewed by Mepham, 1982). Surprisingly, productive responses were considered only occasionally. Advantages and limitations, of the different approaches have been already reviewed (Mepham, 1982; DePeters and Cant, 1992). Generally, methionine and lysine are considered to be limiting. In addition, depending of the approach used, threonine, phenylalanine, tryptophan, histidine (Mepham, 1982), branched chain amino acids (Mäntysaari, Sniffen and O'Connor, 1989), and non-essential amino acids (Bruckental, Ascarelli, Yosif and Alumot, 1991; Meijer, van der Meulen and van Vuuren, 1992) were also implicated. Attention will be focused here on milk response to postruminal supply of individual amino acids.

Methionine and lysine

Methionine and lysine have been frequently studied as potentially limiting amino acids, using either abomasal or duodenal infusions or more recently their ruminally protected preparations. Since 1985, 10 latin-square infusion trials have been run, at the Institut National de la Recherche Agronomique Station in Rennes, on a total of 45 high-producing fistulated cows and 11 feeding trials with protected lysine and/or methionine on some 220 normal cows. From an extensive literature survey, including our own unpublished results, on some 121 comparisons on dairy cows (43 using postruminal infusion and 78 using rumen protected products; Rulquin, 1992) the effects of methionine and/or lysine on productive characteristics appeared somehow inconsistent, at first sight.

Average milk yield, protein yield, protein content and fat content responses were respectively +0.1 kg/d, +29 g/d, +0.9 g/kg and +0.1 g/kg. They ranged from −2.3 to +2.2 kg/d for milk yield, from −131 to +118 g/d for milk protein yield, from −0.6 to +3.6 g/kg for milk protein content and from −4.3 to +5.6 g/kg for milk fat content. Some factors explaining such large variability could be identified (Table 4.5). As mentioned by Schwab, Satter and Clay (1976), Bozak and Schwab (1988), Rulquin (1987), and Schwab, Bozak and Mesbah (1988a; b), methionine

Table 4.5 PRODUCTIVE RESPONSES TO POSTRUMINAL METHIONINE AND/OR LYSINE SUPPLY

Treatments	*Sample number*	*Milk yield* (kg/d)	*Protein yield* (g/d)	*Protein content* (g/kg)	*Fat content* (g/kg)
		Responses to extra amino acid			
Lys	3	0.5	8	−0.2	−0.2
Met	22	−0.2	5	0.4	0.5
Lys + Met					
All diets	96	0.2	35	1.1	−0.2
Low maize diets[a]	9	−0.2	10	0.7	1.5
High maize diets	87	0.2	38	1.1	−0.2
		High maize diets only			
Low CP diets[b]	25	0.5	38	0.9	−0.2
High CP diets	62	0.3	49	1.3	−0.3
Early lactation[c]	16	0.7	56	1.2	−0.5
Mid lactation	71	0.1	31	1.0	0.0

[a] Dietary protein from maize < or = to 32% of CP diet; [b] CP content of the diet < or = to 14.2% DM; [c] Lactation stage < or = to 10 weeks. Literature review on 121 trials using rumen protected amino acids or postruminal infusions (After Rulquin, 1992)

and lysine often seem to be co-limiting as the responses are higher when they are associated. Association of methionine and lysine gave better responses with maize based diets than with other diets (Table 4.5). On grass silage diets or hay, cow performance was not really improved by increasing methionine and lysine supply, in spite of their apparent shortage (Girdler, Thomas and Chamberlain, 1988a; b; Le Henaff, Rulquin and Vérité, 1990).

With maize based diets, the responses to extra amino acids appear to be higher on high crude protein than on low crude protein diets (Table 4.5). Such an effect has been noticed when increasing the crude protein content by supplemental corn gluten meal (Rulquin *et al.*, unpublished) but not with extra soya bean meal (Robert, Sloan, Saby, Mathé, Dumont, Duron and Dzyzcko, 1989; Sloan, Robert and Mathé, 1989; Christensen, Cameron, Clark and Drackley, 1992). Responses to extra amino acids would also depend on the methionine and lysine content of the protein concentrate (Rulquin *et al.*, unpublished see Figure 4.3); they were lower on fish meal (high lysine and methionine content) than on groundnut meal (low lysine and methionine content). Intermediate responses were obtained with corn gluten meal and blood meal suggesting that the response originated, respectively, either from lysine or from methionine.

The physiological state of cows could also affect the responses: in early lactation methionine and lysine supplements improve milk yield and sometimes protein content whereas in mid lactation mainly the protein content is increased (Bozak and Schwab, 1988; Schwab *et al.*, 1988a; b; Rulquin, 1992).

To summarize, methionine and lysine appear to be the first limiting amino acids, at least with maize-based diets. Correction of this limitation led to an increase in the true protein content of milk and especially milk casein content, without any effect on butterfat (Donkin, Varga, Sweeney and Muller, 1989; Le Henaff *et al.*, 1990; Rulquin, Le Henaff and Vérité, 1990). This change in milk composition fits fairly well with the constraints of the dairy industry. On grass diets, other amino acids or other factors such as energy supply could be also limiting.

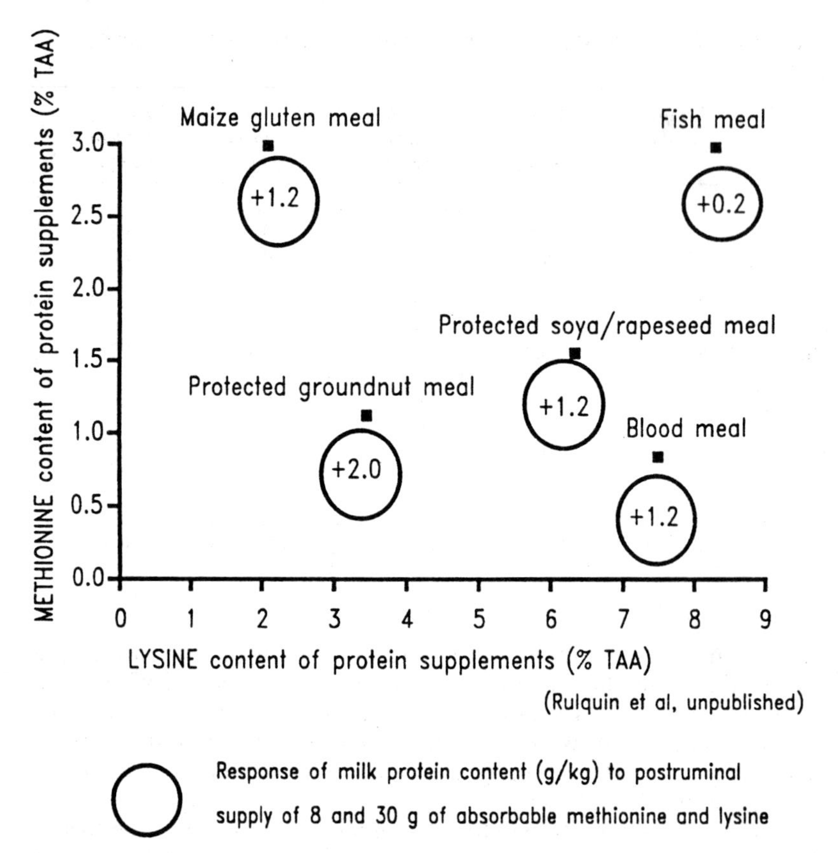

Figure 4.3 Effects of protein sources used with maize- silage based diets on the milk protein response to rumen protected methionine and lysine supply

Other essential amino acids

The most extensive studies with essential amino acids, other than methionine and lysine, were made by Schwab *et al.*, (1976) and Fraser, Ørskov, Whitelaw and Franklin, (1991), who infused nearly all of the essential amino acids. Using the total intragastric nutrition technique, Fraser *et al.* (1991) concluded that histidine is limiting just after methionine and lysine. Such a conclusion could not be drawn from the trials of Schwab *et al.* (1976) and Rulquin (1987); this discrepancy could arise from possible differences in body protein reserves, as carnosine is an endogenous source of histidine. Positive effects of increasing intestinal arginine supply on milk yield were reported by Schwab *et al.* (1976), but not by Vicini, Clark, Hurley and Bahr (1988). None of the other essential amino acids appeared to be as clearly limiting as in the experiments of Schwab *et al.* (1976) or Fraser *et al.* (1991), but conclusions can obviously differ within various feeding situations.

Non-essential amino acids

Infusion data of Oldham and Bines (1984) using aspartic and glutamic acid or of Fraser *et al.* (1991) with several non-essential amino acids did not support the idea that non essential amino acids could be limiting. However, increasing proline supply has been recently reported to increase fat yield (Bruckental *et al.*, 1991); it could have been associated with a modification of the mammary metabolism of arginine but the exact mechanism is not yet understood. Glutamine has been suggested recently because of its low blood level in early lactation (Meijer *et al.*, 1992). In humans, Lacey and Wilmore (1990) suggested that glutamine may become a 'conditionally essential' amino acid during critical illness, but it has no relevance to healthy dairy cows. Thus, until now, there is no clear evidence of any limitation from non essential amino acids for normally-fed dairy cows.

Manipulating the profile of duodenal amino acids not only improves milk protein secretion, as discussed previously, but also tends to decrease urinary nitrogen losses (up to 6%) by increasing the efficiency of nitrogen utilisation (Rogers *et al.*, 1984; Choung and Chamberlain, 1991; Fraser *et al.*, 1991). However, it is obvious that benefits of such manipulation depend on the feeding situation i.e. amino acid composition of dietary protein, its degradability and the amount of absorbable protein supplied. Since the two last items are still integrated in existing protein feeding systems, it was tempting to integrate present knowledge on amino acid digestion and metabolism in these systems to move towards amino acids feeding systems.

Estimation of lysine and methionine supply and requirements from the PDI system

As with non-ruminants, much effort has been made by ruminant nutritionists to develop methods for valuable assessment of individual amino acid feeding in ruminants, at least for the most limiting ones. Most of the new protein systems and models were designed so that they could be extended to account for individual amino acid nutrition. However the implementation of such a 'function' has usually been delayed due to insufficient relevant basic information and lack of validation. Thus it still remains only a future possibility. Indeed, it is necessary to estimate correctly both the nutritive value of feeds and animal requirements. The significance of productive responses to improved amino acid balance, as observed above, along with the existence of new possibilities to manipulate the intestinal amino acid profile (adequate undegraded protein, synthetic protected amino acid, etc.) gives further practical support to those efforts.

Therefore, in addition to the experimental work reported above, we have developed a project aimed at incorporation of the individual amino acid approach into the PDI system; starting with lysine and methionine. This work is now being finished; it was proposed for discussion with potential users in early 1992 and will soon be published. The main aspects and basis of this project are as follows.

The absorbable supply of individual amino acid is assessed from the three parameters for each of the different duodenal protein fractions (microbial, dietary, endogenous):

1 duodenal protein flow (as assessed in the PDI system);
2 amino acid profile;
3 true digestibility of amino acid in the small intestine.

A specific feature of our approach is that its accuracy was validated against a large set of data on measured duodenal amino acid flow; as a consequence, an adjustment step was introduced to get the final evaluation. Finally, tables of feedstuffs and corresponding relative concentrations of lysine and methionine, digestible in the small intestine (LysDI and MetDI), will be given.

Lysine and methionine requirements for dairy cows are not derived from theoretical assumptions but are based on the relationships between the amino acid dose and the productive response of dairy cows, determined from an extensive literature review. Moreover, requirements for LysDI and MetDI are not expressed in terms of daily amounts (g/d) but as the percentage of 'metabolizable' protein supply (i.e. LysDI or MetDI% PDI intake), to be in line with the concept of 'ideal protein' as proposed in non-ruminant nutrition (Agricultural Research Council, 1981; Institut National de la Recherche Agronomique, 1989b).

ESTIMATION OF ABSORBABLE AMINO ACID VALUE OF FEEDS

Reliable absorbable amino acid values of feeds could be given from prediction of duodenal amino acid flux and true intestinal digestibility. Quantitative estimates of intestinal amino acid profiles or flux based on dietary composition characteristics (amino acid profile and apparent digestible organic matter, or digested carbohydrates and undegraded nitrogen) have been obtained by regression by Gabel and Poppe (1986) and Hvelplund and Madsen (1989) respectively. Since variation in intestinal amino acid profile is related to the difference in amino acid composition of the diet and the amount of undegradable protein (see above), another way is to calculate, for each feedstuff, amino acids from undegraded proteins and those coming from the microbial fraction. This approach has been used by Cornell University (Mäntysaari *et al.*, 1989) and Institut National de la Recherche Agronomique (Guinard *et al.*, unpublished). However, in addition to the limitations of systems predicting the relative fractions of microbial and undegraded dietary proteins, the principal limitations of this method are related to the choice of amino acid profile of the by-pass and microbial proteins.

Amino acid profile of rumen non-degraded protein

Only a few data are available on amino acid composition of rumen unfermented feeds. They have been obtained either from *in vitro* studies (Chalupa, 1976; MacGregor, Sniffen and Hoover, 1978; Mäntysaari *et al.*, 1989) or, most often from the *in situ* nylon bag technique. Further, the modifications brought about by rumen fermentation on dietary amino acid profile appear to vary rather broadly. Thus, when compared to the corresponding feedstuffs or diet, the amino acid profiles of the *in situ* by-passed fraction were claimed to be either rather different (Rooke, Greife and Armstrong, 1984; Varvikko, 1986; Crooker, Shanks, Clark and Fahey, 1987; Hvelplund, 1987; Le Henaff, 1991; Puchala, Shelford, Vera and Pior, 1991; Schwab, personal communication), or rather similar (Ganev, Ørskov and Smart, 1979; Varvikko, Lindberg, Setälä and Syrjälä-Qvist, 1983; Weakley, Stern and Satter, 1983; Teller, Godeau, van Navel and Demeyer, 1985; Susmel, Candido and Stefano, 1988).

Indeed, the extent of these modifications could really differ according to feedstuffs; it may be greater for high degradable sources, particularly when they include heterogeneous proteins (such as in by-products enriched with 'solubles') or for treated proteins in which a particular amino acid may become unavailable (Ganev *et al.*, 1979). It may also arise from individual subjective interpretation and methodological differences as reviewed by Setälä (1983) (duration of incubation, diet of test animal, grinding, microbial contamination, etc.). Correction for microbial contamination, though considered necessary (Rooke *et al.*, 1984; Crooker, Clark, Shanks and Hatfield, 1986; Varvikko, 1986; Crooker *et al.*, 1987) increases the discrepancy between results and further its validity is questionable because of uncertainty surrounding the composition of microbial protein (see below).

When considering all available data (n = 90) on amino acid profiles in residues of rumen fermented (9 to 16 h) feeds (Le Henaff, 1991), it appears that arginine tends to decrease and branched-chained amino acids tend to increase after rumen incubation (Figure 4.4). Arginine is known to be very sensitive to fermentation (Makoni, von Keyserlingh, Puchala, Shelford and Fisher, 1990) whereas peptidic bounds with branched-chain amino acids are very resistant to hydrolysis (Hunt, 1985). However, the differences in amino acid profiles between feed protein and the corresponding undegraded protein can obviously be considered as rather small in comparison with the large differences which prevail between different feeds (Figure 4.4). Therefore, due to methodological limitations and the small number of available data, feed amino acid profiles can be taken as a first guide to estimate undegraded protein amino acid profiles.

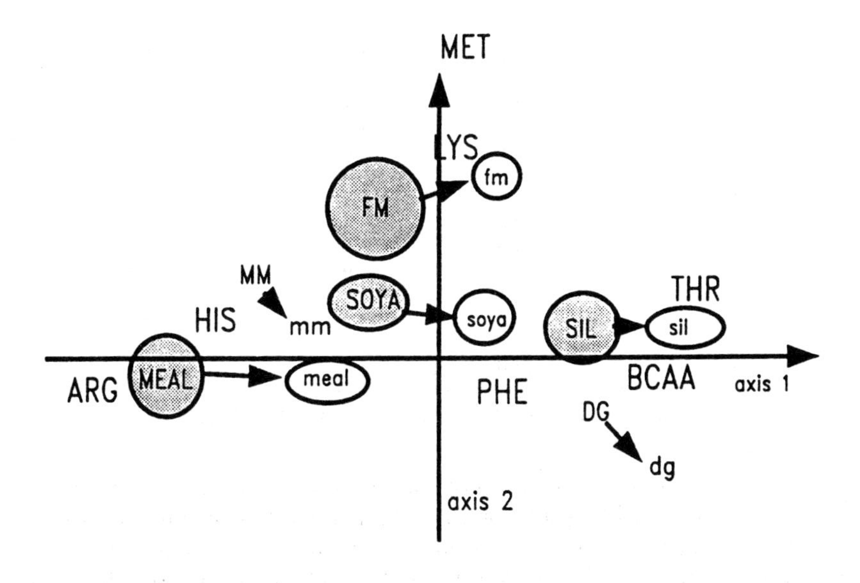

dispersion area of feedstuffs → dispersion area fo rumen residues

FM = fish meal; MM = meat meal; SOYA = soya bean meal; DG = distiller's grains

SIL = silages; MEAL = meals other than soya; BCAA = branched chain amino acids

Figure 4.4 Effects of rumen incubation on amino acid profile of some feeds

Profile of microbial proteins

The ruminal population of microorganisms includes numerous species of bacteria, protozoa and fungi. There are apparent differences in amino acid composition between bacteria (higher level of methionine) and protozoa (higher level of lysine) (see Buttery and Foulds, 1985) and between species, for example between 'liquid associated bacteria' and 'solid associated bacteria' (Gun, Schwab, Bozak, Whitehouse and Olson, 1989; Lallès, Poncet and Toullec, 1992). Extensive literature reviews (Le Henaff, 1991; Clark *et al.*, 1992) provide evidence of fairly large variations in amino acid profiles of bacteria samples from the rumen (Table 4.6). This could partly arise from differences in isolation and analytical procedures, but such variations are also found within laboratories and their origins are still unknown (Clark *et al.*, 1992). Furthermore, in the rumen, the relative importance of bacteria versus protozoa (or of different species) varies fairly broadly between diets. This contribution can be altered in the duodenum due to rumen sequestration of protozoa (Ushida, Jouany and Demeyer, 1991).

Table 4.6 VARIABILITY OF RUMEN BACTERIA AMINO ACID COMPOSITION

Amino acid	*Mean*[a] (g/100 g)	*CV* (%)	*Minimum*[a] (g/100 g)	*Maximum*[a] (g/100 g)
Lys	8.0	12.0	5.0	11.4
His	1.8	18.0	1.0	2.7
Arg	4.9	11.0	4.1	7.5
Thr	5.8	9.0	4.7	7.8
Met	2.5	24.0	1.1	3.9
Val	6.2	10.0	4.7	7.5
Ile	5.9	10.0	4.6	7.6
Leu	7.7	7.0	5.3	8.8
Phe	5.3	12.0	4.2	8.2
Asp	12.1	7.0	7.9	13.8
Ser	4.6	9.0	3.6	5.5
Glu	13.3	7.0	8.9	15.1
Gly	5.7	10.5	4.0	7.6
Ala	7.8	8.0	6.3	9.8
Tyr	4.7	12.0	2.5	5.6
Pro	3.7	14.0	2.3	5.3

[a] % of the sum of analyzed amino acids. After 66 published data (Le Henaff, 1991)

It appears difficult to develop a simple method to estimate accurately the variations in the amino acid profile of the whole microbial protein flowing to duodenum. Therefore, the amino acid profile of rumen micro-organisms has been generally considered to be that of bacterial samples. The reference profile chosen for our approach is the average of 66 data reviewed by Le Henaff (1991).

Estimation of whole intestinal protein amino acid profile

The direct estimation of the amino acid profile of the whole absorbable protein derived from the above simple assumptions could introduce some bias or have only limited accuracy. Therefore, as required for any model, it was checked against real measurements of duodenal amino acid profile, using a large data base of dairy and growing cattle (Guinard *et al.*, unpublished). These direct estimates account not only for microbial and dietary but also for endogenous protein fraction (Institut National de la Recherche Agronomique, 1989a; Ørskov, McLeod and Kyle, 1986).

For each individual amino acid, the estimated value was on average, very close to the measured one (Figure 4.5) and the correlation between both values was either significant (except for Thr) or fairly good at least for Lys, Leu, Asp, Glu, Pro, Ser ($r > 0.60$). However, a slight but significant bias was noticed except for Phe and Gly; the initial estimates tended to increase the range of variations between diets. Therefore an adjustment, based on the regression equations, was introduced in the procedure of estimation. As a result, the intestinal amino acid contents were estimated after adjustment with an acceptable accuracy: standard deviations ranged from 3.5 to 6% for most essential amino acids, except histidine and methionine (9.5 and 11.7%).

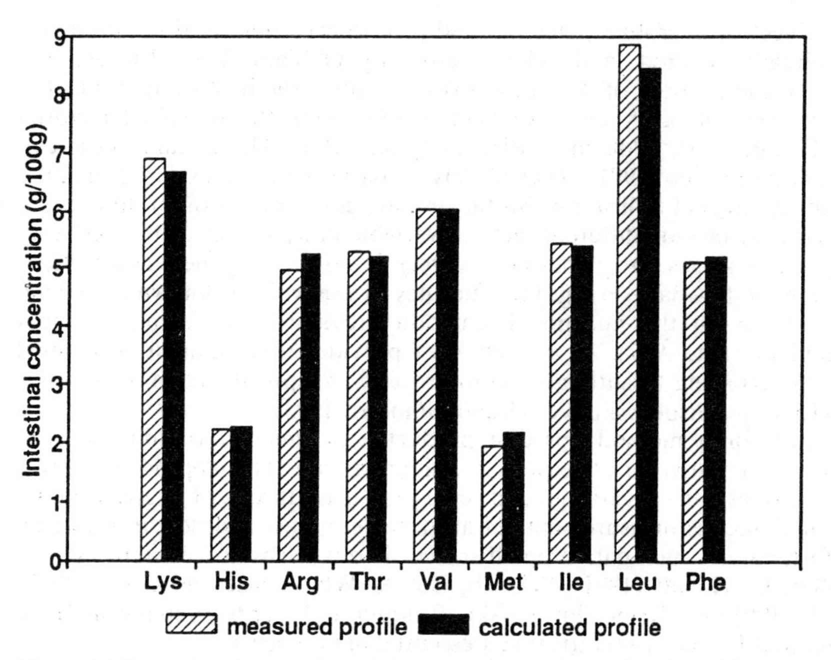

Figure 4.5 Comparison between intestinal measured and calculated amino acid profiles. After a literature review on 133 data (Guinard *et al.*, unpublished)

Digestible lysine and methionine content of feeds

The true digestibility of individual microbial amino acids in the small intestine is assumed to be constant (0.80). True digestibility is also assumed to be a constant feed (residue) characteristic, varying substantially between feeds (0.6 to 0.95; Institut National de la Recherche Agronomique 1989a). Indeed little information is available from post-ruminal infusion or 'mobile nylon bags' studies on by-pass protein. They do not however indicate substantial variations in intestinal digestibility between different amino acids, except cysteine (Hvelplund and Hesselholt, 1987; Le Henaff, 1991).

The digestible lysine and methionine content of feeds (LysDI and MetDI) is therefore calculated as the sum of microbial and dietary fractions, and expressed as a percentage of the PDI content (representing the sum of 18 main amino acids).

The present proposal, designed for estimating individual amino acid supply, may be considered to be a logical development of the PDI system. For this development an extensive literature survey was performed to obtain the new parameters required. Therefore, LysDI and MetDI values can be calculated for nearly all feedstuffs with acceptable accuracy as shown by the validation approach. It is an intermediate step, with obvious limitations, which needs future refinements. Nevertheless, our proposal provides basic working guide-lines on amino acid nutrition of dairy cows for both nutritionists and advisers.

AMINO ACID REQUIREMENTS OF DAIRY COWS

Since the subject of microbial (bacterial and protozoal) amino acid requirements has been already covered in detail (reviewed by Oldham, 1981; Buttery and Foulds, 1988), the purpose of this section is to examine the host's requirements.

Until now, several proposals have been made, using the so called factorial approach (Smith, 1980; Oldham, 1981; Mäntysaari 1989; Owens and Pettigrew, 1989; Rohr and Liebzien, 1991). They all rely on basic assumptions concerning the efficiency of amino acid utilization and the amino acid composition of different N products and they include different factors (variable efficiency of amino acid use, different lactation stages, pregnancy needs, etc.). Unfortunately these assumptions are based only on few data on e.g. the efficiency of use which is known to depend to a large extent on the nutritional situation prevailing during measurement (Rulquin and Journet, 1987). As a result, they provide either optimal or minimal requirements according to authors, and the results are spread over a large range of values (Rulquin, Pisulewski, Vérité and Guinard, 1993).

The 'dose-response' method, based on productive parameters, could be of great interest as an alternative or at least as a validating step. This approach involves measuring responses to several graded levels (at least three) of a given amino acid and simultaneous measurements of absorbable amino acid flow in fistulated cows. Unfortunately, such data are still scarce: Bozak, Schwab, Whitehouse and Olson (1989); Le Henaff (1991); Schwab, Bozak, Whitehouse and Olson (1989) and Schwab (1991) for lysine, Pisulewski, Rulquin and Vérité (unpublished) for methionine, and Fraser (1991) for seven essential amino acids.

To provide a larger experimental base (164 diets from 57 trials), an attempt was made to use also all other literature data (i.e. using only two levels of infused or protected amino acids) (Rulquin *et al.*, 1993). The amounts of LysDI and MetDI from the basal diets were predicted from the equations of Guinard *et al.* (unpublished). To be close to the 'ideal protein' concept, they were expressed as a percentage of the total supply of 'absorbable' amino acid (i.e. PDI). Productive responses were calculated within experiments, against the control treatment (adjusted to a common reference amino acid concentration).

The responses in milk protein yield and content appeared to be clearly related to 'absorbable' amino acid (LysDI and MetDI), following the law of diminishing returns (Figure 4.6 and Figure 4.7). From such nutrient-response relationships, the recommended allowances of lysine and methionine (LysDI or MetDI% PDI)

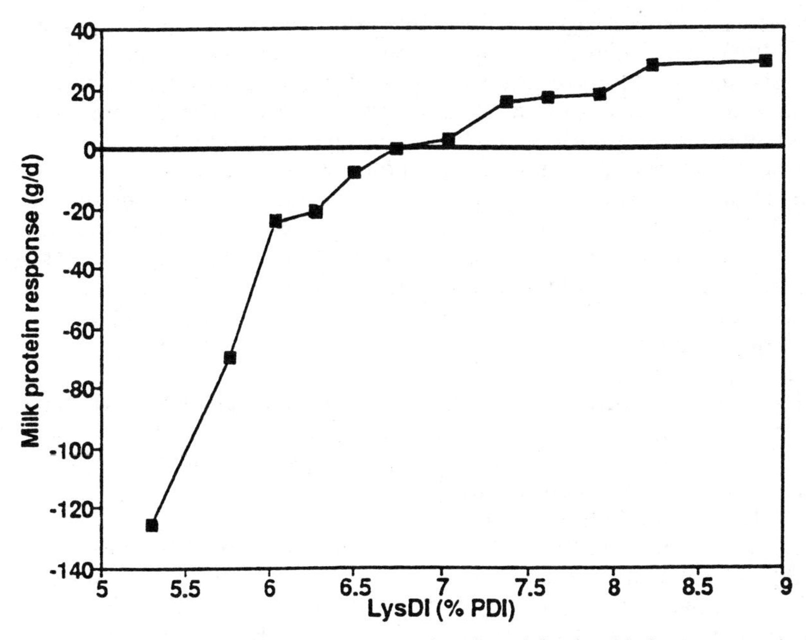

Figure 4.6 Milk protein yield responses as a function of duodenal lysine concentration (LysDI% PDI). The responses were related to the reference yield, calculated at absorbable lysine concentration of 7.0%. After a literature review on 126 data (Rulquin *et al.*, 1992)

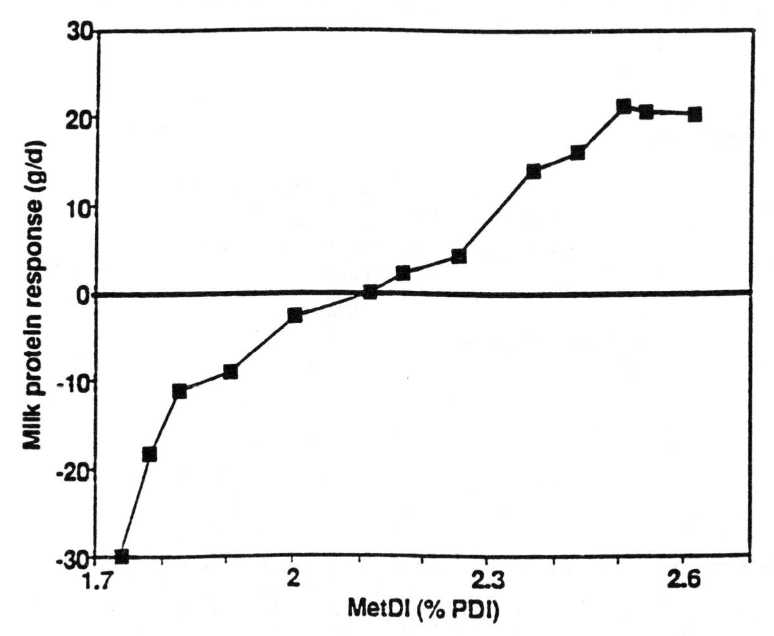

Figure 4.7 Milk protein yield responses as a function of duodenal methionine concentration (MetDI% PDI). The responses were related to the reference yield, calculated at absorbable lysine concentration of 2.1%. After a literature review on 113 data (Rulquin *et al.*, 1992)

for milking cows were estimated to be 7.3% for lysine and 2.5% for methionine (Rulquin *et al.*, 1993). For technical reasons, these values were fixed slightly below asymptotic levels and would correspond to optimal practical recommendations. There is some evidence that protein yield is dramatically lowered with diets providing less than 6.8% lysine or 2.0% methionine. Therefore these latter values should be considered as minimum thresholds. In this context, for dietary supplies of lysine and methionine ranging between the minimum thresholds and the recommended allowances, the use of extra amino acids will be governed by economic considerations. It should be noticed that the above requirements depend directly on the method used to estimate amino acid feed values and only then would they ensure unbiased and realistic rationing. If any other feed evaluation method is to be used with our requirement values such good agreement would disappear and significant discrepancies could then arise.

These requirements obtained 'empirically' on a large number of high yielding cows (average milk yield of 30 kg) are higher than the previous factorial estimations. For instance the lysine requirement for 35 kg of milk is 151 g/d instead of 115, 130 and 89 g/d (respectively for Oldham 1981; Kalnitsky and Aitova, 1991; Owens and Pettigrew, 1989). For methionine the requirement (2.5% or 52 g/d for 35 kg of milk) is close to that of Smith (1980) (2.60%) and higher than that of Oldham (1981); Kalnitsky and Aitova (1991); Owens and Pettigrew (1989) (38, 44 and 31 g/d, respectively). When expressed on the same basis (% of essential amino acids) these requirements (15.0 and 5.1% for lysine and methionine) are close to those obtained experimentally on a limited number of cows by Fraser (1991) (15.6 and 5.8%) and Schwab (1991) (15.4 and 5.2%).

In practice the first step is to formulate a ration to meet the absorbable protein (PDI) requirements. Then, its lysine and methionine content (LysDI and

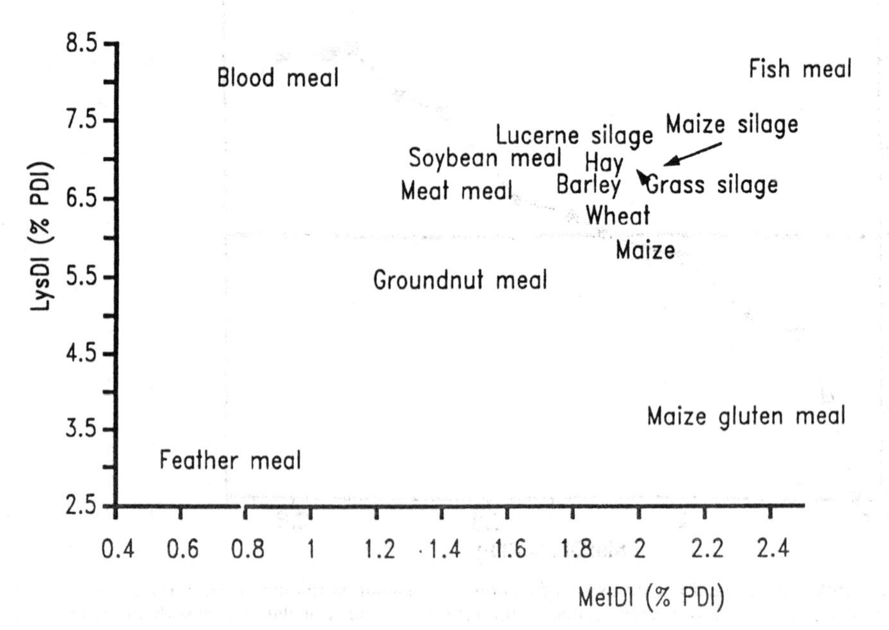

Figure 4.8 Absorbable methionine and lysine values (MetDI and LysDI% PDI) of some feeds

MetDI from feed tables as cited above) are compared against the recommended allowances. If necessary the diet is reformulated using more appropriate feedstuffs (Figure 4.8) or supplied with protected synthetic amino acids. Typical grass silage or maize silage diets complemented with cereal and soya bean meal provide 7.07–7.03% LysDI and 1.80–1.82% MetDI. The use of fish meal increases these concentrations to 7.24–7.19% and 2.08–2.1%. For these diets the amino acid requirements will be met by providing 15–25 g and 5–10 g of absorbable lysine and methionine, respectively. For this purpose, rumen protected amino acids can be effectively used.

Various technical procedures have been used to protect amino acids against ruminal degradation (Loerch and Oke, 1989): chemical analogues, metal chelates, lipid encapsulation, zein coating, polymeric encapsulation, gypsum encapsulation (Kojima, Takuwa, Ushida and Nakanishi, 1989), acetylation of peptides (Wallace, 1992). With regard to the effectiveness and the cost of these products, their required doses can be calculated from the dose-response equations given by Rulquin (1992).

Conclusions

Quantitative prediction of amino acid outflow from the rumen has been the most limiting step in the determination of ruminant amino acid requirements. Extensive utilization of literature data on amino acid digestion in ruminants resulted in a system able to tackle this problem.

Of course for the present, our proposal is far from being perfect. However, it will be improved as the knowledge of amino acid digestion progresses. It has the advantage of simplicity and accuracy of the dose response relationship. Moreover it may be already recommended for practical implementation (amino acid tables for all feedstuffs).

This step in the direction suggested by Buttery (1977) to feed ruminants with the same precision as non-ruminants is small because it concerns only methionine and lysine. However, some obscurities surrounding the subject of amino acid utilization in dairy cows have been already clarified. Finally, and above all, protein yield improvements obtained by correcting lysine and methionine deficits are encouraging for further studies on amino acid utilization in ruminants.

Acknowledgements

The authors wish to thank Rhône Poulenc Animal Nutrition and P.M. Pisulewski for their collaboration.

References

Agricultural Research Council (1981) *The Nutrient Requirements of Ruminant Livestock*. Farnham Royal: Commonwealth Agricultural Bureaux

Beever, D.E., Gill, M., Dawson, J.M. and Buttery, P.J. (1990) *The British Journal of Nutrition*, **63**, 489–502

Bozak, C.K. and Schwab, C.G. (1988) *Journal of Dairy Science*, **71 (1)**, 187 (abstr)
Bozak, C.K., Schwab, C.G., Whitehouse, N.L. and Olson, V.M. (1989) *Journal of Dairy Science*, **72 (1)**, 496 (abstr)
Bruckental, I., Ascarelli, I., Yosif, B. and Alumot, E. (1991) *Animal Production*, **53**, 299–303
Buttery, P.J. (1977) In *Recent Advances in Animal Nutrition*, pp. 8–24. Edited by W. Haresign and D. Lewis. London: Butterworths
Buttery, P.J. and Foulds, A.N. (1985) In *Recent Advances in Animal Nutrition*, pp. 257–271. Edited by W. Haresign and D.J.A. Cole. London: Butterworths
Cecava, M.J., Merchen, N.R., Berger, L.L. and Fahey, Jr. G.C. (1988) *Journal of Animal Science*, **66**, 961–974
Chalupa, W. (1976) *Journal of Animal Science*, **43 (4)**, 828–827
Choung, J.J. and Chamberlain, D.G. (1992) *Animal Production*, **54**, 502 (abstr)
Christensen, R.A., Cameron, M.R., Clark, J.H. and Drackley, J.K. (1992) *Journal of Dairy Science*, **75 (1)**, 280 (abstr)
Clark, J.H., Klusmeyer, T.H. and Cameron, M.R. (1992) *Journal of Dairy Science*, **75**, 2304–2323
Cody, R.F., Murphy, J.J. and Morgan, D.J. (1990) *Animal Production*, **51**, 235–244
Crooker, B.A., Clark, J.H., Shanks, R.D. and Hatfield, E.E. (1986) *Journal of Dairy Science*, **69 (10)**, 2648–2657
Crooker, B.A., Shanks, R.D., Clark, J.H. and Fahey, G.C. (1987) *Canadian Journal of Animal Science*, **67**, 1143–1148
Cross, P., Vernay, M., Bayourthe, C. and Moncoulon, R. (1992) *Canadian Journal of Animal Science*, **72**, 359–366
DePeters, E.J. and Cant, J.P. (1992) *Journal of Dairy Science*, **75**, 2043–2070
Donkin, S.S., Varga, G.A., Sweeney, T.F. and Muller, L.D. (1989) *Journal of Dairy Science*, **72**, 1484–1491
Faichney, G.J. (1974) *Australian Journal of Agricultural Researches*, **25**, 583–598
Fraser, D.L., Ørskov, E.R., Whitelaw, F.G. and Franklin, M.F. (1991) *Livestock Production Science*, **28**, 235–252
Gabel, M. and Poppe, S. (1986) *Archiv fur Tierernahrung, Berlin*, **36**, 429–454
Ganev, G., Ørskov, E.R. and Smart, R. (1979) *The Journal of Agricultural Science*, **93**, 651–656
Girdler, C.P., Thomas, P.C. and Chamberlain, D.G. (1988a) *The Proceedings of the Nutrition Society*, **47 (1)**, 50A
Girdler, C.P., Thomas, P.C. and Chamberlain, D.G. (1988b) *The Proceedings of the Nuturition Society*, **47**, 82A
Hamilton, B.A. (1987) *Biennial Research Report 1984–86 of North Coast Agricultural Institute Wollonbar*, 18–19
Hunt, S. (1985) In *Chemistry and Biochemistry of the Amino Acids*, pp. 376–398. Edited by C.G. Barret. London: Chapman and Hall
Hvelplund, T. (1987) In *EUR 10657 – Feed Evaluation and Protein Requirement Systems for Ruminants*, pp. 159–169. Edited by R. Jarrige and G. Alderman. Luxembourg: OOPEC
Hvelplund, T. and Madsen, J. (1985) *Acta Agriculturae Scandinavica*, **25**, 20–35
Hvelplund, T. and Hesselholt, M. (1987) *Acta Agriculturae Scandinavica*, **37**, 469–477

Institut National de la Recherche Agronomique (1989a) 389 pp. Edited by R. Jarrige. INRA Paris, J. Libbey Eurotext Paris

Institut National de la Recherche Agronomique (1989b) 282 pp. Paris: INRA

Journet, M., Faverdin, P., Remond, B. and Vérité, R. (1983) *Bulletin Technique C.R.Z.V. Theix. I.N.R.A.*, **51**, 7–17

Kalnitsky, B.D. and Aitova, M.D. (1991) In *Proceedings of the VIth International Symposium on Protein Metabolism and Nutrition*, pp. 303–305. Edited by B.O. Eggum, S. Boisen, Ch. Borsting, A. Danfaer and T. Hvelplund. Folum: NIAS, EAAP No 59

Klusmeyer, P.M., McCarthy, R.D., Clark, J.H. and Nelson, D.R. (1990) *Journal of Dairy Science*, **73**, 3526–3537

Klusmeyer, T.H., Lynch, G.L., Clark, J.H. and Nelson, D.R. (1991) *Journal of Dairy Science*, **74**, 2206–2219

Kojima, Y., Takuwa, M., Ushida, K. and Nakanishi, H. (1989) *Japanese Journal of Zootechnical Science*, **60**, 659–665

Lacey, J.M. and Wilmore, D. (1990) *Nutrition Reviews*, **48**, 297–309

Lallès, J.P., Poncet, C. and Toullec, R. (1992) *Annales de Zootechnie*, **41**, 75–76

Le Henaff, L. (1991) *Thèse de Docteur No 253*, Université de Rennes I: Section Science Biologiques. France

Le Henaff, L., Rulquin H. and Vérité R. (1990) *Reproduction Nutrition Développement*, **suppl 2**, 237s

Leibholz, J. and Hartmann, P.E. (1972) *Australian Journal of Agricultural Research*, **23**, 1073–1083

Loerch, S.C. and Oke, B.O. (1989) In *Absorption and Utilization of Amino Acids, Vol III*, pp. 187–200. Edited by M. Friedman. Boca Raton: CRC Press

Lynch, G.L., Berger, L.L., Merchen, N.R., Fahey, G.C. and Baker, E.C. (1987) *Journal of Animal Science*, **65**, 1617–1625

Makoni, N.F., von Keyserlingh, M., Puchala, J.A., Shelford, J.A. and Fisher, L.J. (1990) In *Proceedings of 70th CSAS Meeting*, p 12 Penticton. BC, Canada

Mäntysaari, P.E., Sniffen, C.J. and O'Connor, J.D. (1989) *Feedstuffs*, **61 (May)**, 13–21

MacRae, J.C. and Ulyatt, M.J. (1974) *The Journal of Agricultural Science*, **82**, 309–319

McGregor, C.A., Sniffen, C.J. and Hoover, W.H. (1978) *Journal of Dairy Science*, **61**, 566–573

McMeniman, N.P. and Armstrong, D.G. (1979) *The Journal of Agricultural Science*, **93**, 181–188

Meijer, G.A.L., van der Meulen, J. and van Vuuren, A.M. (1992) In *EAAP/ASAS Workshop on the Biology of Lactation in Farm Animals*. September 11–12, Madrid, 24 (abstr)

Mepham, T.B. (1982) *Journal of Dairy Science*, **65**, 287–298

Mohamed, O.E., Satter, L.D., Grummer, R.R. and Ehle, F.R. (1988) *Journal of Dairy Science*, **71**, 2677–2688

Oldham, J.D. (1981) In *Recent Advances in Animal Nutrition – 1980*, pp. 53–56. Edited by W. Haresign. London: Butterworths

Oldham, J.D. (1987) In *Feed Evaluation and Protein Requirement Systems for Ruminants*, pp. 171–186. Edited by R. Jarrige and G. Alderman. Luxembourg: Commission of European Communities

Oldham, J.D. and Tamminga, S. (1980) *Livestock Production Science*, **7**, 437–452

Oldham, J.D. and Bines, J.A. (1984) *The Proceeding of the Nutrition Society*, **43**, 65A

Ørskov, E.R., Fraser, C. and MacDonald, I. (1971) *The British Journal of Nutrition*, **25**, 243–252

Ørskov, E.R., McLeod, N.A. and Kyle, D.J. (1986) *The British Journal of Nutrition*, **56**, 241–248

Owens, F.N. and Pettigrew, J.E. (1989) In *Absorption and Utilization of Amino Acids, Vol I*, pp. 15–30. Edited by M. Friedman. Boca Raton: CRC Press

Pena, F., Tagari, H. and Satter, L.D. (1986) *Journal of Animal Science*, **62**, 1423–1433

Puchala, R., Shelford, J.A., Vera, A. and Pior, H. (1991) In *Proceedings of VIth International Symposium on Protein Metabolism and Nutrition*, pp. 101–103. Edited by O. Eggum, S. Boisen, C. Bosting, A. Danfaer and T. Hvelplund. Folum: EAAP pub. 59. NIAS 2

Robert, J.C., Sloan, B., Saby, B., Mathe, J., Dumont, G., Duron, M. and Dzyzcko, E. (1989) *Asian-Australian Journal of Animal Sciences, 2 (3)*, 544–545

Rogers, J.A., Clark, J.H. and Drendel, T.R. (1984) *Journal of Dairy Science*, **67**, 1928–1935

Rohr, K. and Liebzien, P. (1991) In *Proceedings of the VIth International Symposium on Protein Metabolism and Nutrition*, pp. 127–137. Edited by B.O. Eggum, S. Boisen, Ch. Borsting, A. Danfaer and T. Hvelplund. Foulum: EAAP

Rooke, J.A., Greife, H.A. and Armstrong, D.G. (1984) *The Journal of Agricultural Science*, **102**, 695–702

Rulquin, H. (1986) *Reproduction Nutrition Développement*, **26 (1B)**, 347–348

Rulquin, H. (1987) *Reproduction Nutrition Développement*, **27 (1B)**, 299–300

Rulquin, H. (1992) *INRA Productions Animales*, **5**, 29–36

Rulquin, H. and Champredon, C. (1987) *Bulletin Technique CRZV Theix. INRA*, **70**, 99–104

Rulquin, H. and Journet, M. (1987) In *Feed Evaluation and Protein Requirement Systems for Ruminants*, pp. 213–223. Edited by R. Jarrige and G. Alderman. Luxembourg: Commission of European Communities

Rulquin, H., Le Henaff, L. and Vérité, R. (1990) *Reproduction Nutrition Developpement*, **suppl 2**, 238s

Rulquin, H., Pisulewski, P.M., Vérité, R. and Guinard, J. (1993) *Livestock Production Science* (in press)

Schwab, C.G. (1991) In *New England Dairy Feed Conference*. April 10, 1991. Keene, New Hampshire

Schwab, C.G., Satter, L.D. and Clay, A.B. (1976) *Journal of Dairy Science*, **59 (7)**, 1254–1270

Schwab, C.G., Bozak, C.K. and Mesbah, M.M.A. (1988a) *Journal of Dairy Science*, **71 (1)**, 160

Schwab, C.G., Bozak, C.K. and Mesbah, M.M.A. (1988b) *Journal of Dairy Science*, **71 (1)**, 290 (abstr)

Schwab, C.G., Bozak, C.K., Whitehouse, N.L. and Olson, V.M. (1989) *Journal of Dairy Science*, **72 (1)**, 506 (abstr)

Setälä, J. (1983) *Journal of Scientific Agricultural Society of Finland*, **55**, 1–78

Sloan, B.K., Robert, J.C. and Mathé, J. (1989) *Journal of Dairy Science*, **72 (1)**, 506 (abstr)

Sloan, B.K., Rowlinson, P. and Armstrong, D.G. (1988) *Animal Production*, **46**, 13–22

Small, J.C. and Gordon, F.J. (1990) *Animal Production*, **50**, 391–398

Smith, R.H. (1975) In *Digestion and Metabolism in The Ruminant*, pp. 399–415. Edited by I.W. McDonald and A.C.I. Warner. University of New England: Armidale

Smith, R.H. (1980) *The Proceedings of the Nutrition Society*, **39**, 71–78

Smith, R.H. (1984) In *Proceedings of the VIth International Symposium on Amino Acids*, pp. 319–329. Edited by T. Zebrowska, L. Buraczeska, S. Buraczewski, J. Kowalczyk and B. Pastuszewska. Warsaw: Polish Scientific Publisher

Spain, J.N., Alvarado, M.D., Polan, C.E., Miller, C.N. and McGilliard, M.L. (1990) *Journal of Dairy Science*, **73**, 445–452

Stern, M.D., Rode, L.M., Prange, R.W., Stauffacher, R.H. and Satter, L.D. (1983) *Journal of Animal Science*, **56 (1)**, 194–205

Susmel, P., Candido, M. and Stefano, B. (1988) In *Proceedings of The VIth World Conference on Animal Production*, pp. 405 Helsinki, June 27 — July 1 1988. The Finnish Animal Breeding Associations

Tamminga, S. (1973) *Journal of Animal Physiology and Animal Nutrition*, **32**, 185–193

Tamminga, S. and Oldham, J.D. (1980) *Livestock Production Science*, **7**, 453–463

Teller, E., Godeau, J.M., van Navel, C.J. and Demeyer, D.I. (1985) *Journal of Animal Physiology and Animal Nutrition*, **54**, 121–130

Ushida, K., Jouany, J.P. and Demeyer, D.I. (1991) In *Proceedings of VIIth International Symposium on Ruminant Physiology*, pp. 625–654. Edited by T. Tsuda, Y. Sasaki and R. Kawashima. London: Academic Press

Varvikko, T. (1986) *The British Journal of Nutrition*, **56**, 131–140

Varvikko, T. (1987) *Acta Agriculturae Scandinavica*, **37**, 437–448

Varvikko, T., Lindberg, J.E., Setälä, J. and Syrjälä-Qvist, L. (1983) *The Journal of Agricultural Science*, *101*, 603–612

Verite, R., Poncet, C., Chabi, S. and Pion, R. (1977) *Annales de Zootechnie*, **26**, 167–181

Vicini, J.L., Clark, J.H., Hurley, W.L. and Bahr, J.M. (1988) *Journal of Dairy Science*, **71**, 658–665

Wallace, R.J. (1992) *The British Journal of Nutrition*, **68**, 365–372

Waltz, D.M., Stern, M.D. and Illg, D.J. (1989) *Journal of Dairy Science*, **72**, 1509–1518

Weakley, D.C., Stern, M.D. and Satter, C.D. (1983) *Journal of Animal Science*, **56 (2)**, 493–505

Wohlt, J.E., Chmiel, S.L., Zajac, P.K., Backer, L., Blethen, D.B. and Evans, J.L. (1991) *Journal of Dairy Science*, **74**, 1609–1622

Zerbini, E., Polan, C.E. and Herbein, J.H. (1988) *Journal of Dairy Science*, **71 (5)**, 1248–1258

5

SOCIAL CONCERNS AND RESPONSES IN ANIMAL AGRICULTURE

W.V. JAMISON
Oregon State University, College of Agricultural Sciences and College of Science, Corvallis, Oregon, 97331–2202, USA

Introduction

Beginning during the early stages of the Industrial Revolution, an agrarian revolution has largely transmuted agriculture from the pursuit of localized subsistence farming to the specialized pursuit of economic livelihood. Agriculture, particularly since 1950, has been characterized by astonishing technological change. Transformative developments include the application of telecommunications and computers coupled with the development of efficient transportation and the construction of extensive infrastructures. Arguably the most profound agricultural developments have occurred through the dissemination and use of empirical methodology on a widespread basis. From scientific management to the development of hybrid grains to the application of environmentally-conscious farming practices, the intercourse between science and technology in agriculture has been extraordinary. Consequently, the advent of biotechnology and genetic engineering provides the feed industry with the potential to radically redefine previous expectations concerning feed efficiency and productivity.

The contemporary manufacturing of animal feed involves a complex and technologically dependent industry which provides products and services to animal agriculture. The feed industry's growth has been characterized by increasing consolidation, product specialization and market segmentation in response to high-intensity animal production. Even though feed manufacturers are not insulated from cyclical economic downturns, they continue to experience a market for processed feeds and additives. As long as consumers demand readily-available high-quality animal products at low cost, high-intensity production will continue. Similarly, Malthusian predictions regarding world population apparently necessitate the accelerated development of industrialized agriculture. Thus, based upon solely quantitative measures, the feed industry would anticipate continuing demand.

Feed production is affected by factors which are largely outside the control of the industry. These include the weather, availability of raw commodities, variable costs, demand, and government policy. Agriculture is also encountering increasingly direct manifestations of sociological change. Historically, agribusiness has attempted to minimize the impacts of these episodic fluctuations by pursuing stability and predictability, which ultimately maximizes profit. It is within this larger context that modern feed manufacturing is found, which can be resolved into five characteristic variables: *Input, Processing, Output, Customer,* and *Animal.*

The *Input* variable consists of the raw materials and capital necessary to initiate the manufacturing process. It includes nutrient commodities such as soya beans, wheat, vitamins and minerals, and also incorporates feed additives such as probiotics. Also included is initiation and operating capital. The *Processing* variable involves the manufacturing process where the assimilation and reconstitution of raw input into usable feed products occurs. It encompasses the interface between tangibles and abstractions, namely technology and thought. It includes real property such as land, buildings, and machinery; human contact including labour, management, administration and expertise; and the actual raw materials.

Next, the *Output* variable consists of the products that are demanded by high-intensity animal agriculture. These include feeds, vitamin and mineral remixes, and additives.

The subsequent variable is the *Customer*, which superficially completes the cycle from raw material to feed product to consumer. This variable includes farmers, corporations, and others that purchase the output. The profitable feed company has traditionally focused upon the discovery and satisfaction of customer needs as the basis of long-term survival.

However, everything the feed industry does is predicated upon the final variable, the *Animal*. This variable consists of the large-scale production of animals for proteins, fats, and carbohydrates. The uses of animals in modern life have become as limitless as the human imagination. Yet without large-scale animal production the feed industry as it now exists would collapse. Hence, animal utilization is the dependent variable upon which the entire production of feed and feed additives rests, i.e. no animals to feed, no feed industry.

It would follow that the feed industry, whose absolute dependence on animal production is self-evident, would be exceedingly attentive to societal value shifts that challenge the use of animals. Yet agribusiness has historically discounted the effects of normative values upon empirical measurements of success. Time and again the industry has exhibited a chronic failure to anticipate and evaluate social change. Since the 1970's an enigmatic variable has increasingly opposed the production of animals, and consequently threatened the feed industry. This variable is *Animal Protection*. This political movement is a manifestation of social concerns over the human/nature relationship. It apparently threatens agribusiness's quantitative profitability by posing regulatory solutions to qualitative aesthetic evaluations. It also questions the fundamental premise upon which the feed industry operates by opposing the tacit acceptance of the utility of animal production.

The movement's power is often underestimated by those whose livelihoods depend on large-scale animal utilization. By all estimates the movement is growing: e.g. the animal rights movement in the United States has experienced notable successes in recent years including the creation of a large number of organizations, some of which claim hundreds of thousands of members and annual budgets in the tens of millions of dollars. Leaders of the American movement estimate that some 10 million members contribute in excess of 30 million dollars (US) annually to the cause. A movement that only ten years ago was laughingly dismissed as the lunatic fringe today permeates policy discussion. But why has the movement materialized, and who belongs to it?

Little empirical data concerning American animal rights activists has previously been available[1]. In the summer of 1990 the increasing popularity of the American movement was dramatically illustrated when animal rights organizations staged an unprecedented march on Washington, D.C. that attracted approximately 25,000 marchers[2]. The ability of the relatively immature animal rights movement to muster this level of support is illustrative of the movement's continued growth and may be indicative of the its increasing political efficacy. This chapter reports research conducted at the march and provides an initial profile of the attitudes, political behaviour, and demographic characteristics of American animal rights activists. Four sections are included: the historical context of the movement, abbreviated methodology, results, and a discussion of the context for this movement in American social and political life with an assessment of the implications for the feed industry.

Historical context

Our examination of the animal protection movement indicates that the underpinnings for a radical redefinition of the human/non-human relationship existed prior to the 20th century. Rather than a fringe movement of radicals, the contemporary animal protection movement that exists only in liberal democracies such as Great Britain is the result of the convergence of several favourable factors. Among these are the emergence of natural rights theory in England, the protestant reformation with its emphasis on autonomy and individuality, and the sociological alienation that appeared as a result of the Industrial Revolution (Fuyukama, 1992).

Great Britain is acknowledged as the origin of the contemporary animal protection movement. It is from within the British sociopolitical milieu that rights theory arises, from within the British Industrial Revolution that the disruptive displacement from farm to factory arises, and from within British society that the first signs of discontent and alienation coalesce. The Luddite's smashing of industrial equipment and Marx's critique, seemingly disjoined in levels of sophistication and targeting, both were reactions to the profound alienation of the Industrial Revolution. While America was initially spared the brunt of the sociological reactions to industrialization, she nonetheless was profoundly influenced by the British model. Hence, the history of animal rights in America closely resembles that of her progenitor, Great Britain. Rather that differing substantively from her British progenitor, American industrialization, secularization, and alienation merely lags behind England by a few decades. Thus, to understand the contemporary animal protection movement, whether it be in America, Canada, or Australia, (or any other western liberal democracy, for that matter), one must examine the roots of its advent.

---

[1] Susan Sperling provides a striking if limited ethnographic portrayal of San Francisco Bay-Area animal rights advocates in her seminal work, *Animal Liberators: Research and Morality*. (Berkeley: University of California Press, 1988). Additional information regarding the attentive public and attitudes on animal rights is provided by the direct mail survey of subscribers of *The Animal's Agenda* magazine. For further information see Richards and Krannich, *The Proceedings of the North American Wildlife and Natural Resources Conference*, Edmonton, Alberta, March 27, 1991.

[2] J. Mower. (June 10, 1990). Associated Press News Release. These figures were provided by the United States National Park Service Park Police, who are known for their conservative estimates of demonstration size.

Two social movements concerned with human use of animals came into existence in the 19th century. Originating in England, the reformist Humane movement and the radical Anti-vivisection movement both arose out of profound social reactions to increasing technological change, and were concerned with the symbolic position of animals as liaisons between humanity and nature. Each movement was reacting in part to perceptions of the increasing human transformation and exploitation of the natural world (Sperling, 1988). The increasing intensification of agriculture became symbolic of human manipulation of, and intrusion into, all things 'natural'. Simultaneously, literary romanticism personified animals which were previously characterized as brutes and beasts. Literary classics such as *Black Beauty* and *Alice in Wonderland* vested animals with strikingly human qualities. Coinciding with these changes, science contributed to a greater understanding of biological processes and demonstrated that animals were not the unfeeling automatons of Cartesian Dualism. Culminating with Darwin, 18th and 19th century science eroded the previous sanctity with which theology had viewed the separation of human and non-human.

Hence the symbolism of a helpless animal, whether it be under the glaring cruelty of the vivisector's knife or the oppression of the stockyard, was at once synonymous with the intrusiveness of new technologies and the victimization of disenfranchised beings. This symbolism became central to the mobilization of, and value manifestation among, the divergent Humane and Anti-vivisection groups.

Important continuity existed between the early humane movement and other social reform movements (Lansbury, 1985). Lansbury illuminated the symbolic political nexus between animals, feminist suffragettes and labourers that was central to the anti-vivisection riots in Edwardian England. Indeed, many of the same people were involved with both animal and human rights reform. While sincere in their passion, these opponents of the inhumane treatment of animals were in reality opposing something much deeper. During periods of intense technological change and social displacement, there has often been receptivity to criticism of forces in society, such as science, that appear responsible for technological change (Hoffer, 1951). When placed in this context both the anti-vivisection movements of the 19th and 20th century and the animal rights movement reflect increasing social anxiety regarding scientific and technological advancement. They also represent manifestations of a profound estrangement from nature which has proven to be fertile ground for revitalistic, redemptive mass movements[3].

In contemporary liberal democracies the Humane movement has sought reform and moderation in the treatment of animals. The Royal Society for the Prevention of Cruelty to Animals (RSPCA) reflects this outlook. Until recently the Humane movement, which reflects a utilitarian ethical position, accepted animal agriculture. Because of this tacit acceptance, the reformist movement has not posed a significant political or ideological threat to agriculture. Indeed, the Humane movement was historically quite pro-science, and implicitly pro-agriculture (Rowan, 1984). However, the contemporary American animal rights and British animal liberation movements have evolved to question virtually all forms of animal utilization. Sperling (1988) points to many parallels between the ideological views

[3] For an in-depth discussion of the dynamics and composition of redemptive groups, see J.Q. Wilson (1974).

of the anti-vivisection movement and the animal rights movement. Thus, while it is possible to trace the lineage and influence of anti-vivisection thought within contemporary animal rights philosophy, little empirical research has been done on the contemporary animal rights movement. Sperling's work suggests that potential constituencies for the movements are found on both the right and left, including feminists, environmentalists, and the urban/suburban middle class.

To date no large-sample scientific survey information on the active followers of the American animal rights movement has been available. Likewise, to the author's knowledge no data exist on the followers of the British animal liberation movement. Because the causes of animal protection are similar regardless of locale, an examination of the composition of the American movement should provide substantive evidence concerning its British predecessor. Estimates by pro-animal utilization organizations place the number of American animal rights groups at 600, with combined assets of $50–60 million (Kopperud, 1989). People for the Ethical Treatment of Animals (PETA), the largest and most visible of the radical American groups, reports a membership of approximately 350,000 and an annual budget of $10 million (Myers, 1990). Movement leaders place the number of active supporters of animal rights at approximately 8–10 million. Whatever the actual number of supporters may be, the movement is substantial and continues to grow.

Methodology

During the summer of 1990 The March for the Animals in Washington, D.C. offered an unprecedented and convenient venue for large-sample survey research. Never before had such a large gathering of the previously nebulous movement occurred at one time in one place. The research project entailed interviewing march participants through the administration of a survey. The survey used both open-ended and closed-end questions concerning political and movement experience, attitudes and demographic characteristics. Assistance in drafting the survey was provided by the Oregon State University Survey Research Center (OSU SRC). The survey and subsequent revisions were pre-tested among OSU graduate students. The questionnaire was eight pages in length and took approximately twelve to 15 minutes to complete. A modified stratified systematic sampling technique developed in consultation with the OSU SRC was used for randomization. At the march questions were posed by interviewers, who read from pre-printed copies of the questionnaire. Respondents were given response cards for most questions. They were instructed to give the letter corresponding to their response for each question. Interviewers recorded all responses directly on a copy of the questionnaire, so we have separate questionnaires for each of our respondents. Four hundred and 26 interviews were initiated; there were seven refusals and seven of the interviews were interrupted or terminated prior to completion (n = 412, a 97% response rate).

Indirect evidence of successful randomization is provided by a videotape of the marchers. A count of activists on the videotape reveals that the march included 71% female and 29% male participants. Race was similarly verified; 90% were white activists and 10% were minorities.

Results

DEMOGRAPHIC CHARACTERISTICS

Previous accounts of the American movement have characterized its activists as educated, but the extent of their education has been unknown. Contrary to the misconceptions of some sectors of the agricultural community, the respondents were highly educated. Indeed, 79% had received some college or university education, while fully 19% had an advanced graduate or professional degree, such as an M.S., Ph.D., or law degree (Table 5.1). These figures are contrasted against the American general public, where according to the 1990 census less than 15% have undergraduate degrees and less that 7% have graduate degrees.

Table 5.1 HIGHEST EDUCATIONAL LEVEL COMPLETED (n = 407)

Level	%
Eighth grade or less	< 1
Grade 9–11	4
High school graduate or GED	13
Technical or trade school beyond high shool	4
Some community college	6
Two year community college degree	7
Some four year college or university	18
College or university degree (B.S., B.A., etc)	22
Some graduate school	7
Graduate or professional degree (Ph.D., Law, M.S., M.A., etc)	19

Anecdotal descriptions concerning the movement's racial composition had suggested it was overwhelmingly caucasian. This was confirmed. Ninety-three percent of respondents were white and 2% were black, while American Indians, hispanic Americans and Asians each accounted for 1%. Two percent of the respondents reported their race/ethnicity as 'other'. One explanation for the predominantly caucasian composition of the American animal rights movement is the socio-economic differences in median income, mobility, education, and time availability between the white activists and minorities. Indeed, the racial composition of the march closely mirrors the racial composition of various other social movements in this country. However, the march was held in Washington, D.C., which has a majority African-American population; likewise the Washington metropolitan area has a large percentage of African-Americans proximate to the march location. Yet very little participation by this group was revealed. Thus, the lack of participation of black Americans in the contemporary animal rights movement cannot be explained solely in terms of socio-economic factors.

There has been speculation on the income level, gender and age of the activists. Respondents generally had annual household incomes of between $20,000 and

Table 5.2 TOTAL ANNUAL HOUSEHOLD INCOME BEFORE TAXES IN 1989 (n = 407)

Income (US $)	%
< 10,000	10
10,000 – 19,999	9
20,000 – 29,999	18
30,000 – 39,999	17
40,000 – 49,999	12
50,000 – 59,999	11
60,000 – 69,999	6
70,000 – 79,999	6
≥ 80,000	8
Do not know/no answer	3

$40,0000, with a mean income of $37,400 and a median income of $33,000 (Table 5.2). Evidence also suggested that most animal rights activists are female, and this was confirmed as well. Sixty-eight percent of those interviewed were female while 32% were male. The age of most respondents was between 20 and 50 years old with a mean age of 29. Interestingly, a large fraction of the marchers were younger. For example, over a third of the sample consisted of people under 30 years of age. (Figure 5.1).

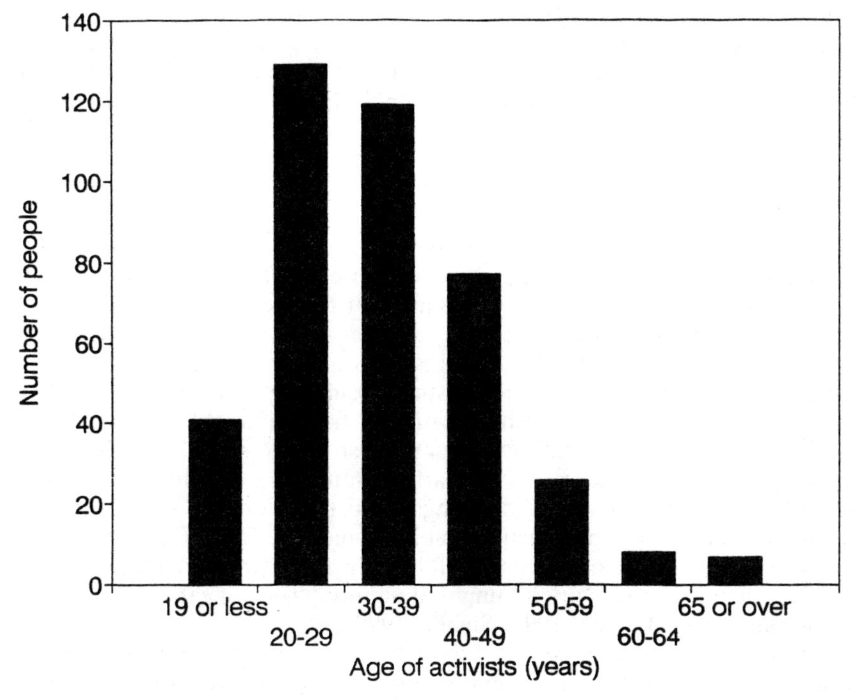

Figure 5.1 Activist age in years (n = 407)

The activists frequently reported being employed in professional jobs. Forty-four percent were professionals such as nurses, doctors, architects, lawyers, engineers, professors, administrators and so on. Respondents were employed in a wide variety of other occupations as well. The various sectors of the economy were well represented; however, Government, Health Care and Education had the greatest representation. It has also been speculated that many activists do not work outside the home, but the data do not support this. Sixty-nine percent classified their current job status as 'working for pay' while 14% were full-time students. Only 4% reported their job status as not working outside the home, and of this 4%, only 15 were housewives.

The data confirmed speculation that animal rights activists are predominantly from urban areas. Sixty-six percent of the activists lived in metropolitan areas, suburbs or cities with populations of more than 50,000 people, while 19% lived in towns of between 10,000 and 50,000 people. Ten percent of the respondents lived in towns of less than 10,000, while 5% did not know or did not answer.

Some agriculturalists may believe that animal rights activists are acting out of ignorance, but this was not substantiated by the data. The respondents get their information about important public issues from a variety of sources. When asked for their most important source of information about public issues, 27% identified newspapers, 22% identified television, and 19% identified magazines. The agricultural community may believe that activists rely upon movement magazines or direct mail distributed by the animal rights groups for their information. The data does not support this contention. Nineteen percent of the respondents reported that magazines were their most important source of information, 15% reported direct mail, and 6% reported other sources such as word-of-mouth. Over 90% reported routine use of more than one source. Use of more than one source of information about public affairs is unusual and indicates a high level of interest in public affairs. Compared with the general public these activists are well-informed and notably less reliant on television for political information (Ranney, 1983).

ATTITUDINAL CHARACTERISTICS

Central to much of the literature within the movement is a rejection of the view that humanity should have dominion over the environment. In eurocentric cosmology 'dominion' is interpreted as meaning that people have received divine authorization to use animals for food, fibre, and as beasts of burden. The word 'dominion' appears in the first book of the Bible, and is often cited as the ethical cornerstone upon which Judeo-Christian cosmology, and indeed western epistemology, constructs its relationships with the natural world (White, 1967; Singer, 1990): 'You shall have dominion over the fish of the sea, the birds of the heaven, and all creatures that move on the earth' (Genesis 1:28). This verse is one of the most often cited in defense of the religious justifications for the utilization of animals. And while identifying a distinct 'biblical' mandate for the unconstrained use of animals is problematic, animal rights philosophy clearly rejects both religious and secular arguments that rely upon human 'dominion' as justification for animal utilization (Regan, 1986; Linzey, 1991; Singer, 1990).

The research sought to measure the extent to which this rejection of human dominion has moved from the movement's opinion leaders outward to activists in general. The survey data, reinforced by additional informal interviews with movement activists, indicates that rejection of human dominion over the environment is a central unifying theme among the divergent animal rights groups. Animal rights supporters at the march also rejected hierarchical structures in society. When asked if the main cause of animal exploitation is belief in human dominion over the environment, 87% of the respondents either agreed or strongly agreed.

The research also identified the feelings of activists towards various occupations and groups by using feeling thermometers. Feeling thermometers are survey instruments that gauge respondent perceptions of phenomena by positing a scale ranging from point 'A' to point 'B'; this allows respondents to indicate their responses by self-selecting a value on the scale. Our scale ranged from 0 to 100, with a 0 response indicating a very cool or negative response and 100 indicating a very warm or positive response. Animal rights activists felt most positively towards environmentalists. Activists also felt favourably disposed towards feminists and veterinarians. We anticipated a gradual decline in feeling thermometer scores from high to low, but found instead a dramatic drop to disfavoured groups, including farmers. There simply were no middle scores. The activists felt negatively towards farmers and ranchers: farmers and ranchers ranked with scientists, politicians and businessmen as the groups that solicited the most negative responses (Table 5.3).

Table 5.3 VIEWS CONCERNING DIFFERING OCCUPATIONS AND GROUPS (100 = the most positive, 0 = the most negative)

Occupation	*Sample size*	*Mean*	*Median*	*Standard deviation*
Animal rights advocates	408	93	> 99	14
Environmentalists	403	88	97	18
Feminists	380	70	76	26
Veterinarians	401	70	74	24
Farmers/ranchers	392	30	25	26
Scientists	394	28	21	27
Politicians	401	27	23	22
Businessmen	396	25	20	22

The research utilized additional feeling thermometers to examine activist perceptions of the farm treatment of various farm animals. Ten animals were posited to the respondents, who in turn scored them. It should be noted that the composite perceptions of agriculture among respondents was negative. Horses scored the highest median score (39) and were the only animals with a positive skew (Table 5.4). Once again a gradual decline in thermometer scores was anticipated. Instead, all scores were below the neutral/ambivalence score of 50. Analysis indicates that there is an inverse relationship between the perception of highly-intense production and activist approval, e.g. the greater the perceived intensity and industrialization, the lower the score. Likewise, it is believed that terminal production practices (mink, veal, etc.) may be the locus of negative activist perceptions. However, evidence is indirect and highly speculative.

Protests against animal-based research are increasingly directed at the scientific community. During much of this century, research practices were accepted by the general public as a necessary tool of science (Bronowski, 1978). In the past decade however, acceptance of such animal research has repeatedly been called

Table 5.4 VIEWS CONCERNING THE FARM TREATMENT OF VARIOUS FARM ANIMALS (100 = the most positive, 0 = the most negative)

Occupation	*Sample size*	*Mean*	*Median*	*Standard deviation*
Horses	370	36	39	30
Sheep	347	29	24	27
Dairy cows	389	22	11	25
Beef cows	389	15	1	21
Pigs	376	13	0	21
Turkeys	376	13	0	18
Laying hens	385	11	0	20
Broilers	390	7	0	15
Mink	388	4	0	12
Veal calves	400	2	0	9

into question. The survey measured the reaction of the activists to different levels of animal-based research. The activists were generally opposed to research that utilizes animals, regardless of the level of harm to the animal or benefit to human beings. However, not **all** animal rights activists are opposed to **all** animal experimentation. The respondents were asked a series of questions that measured their approval level regarding scientific research that was beneficial for humans and incorporates animal experimentation. Fifty-six percent of the respondents either disapproved or strongly disapproved of scientific research that uses animals but does not harm them. Interestingly, fully 26% of the animal rights activists approved or strongly approved of such research. This level of approval suggests that some common ground may exist between animal rights activists and the scientific community regarding utilization of animals for laboratorial research. Yet, physiological and psychological harm seems to be the turning point for animal activists that may otherwise support animal research. Fully 84% either disapproved or strongly disapproved of animal research that causes harm to the animals. Indeed, in follow-up interviews conducted after the march, those invasive procedures that appear to the activists to be responsible for causing harm to the animal are the focus point of activist grievances against biomedical research.

In order to approximate the activists' attitudes towards science the survey asked if they felt science does more good than harm, or more harm than good. Twenty-six percent of respondents felt that science does more good than harm. Fifty-two percent of the activists believed that science does more harm than good. These data are in stark contrast to the general public. In polling conducted by the National Science Foundation in 1985, and in keeping with a consistent pattern over time, 58% of the public felt that science does more good than harm, while only 5% felt that science does more harm than good (Barke, 1986).

The research also tested the hypothesis that animal rights activism is in part motivated by concern for pets and mobilized by highly personal experiences with pets. The survey sought to measure the activists' emotional attachment to pets. When asked if they approved of keeping pets at home, fully 87% either strongly approved or approved of doing so. Nine percent of the respondents were neutral about pet ownership, while only 4% opposed or strongly opposed keeping pets at home. Likewise, responses to the open-ended questions indicated that intensely emotional experiences with pets were a significant mobilizational force in the activists' lives. Yet these views are in direct conflict with statements of animal rights opinion leaders. Ingrid Newkirk, a prominent leader of PETA, states that pet ownership is an 'absolutely abysmal situation brought about by human manipulation' (McCabe, 1986). And Peter Singer (1990), in the original introduction to *Animal Liberation*, takes pains to disclaim any sentimentality towards pets, or interest in keeping them. While this discrepancy regarding the status of pet ownership is significant, it is by no means atypical of the internal ideological inconsistencies that characterize interest groups. Among ideologically motivated groups, it is not unusual for leaders to be conceptually purist and ideologically mature when compared to supporters or even occasional activists. For example, samples of leaders of ideologically motivated groups have shown them to be less willing to compromise than their membership, and considerably more 'purist' than voters who support their agenda[4].

POLITICAL BEHAVIOUR

Activists tend to be moderately liberal or liberal (Figure 5.2). The survey also identified their political party affiliation. Thirty-seven percent of the respondents were independents, 35% were Democrats, 14% were Republicans, 11% indicated 'other', and 3% did not know or did not answer. Party preference was further clarified by asking those respondents who described themselves as independents if

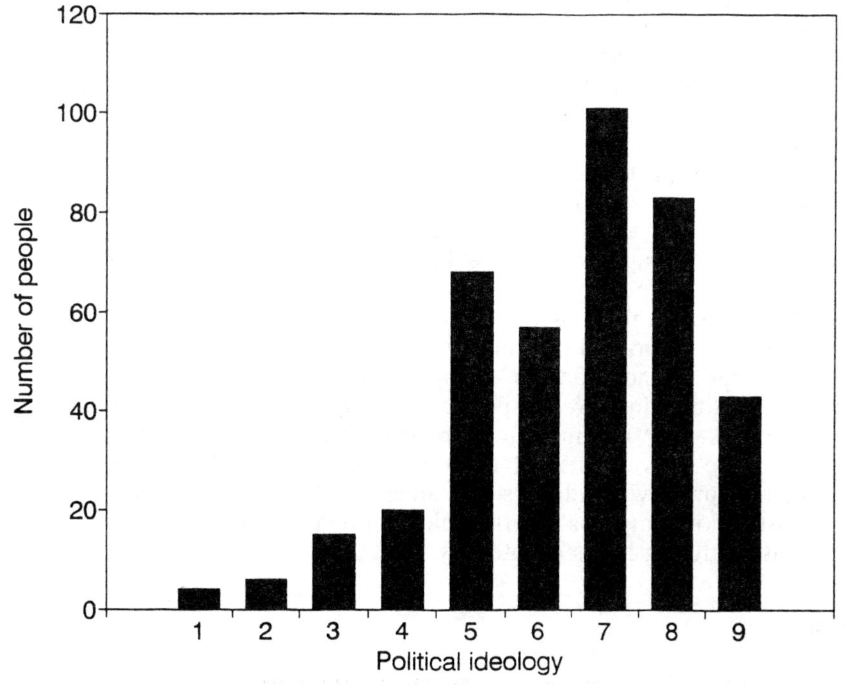

Figure 5.2 Political ideology (n = 397; 1 = most conservative, 9 = most liberal)

[4] For additional exploration on the internal life of interest groups and the ideological inconsistencies implicit in their organizational dynamics, see Berry (1989), Wilson (1974), Wilson (1989) and Lunch (1987).

they 'lean' towards Democrats or Republicans. Thirty-three percent lean towards the Democratic party, 8% lean towards the Republican party, and 59% did not lean towards either party (e.g. 'true' independents). It must be noted that voters who register as independents and then clarify that they 'lean' towards one party consistently vote for that party more than voters who identified themselves with that party to begin with. Thus, data indicate that the activists are self-defined liberals with democratic or independent affiliations and democratic voting records.

If the animal protection debate can be described as a value-laden political confrontation over the relationships between animals and humans, then the political activities of the animal protection movement are the manifestations of values that are, at least on the surface, contradictory to those held by industrial agriculturalists. Viewed in this context, it follows that this debate is indeed ultimately a political contest between the unreconciled perspectives of opposing interests. Thus, the most significant potential impact of the animal rights movement on agricultural policy is reflected in the high level of political activity of the activists. They are politically active and have both the time and inclination to be involved in politics and social movements. This was dramatically illustrated by their participation in the march.

The activists were asked if they approved or disapproved of various political activities sometimes used to advance the cause of animal rights. The survey then asked our respondents in which of these activities they had participated. Ninety-eight percent of the activists strongly approved or approved of contributing money to animal rights groups, and fully 90% of the activists had already done so. Ninety-nine percent strongly approved or approved of writing to elected representatives about animal rights, and 74% had previously contacted their elected representatives on this subject.

Regarding levels of protest directly related to agriculture, 77% of the activists favoured boycotting businesses that sell meat while 45% had already done so. Sixty-six percent of the activists approved or strongly approved of protesting supermarkets that sell meat, while 15% had already done so. But perhaps the most significant level of political activity was the proportion of the activists that had campaigned for pro-animal rights candidates. Ninety-six percent of the activists strongly approved or approved of campaigning for candidates who favour animal rights, and 38% had already done so. Compared with the general public, or even campaign contributors, this level of political activity is truly extraordinary (Nie, Verba and Petrocik, 1979). Thus, the marchers were characterized by profound commitment to the movement and to continued action within the legitimate political system. Our survey indicates that animal rights activists possess marked political sophistication, participate in the political process in many ways, and affect the process with differing levels of intensity.

Conclusion

Dulce et decorum est pro patria mori
(It is sweet and fitting to die for one's country)
– Horace

The data parallel those from studies of other political causes and movements which have found American political activists to be middle class, well educated people with strong views and a sense of obligation about expressing them (Luker, 1984; Wilson, 1962; Buttel and Flinn, 1974; Van Liere and Dunlap, 1980). Another parallel is that animal rights activists, like political activists from all persuasions in the political spectrum, are motivated by concerns which run deeper than their surface sympathies for the political symbols around which the debate revolves. Kristin Luker discovered in *Abortion and the Politics of Motherhood* that often the issue in question is merely a convenient lynch-pin upon which to hang strongly felt political beliefs. The research indicates that this may be the case with the animal rights debate. It is true that respondents demonstrated a marked compassion for animals, and this cannot be dismissed as misguided sentimentality. Yet their intensity suggests motives which reach beyond feelings for animals. While open-ended survey questions and conversations with movement leaders suggest that feelings about pets are closely associated with mobilizing most activists, it is by no means clear whether concern for animals is the motivation for the activists' continued allegiance.

A connection between the animal rights movement and reactions to broader developments in social and political life is indicated. In this, the works of Sperling (1988) and Lansbury (1985) are intriguing. Sperling has illustrated the symbolic roles of animals as liaisons between people and nature. She has pointed out that in times of rapid technological change and social displacement technologically dependent animal practices (including animal production) become symbolic of humanity's estrangement from, and corruption of, nature. Lansbury's illumination of the importance of political symbolism to the anti-vivisection riots in Edwardian England supports the contention that the animal rights debate is largely symbolic. Following this work, the research project indicates that animal rights may be a mechanism for the promotion of egalitarian political values which are in turn exacerbated by a profound estrangement from the non-human world.

Indeed, America's (and Great Britain's) urbanized population no longer has direct experience with animal production practices. In the U.S., only 3% of the population participates in food production. Hence one might expect a lack of basic agricultural knowledge among the general public. On the surface, the data support this expectation: 88% of the respondents reported an urban orientation. Yet the data also indicate that activists are well-informed by a variety of sources. Likewise, subsequent research indicates that animal rights activists are several generations removed from the farm. This leads to the intuitive conclusion that the distance of the consumer from the production and slaughter of animals must have some effect on consumer perceptions of what constitutes good and bad farming. While no causal relationship is implied, the non-agrarian configuration of the movement strongly suggests that the segmentation between producer and consumer that is characteristic of late 20th century western societies may exacerbate the manifestation of animal rights ideology. Yet this analysis is incomplete; being isolated from the production of foodstuffs does not necessitate opposition to the practices which produce them. Instead, it is believed that this isolation is one of several factors which may serve to convert latent political beliefs into manifest political activism.

In conjunction with the urbanization of society it is believed that the contemporary emergence of animal rights ideology is an anthropocentric manifestation of core political values, namely the projection of human egalitarianism upon the non-human world. In *The Rights of Nature*, Nash (1989) illustrates the evolution and progression of 'rights' theory outward from its anthropocentric patriarchal beginnings. Nash provides a theoretical over-view of the emergence of animal/ecological rights, and suggests that animal rights is the logical extension of egalitarianism to the natural world. In effect, he establishes linkage between eco-liberation movements and classic American liberalism. Indeed, in American politics egalitarianism is the central core political value of the left (McClosky and Zaller, 1984). Following Nash it is believed that animal rights finds currency on the left[5]. This was indirectly supported by the self-reported liberal ideology of the activists as well as their democratic party affiliation.

The scope of this discussion is not adequate to allow review of the origins of egalitarianism. However, a unique synthesis that has recently received attention provides the reader with a concise and textualized analysis. In *The End of History and the Last Man*, Fuyukama (1992) argues that the emergence of liberal democracies, with their emphasis on universal suffrage, individual liberty and freedom, represent the end point of the political/philosophical dialectic (Fuyukama, 1992). The purpose of this paper is not to recapitulate Fuyukama's premise. Rather, Fuyukama posits a thesis that helps to explain the emergence of animal rights in the contemporary era. He argues that egalitarianism as manifested in western industrialized democracies is the point beyond which political theory cannot progress. He likewise argues that, given the absence of a delineation between moral agent and amoral agent, the extension of rights to animals is inevitable. Hence, animal protection is a peculiar manifestation of expanded ecological consciousness coupled to rights ideology, which in turn is exacerbated by scientific evidence of the interrelated nature of all mammalian species.

It is posited that the movement's opposition to modern industrialized agriculture is a profound reaction to the commodification and objectification of sentient beings. It is a given that modern agriculture and the feed industry is critically dependent upon technology. Yet, as Sperling (1988) has illustrated, animal rights and animal liberation advocates have a deep aversion to the technological manipulation of the non-human world. The central core political belief encompassed by animal rights ideology is equality, thus it is logical that the resultant manifestation of this belief are activist calls for the protection of the autonomous individual animals. Mass-production techno-industrial agriculture not only alters what are perceived to be natural processes, it is fundamentally polar to the activists' anthropomorphic egalitarianism.

Agriculturalists understand that modern animal production is dependent on the economies of size and scale to maximize productivity. It is similarly evident that econometric measures of efficiency quantify abstractions such as welfare, time and speed. What agriculturalists fail to understand is that the process of evaluating animal life in solely economic terms commodifies the individual animal. In effect, the animal is no longer the subject of its own individual existence; it no longer has intrinsic worth separate from its financial value to the farmer. Under high-intensity agriculture, the animal becomes the object of the farmer's livelihood. Indeed, activists perceive that the **individual** well-being of animals is not the concern of industrialized agriculture. Rather, they believe that agribusiness views the sum of overall flock/herd statistics such as feed efficiency, mortality rate, and growth

[5] An emerging body of evidence indicates that this indeed may be the case. For description of the political ideologies of animal rights activists, see Sperling (1988) and Jasper and Poulsen (1989).

rate as the determinant of success. It is this objectification of the individual, this 'dehumanization' of the farm animal, that offends the animal rights activist.

Thus the feed industry confronts a threatening movement that it poorly understands. Activists, whether they be from New York or London, Los Angeles or Nottingham, are demographically much more mainstream than previously anticipated. They are highly educated, well informed and politically sophisticated. They are not marginal to the political system, and their political values are based on classic liberal ideals of equality. Similarly, they are urban dwellers whose experience with the life and death processes inherent to animal production are severely limited. Ultimately, the debate over the rights of farm animals has little to do with the reality of their treatment. Instead the debate is about the perception of what is real, and in public policy, perception becomes reality. Agriculturalists and animal rights activists have different realities. The farmer sees animal production through an economic reality which pictures the animal as a profit/loss commodity. The activist sees animal production through a political reality which pictures the animal as an individual worthy of rights. Whereas the farmer views highly-intensive, industrialized agriculture as a necessary response to economic realities, the activist views these practices as a corruption of 'natural' processes.

In response, animal rights activists extend the ethical circle to farm animals, much as abolitionists extended political equality to slaves. The activists in effect project their own anthropocentric political beliefs into the farmyard. To a feed industry whose existence is predicated on the social legitimacy of animal production, the animal rights movement is indeed a deeply disturbing development. Neither agriculture nor agriculturalists are well regarded among animal rights advocates. Viewed from a political perspective, the critique of industrialized agriculture has little to do with the status of farm animals. Rather, the debate over the rights of animals is symbolic of a much larger value conflict; it is ultimately a manifestation of deeply rooted sociopolitical beliefs which have the potential to transform agriculture.

References

Barke, R. (1986) *Science, Technology, and Public Policy*, pp. 127–128. Washington: CQ Press

Berry, J. (1989) *The Interest Group Society. 2nd Edition*, pp. 66–76. Boston: Tufts University Press

Bronowski, J. (1978) *The Common Sense of Science*, pp. 133, 144–150. Cambridge: Harvard University Press

Buttel, F.H. and Flinn, W.L. (1974) The structure of support for the environmental movement: From recreation to politics. *Pacific Sociological Review*, **14**, 270–287

Fuyukama, F. (1992) *The End of History and the Last Man*. New York: The Free Press

Hoffer, E. (1951) *The True Believer*. New York: Harper and Row

Jasper, J. and Poulsen, J. (1989) Animal Rights and anti-nuclear protest: Condensing symbols and the critique of instrumental reason. In *Proceedings of The Annual Meeting of the American Sociological Association, San Francisco*. **(August)**

Kopperud, S. (1989) Out of the shadows; the animal rights movement exposed. In *The National Cattlemen's Association Annual Report*. **Summer**, 27–42

Lansbury, C. (1985) *The Old Brown Dog: Women, Workers, and Vivisection in Edwardian England*. Madison: The University of Wisconsin Press

Lichter, R. and Rothman, S. What interests the public and what interests the public interests. *Public Opinion*. **(April/May)**, **6**, 44–48

Linzey, A. (1991) *Christianity and the Rights of Animals*, pp. 22–38. New York: Crossroad

Luker, K. (1984) *Abortion and the Politics of Motherhood*. Berkeley: The University of California Press

Lunch, W. (1987) *The Nationalization of American Politics*, pp. 214–222. Berkeley: The University of California Press

McCabe, K. (1986) Who will live, who will die? *The Washingtonian*. **(August)**

McClosky, H. and Zaller, J. (1984) *The American Ethos: Public Attitudes toward Capitalism and Democracy*, pp. 1–17, 62–100. Cambridge: Harvard University Press

Myers, C. (1990) A life affirming ethic . . . In *The Chronicle of Higher Education*. **(October 10, 1990)**, A21–A28.

Nash, R. (1989) *The Rights of Nature,* pp. 3–12, chapters 1–2. Madison: The University of Wisconsin Press

Nie, N., Verba, S. and Petrocik, J. (1979) *The Changing American Voter*. Cambridge: Harvard University Press

Ranney, A. (1983) *Channels of Power*. Washington: American Enterprise Institute

Regan, T. (1986) *Animal Sacrifices: Religious Perspective on the Use of Animals in Science*. Philadelphia: Temple University Press

Rowan, A. (1984) *Of Mice, Models, and Men*. Albany: State University of New York Press

Singer, P. (1990) *Animal Liberation. 2nd Edition*, pp. 185–211. New York: A New York Review Book

Sperling, S. (1988) *Animal Liberators: Research and Morality*, pp. 51–130. Berkeley: University of California Press

Van Liere, K. and Dunlap, R. (1980) The social bases of environmental concern: A review of hypotheses, explanations, and empirical evidence. *Public Opinion Quarterly*. 181–197

White, L. (1967) The Historical Roots of Our Ecological Crisis. *Science*. **(March 10, 1967)**, 1203–1207

Wilson, J.Q. (1962) *The Amateur Democrat*. Chicago: The University of Chicago Press

Wilson, J.Q. (1974) *Political Organizations*, pp. 45–50, chapters 1–3. New York: Basic Books

Wilson, J.Q. (1989) *Bureaucracy: What Government Agencies Do and Why They Do It*, pp. 83–84. New York: Basic Books

II

Poultry Nutrition

6

NUTRITIONAL MANAGEMENT OF BROILER PROGRAMMES

C.G. BELYAVIN
Chris Belyavin (Technical) Limited, 2 Pinewoods, Church Aston, Newport, Shropshire TF10 9LN, UK.

Introduction

In the past, the major criteria for assessing the performance of a flock of broilers would have been growth rate and feed conversion ratio (FCR). Diet specifications and feeding programmes would have been produced in order to maximise these two parameters and overall flock performance would have been assessed at the end of the growing period by calculating the total weight of chicken produced from the factory weight and combining that with the total feed deliveries to give an overall FCR.

By today's standards this is a fairly crude approach to chicken production and does not incorporate any steps during the growing period to correct any factors causing growth to be below the expected target. Because chicken markets generally want birds within fairly specific weight bands, if a crop is off target during the growing period, this would normally be overcome by either killing the birds early if they are overweight, or keeping them longer if the reverse is the case. Because this is a fairly imprecise approach, the net effect may be considerable disruption to any stocking schedule with consequential effects on the profitability per unit of growing space.

Development of our understanding of nutritional factors affecting broiler growth and carcass composition, together with the availability of equipment for manipulating feeding programmes on site, means that it is now possible to implement sophisticated approaches to feeding broilers.

Free choice feeding

Traditionally, poultry feeds in the UK are cereal based. For over 20 years the concept of offering chickens or turkeys a choice between a cereal and a balancer meal or pelleted feed has been discussed. The theory is based on the belief that the chicken knows best and by feeding a complete feed on its own, not every animal is being provided with its specific nutritional requirements. By offering the choice, each individual animal theoretically can choose a blend between the cereal and the balancer to suit its daily needs for protein and energy and possibly some other nutrients. Therefore, it could be said that with this approach the initiative for getting the final feed correct is with the animal and not the nutritionist.

With such an approach as this, relatively large savings in food costs may occur because the whole cereal does not need to be milled, mixed and pelleted with the other dietary ingredients.

While the practice has been researched in layers, broilers and turkeys, it is beyond the scope of this chapter to consider laying hens. Work involving broilers will be reviewed in detail and some consideration will be given to the work done with turkeys as a comparison with the broiler.

In one study reported by Cowan and Michie (1978a) male and female broilers were fed either a complete diet or given a choice of whole wheat and one of two higher protein feeds formulated by omitting some or all of the cereal from the complete diet. For males receiving the choice treatments, on average 44.7% and 73.1% by weight respectively of the food consumed from 21 days consisted of whole wheat. The corresponding figures for the females were 49.9% and 77.3% respectively suggesting that the female has a lower protein requirement. However, growth rate from 21 days was significantly lower for the males which received wheat and the cereal-free, higher-protein feed and for the females on either choice treatment. Differences in the treatment means were small. These findings suggest that, in fact, the female birds were not as capable as the males in controlling their daily protein intake.

The composition of the balancer diet affects the whole cereal intakes of choice fed broilers (Rose and Michie, 1984). Rose, Burnett and Elmajeed (1986) working with broilers from 21 days of age onwards confirmed that food form had a large effect on diet selection although there was no effect on total food intakes or weight gains. If a mash balancer was provided, which is unlikely to be the case in commercial practice, the broilers selected a greater proportion of wheat when it was ground rather than whole, but if the balancer was pelleted, a greater proportion of wheat tended to be selected when whole wheat was provided.

Research with turkeys started some 30 years ago. Chamberlin *et al.* (1962) cited by Cowan and Michie (1978b) allowed male large white turkeys in confinement a free choice of maize and a pelleted protein concentrate (320 g/kg crude protein) from eight to 24 weeks of age. On average, 57% of food consumed was the maize. Final body weights on average were similar to those of other males fed on a pelleted complete diet. In their own work Cowan and Michie investigated choice feeding of male turkeys from 50 days of age using pelleted barley and a turkey starter diet, whole barley and the starter diet, a ground barley based turkey finisher diet and the starter diet, pelleted barley and a series of complete diets compared with feeding the series of complete diets. For turkeys on the first four treatments 21%, 21%, 42% and 5% by weight respectively, of food consumed was the barley alternative. There were no significant differences in final bodyweight between turkeys on choice treatments and those receiving the choice of complete diets alone.

Rose and Michie (1982) offered choice fed growing turkeys whole wheat and each of six balancer diets the composition of which were identical except for the content of ground cereals. In a second experiment turkeys were offered whole wheat and each of four balancers which varied only in their calculated metabolizable energy content and the type of protein concentrate (white fish meal or meat and bone meal). The turkeys which were fed on balancers with a high white fish meal content ate more whole wheat and correspondingly less balancer than the turkeys offered balancers with a high meat and bone meal content. The energy content of the balancer did not affect food intake of the turkeys in the total feeding period. A high proportion of barley in the balancer hastened the increase in whole wheat intake after the introduction of the choice feeding regime.

The management of broiler feeding programmes

When a chicken is young it has a small maintenance requirement (Filmer, 1991) but its growth potential is enormous. The older bird tends to have high maintenance and no growth. The consequence of this is that the young bird has a relatively high requirement for protein and essential amino acids and the requirement of the older animal is comparatively low.

These gradual changes in requirements are usually catered for by using feeds with lower protein content as the animal gets older. The number of diets used will depend on the type of stock and the length of the growing period. This approach can lead to periods of under and over feeding of key nutrients. When nutrients are underfed the maximum genetic potential for growth may not be achieved and efficiency of feed conversion will be poor. When they are overfed, excess protein, for example, has to be deaminated and excreted with the inevitable consequences of this.

Two approaches have been taken by nutritionists to overcome the problems. Firstly, by using more feeds to reduce the size of the steps but this has practical difficulties of stock control both at the mill and on the farm and, secondly, by equalising the amount of under and overfeeding of nutrients and specifically of protein.

In recent years a number of additional performance criteria have become important which have exacerbated the situation. In addition to weight for age and feed conversion ratio, other criteria of increasing commercial importance are now carcass yield, carcass composition and breast meat yield. The applicable criteria may vary from flock to flock and a further complicating factor may be the sex of the birds in the flock. Combinations of research and field results have identified the nutrient and specifically amino acid requirements and intake profiles that the average bird needs in order to achieve the criteria. If the requirements have been identified correctly and are met in practice they enable producers to consistently achieve their performance objectives. Generally speaking, if the requirements are met which maximise breast meat yield, then the requirements for the other production criteria will also be met.

In practice, it is simpler to use one dietary criterion rather than several. Lysine is an essential nutrient for optimising broiler performance and carcass characteristics. By using lysine as the nutritional variable it is assumed that all other amino acids are in balance, i.e. the birds are being fed a balanced protein. Figures 6.1 and 6.2 show the lysine requirements in terms of daily intakes for male broilers with a view to maximising breast meat yield and for females maximising weight for age. They clearly illustrate the differences in requirement as influenced by performance objective and sex.

In order to supply these target intakes accurately to a flock of birds, which is essential if maximum benefit is to be obtained, feed intake data have to be available when making the calculations as to what must be provided in the final feed delivered to the birds. This can be arrived at in a number of ways. Firstly, the feed intake targets specified by the relevant breeding company could be used but this has obvious limitations. Alternatively, the actual figures for the

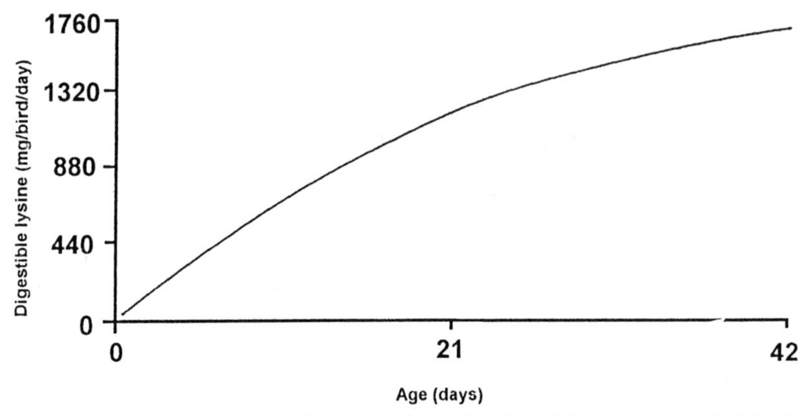

Figure 6.1 Digestible lysine requirements of male broilers with a view to maximising breast meat yield over the growing period

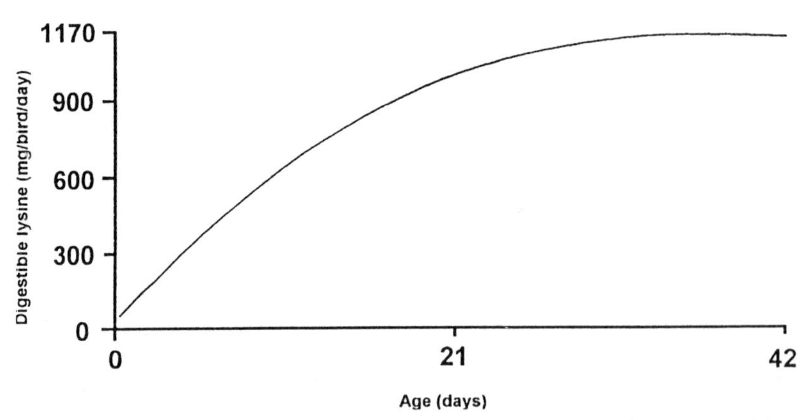

Figure 6.2 Digestible lysine requirements of female broilers with a view to maximising weight for age over the growing period

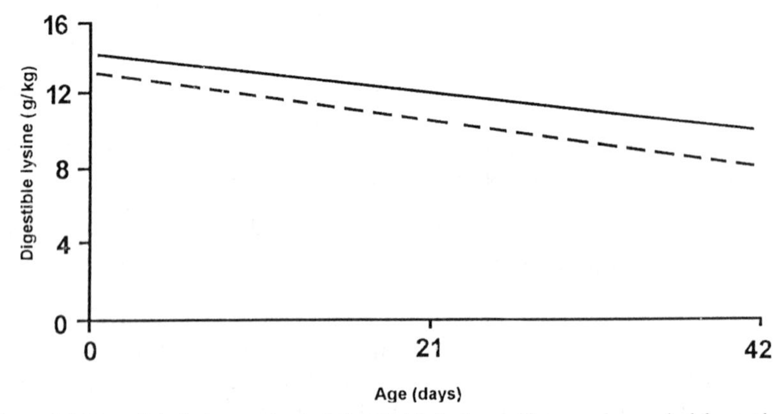

Figure 6.3 Digestible lysine content of the diet (g/kg) over the growing period for male and female broilers with a view to maximising breast meat yield (——) and weight for age (– – –) respectively

previous flock through the building could be used. These could be derived from details of feed deliveries or usage and bird numbers. The dates of feed changes would provide the points for the graph. By combining these data and the daily lysine intake requirement (dependent on the performance objective; see above) the required dietary lysine content for any part of the growing period can be calculated (Figure 6.3).

The lysine content of the feeds available for the respective stages of the growing period can be superimposed on to the graph of the calculated requirement and the compatibility of the two assessed (Figure 6.4). The content of any of the feeds, or all of them, and the period over which they are fed, can then be adjusted if necessary to achieve as near to 100% fit to the requirement curve as possible. This then gives the lysine contents to which each diet should be formulated and when it should be fed in order to achieve the required performance objective. The more stages in the feeding programme over the growing period, the greater the chance of a perfect fit.

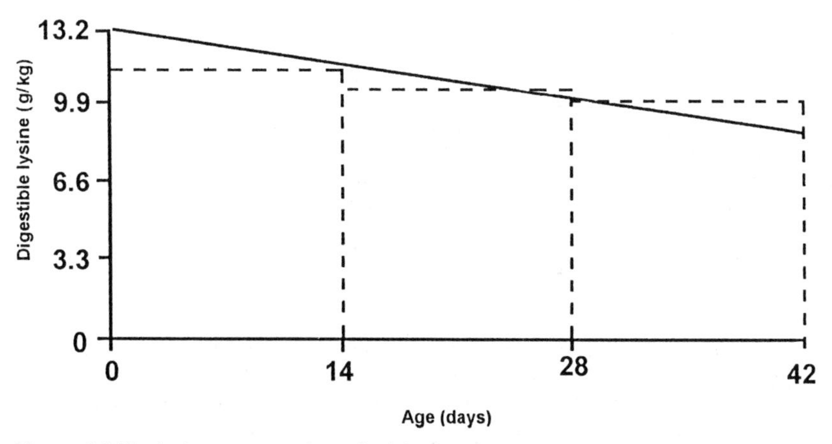

Figure 6.4 The lysine content of standard feeds available, superimposed onto the graph of the calculated requirement (female broilers (maximum weight for age); —— ideal feed composition, - - - standard diets)

Even with this approach it can be seen that there are periods when the birds are 'overfed' nutrients and at these stages some means of nutrient dilution is necessary in order to optimise feed utilization.

It is important to note that more stages need not mean more diets as two existing diets could be combined to produce an infinite number of feeds of intermediate composition.

PRACTICAL APPLICATION

Some companies have taken a very simple approach by literally putting 100 or 150 g/kg wheat on top of each load of broiler grower or finisher feed as it leaves the mill. In Denmark, whole wheat addition to broiler diets has been practised since 1984 (Anon, 1992). Wheat is introduced into the ration from day 12 at

50 g/kg inclusion rising to 300 g/kg at 35 days at which point the maximum recommended inclusion is reached and this inclusion rate is held until the birds go for processing. This is not sufficiently scientific if the aim is to provide the optimum nutrient profile for a flock. However, it does give individual birds the opportunity to select feed or wheat as they wish. This is 'free choice feeding' in commercial practice.

In a controlled experiment using 3750 Cobb broilers undertaken at the Harper Adams Poultry Research Unit, the feeding of standard starter, grower and finisher feeds at 0.75, 1.0 and 3.0 tonnes/1000 birds respectively was compared with feeding the same diets but with 150 g/kg whole wheat mixed in with the grower and finisher feeds. Supplement inclusion rates were increased to compensate for the wheat dilution effect. Results from the study are summarised in Table 6.1.

Table 6.1 OVERALL PERFORMANCE OF BROILER CHICKENS WITH AND WITHOUT THE ADDITION OF WHOLE WHEAT TO THE GROWER AND FINISHER DIETS

	Whole wheat used	
	Nil	*150 g/kg*
Mortality (% of birds housed)	12.48	9.49
Total culls (% of birds housed)	4.75	4.05
Hock lesions (% of birds inspected)	6.32	3.68
Average feed fed/survivor (kg)	4.79	4.71
Final average bird weight (kg)	2.45	2.50
Feed conversion (kg feed/kg LW)	1.96	1.89

In general, the effects of feeding the whole wheat were as follows. Mortality and total culls were reduced by 21% and hock lesions by 41%. Average liveweight increased significantly from 2.45 kg to 2.50 kg or 2.1% (P = 0.022) and this led to a 6.7% increase in yield per square foot of growing space (4.05 kg v 3.79 kg). Feed conversion ratio improved from 1.96 to 1.89 (3.75%).

The ultimate practical solution to the problem of meeting precisely the nutrient requirement profile is to set up two feed bins at each growing shed and put a high specification feed in one and a low specification feed in the other. The first feed would be higher in key nutrients than the highest specification required in any mixture and would probably resemble a starter diet. The second must be lower in these nutrients. The low specification feed may be a complete diet or whole wheat which can be more practical. In Denmark, growers have moved from the practice of spreading wheat on the top of deliveries to a system where wheat is included on a gradual daily increment through an accurate feed weighing system and this has led to better results. It requires a second feed bin at the shed in which to store the wheat. With this approach an average wheat inclusion of up to 250 g/kg is achieved over the growing period which is higher than that achieved in the UK. A partial explanation for this is that, after the starter diet, the Danes only feed one diet for the remainder of the growing period.

The availability of near infra-red equipment for the rapid analysis of complete feeds and feed raw materials means that an actual analysis could accompany each delivery of feed to a growing site meaning that any deviations between the theoretical and actual composition of the feed could be taken into account. Between batch variability could also be taken into account.

A mixture can then be fed to the flock and the blend ratio changed gradually and progressively each day so as to meet exactly the changing needs of the birds as they get older or feed intake changes. This is probably an over simplification of the approach but it removes the emphasis from the animal back to the nutritionist.

There has been a surge of equipment coming on to the UK market with the ability to blend wheat into feed on-site giving the combined arable and poultry farmer the opportunity to use home grown wheat available on the farm. The equipment now available ranges from sophisticated computer based systems working in real time which can blend in wheat on a daily basis so taking actual feed intake into account instead of forecast appetite, to the less sophisticated equipment designed to simply 'trickle' wheat on to the feed as it passes into the poultry house. A further alternative advocated by some is to simply supply whole wheat on alternate feeds.

A sophisticated approach to the concept described above has been developed beyond the theoretical stage and is in use on commercial farms. One such example is an independent broiler grower in the North of England who has installed the Flockman system into all of his poultry houses. With such equipment, which includes accurate weighing of feed delivered to the birds each day, daily feed consumption is accurately recorded and the birds precise requirements in terms of feed can be delivered by the system each day by blending the feeds available in the storage bins outside the shed. By using the system, a house of 29,000 Ross mixed sex broilers achieved an average factory weight of 2.78 kg at 49.4 days with a FCR of 1.97 and an EPEF (European Performance Efficiency Factor) of 270.

Feed was on average 115 g/kg whole wheat and was fed over the growing period. A second crop of male birds through the house achieved an average body weight of 3.18 kg at 49 days against a breeder's target of 2.61 kg with an overall FCR of 1.898. This level of performance is 22% above the breed target and is illustrated in Figure 6.5.

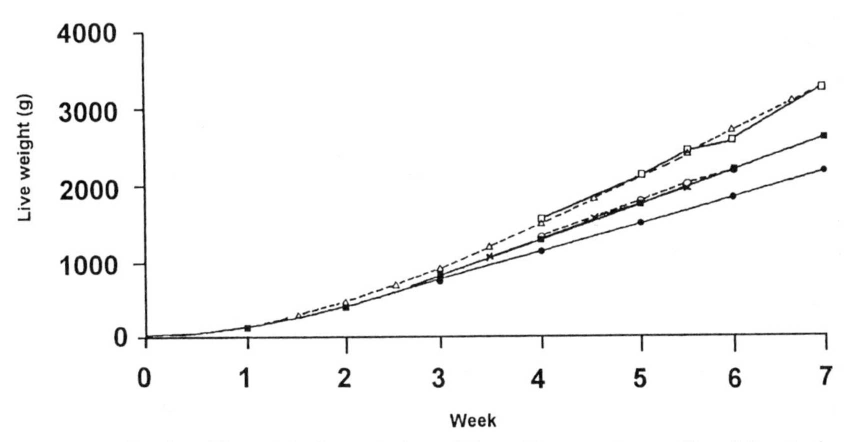

Figure 6.5 Graphs of liveweight for male (—— □ crop 1; - - - △ crop 2) and female (- - - ○ crop 1; —— × crop 2) broilers from a commercial unit where wheat blending is practised compared with breed targets (—— ■ Ross male standards; —— ● Ross female standards)

Table 6.2 OVERALL PERFORMANCE OF COMMERCIAL BROILERS GROWN WITH AND WITHOUT CONTROLLED WHEAT BLENDING

	Whole wheat used	
	Nil	*150 g/kg*
Slaughter age (days)	42.0	42.1
Final average bird weight (kg)	1.81	1.82
Feed conversion (kg feed/kg LW)	1.88	1.78
Margin per bird (pence)	2.63	8.45

Results from a second commercial unit are summarised in Table 6.2 and show an improvement of 0.10 in FCR where on average 155 g/kg whole wheat was blended into the feed. An improvement of 5.82 pence/bird in margin resulted from lower feed costs as a result of incorporating the wheat.

Carcass yield

Some concern has been expressed about possible loss of meat yield and in particular breast meat yield when whole wheat is blended into broiler feeds. Figure 6.6 shows total carcass yield as a percentage of liveweight and breast meat yield as a percentage of the eviscerated carcass for male birds sampled from the flock described above. As a comparison, the Ross targets for the parameters are included in the figure. No control birds were available but it can be seen that yields from birds fed wheat in a scientific manner were on or close to expected targets.

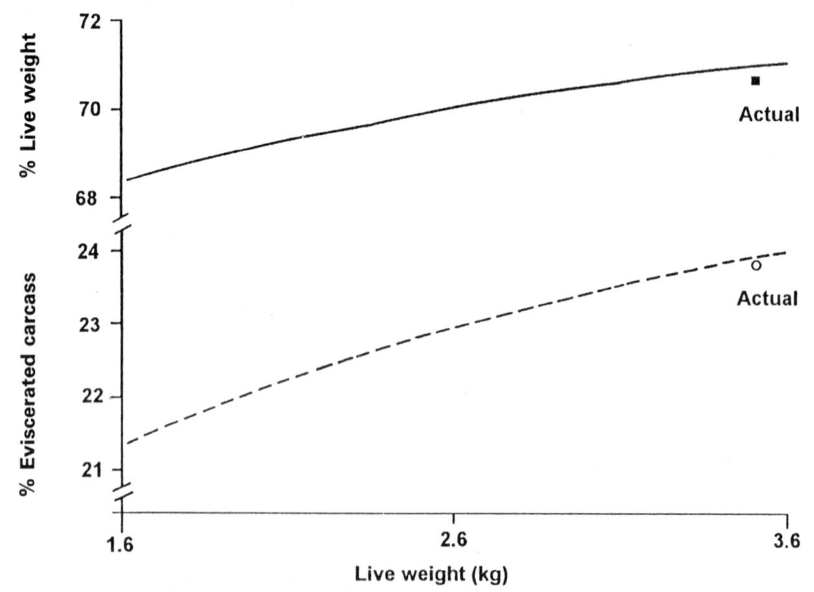

Figure 6.6 Carcass (– – –) and breast meat (——) yield of male broilers grown on a commercial unit where wheat blending is practised compared with breed targets

Safety

Concerns about safety of using whole wheat on farms relate to *Salmonella* and coccidiosis. *Salmonellae* have been isolated from most raw materials used in broiler feed but the spread of the organism can be controlled by heat treating the feed. This of course does not take account of any subsequent contamination after manufacture which is possible and does happen. It is felt that the introduction of whole wheat on to the growing site and into the poultry shed increases the chances of contaminating a flock with the inevitable risks to consumers.

In reality, very little wheat is contaminated, particularly if it is bought from a reliable source and has been stored in bird proof stores. Also, treatment of the wheat with salmonella inhibitor on delivery to the bin on site is now possible and should greatly reduce further what is already a small risk.

Table 6.3 shows the efficacy of a commercial liquid salmonella inhibitor, based on organic salts, in reducing the count of *Salmonella enteritidis* organisms on artificially infected wheat samples (G. Hall, personal communication). The initial inoculum of salmonella bacteria on the whole wheat was 5 ml of a 10^{-4} dilution of a suspension calculated by standard plate count assay to contain 3.16×10^6 organisms/ml, i.e. an initial theoretical dose of 316 organisms/100 g using a bulk initial sample of 500 g. This is approximately 30 times the amount found in contaminated wheat samples.

Table 6.3 THE EFFICACY OF A COMMERCIAL LIQUID SALMONELLA INHIBITOR IN REDUCING THE COUNT (MPN/100 g) OF *Salmonella enteritidis* ORGANISMS ON ARTIFICIALLY INFECTED WHEAT SAMPLES

	Organisms (MPN/100 g)	
	Control	*Treated*
After six hours	267	223
After twenty-four hours	147	78
After forty-eight hours	100	<1

After G. Hall, personal communication

The inoculum was administered using a fine aerosol jet on a wheat sample laid out in a large tray and constantly agitated to ensure an even mix. The test was carried out in triplicate for enumeration of bacteria after dosing. The control counts were made in infected wheat with no product added.

It can be seen from Table 6.3 that the application of the product at a rate of 4 kg/tonne led to a substantial reduction in organism counts expressed as MPN (Most Probable Number)/100 g some 48 hours after treatment.

The concerns about coccidiosis are probably equally unfounded. They arise from the fact that the addition of whole cereal to a compound feed containing an anti-coccidial will dilute the product below its recommended level of intake.

Work undertaken in Australia (Cumming, 1991) showed that the gizzards of birds fed only compound feed are atrophied and did not develop normally as they had no hard particles to grind down. Male broilers were fed either a standard feed or whole wheat offered free choice with pelleted feed (Table 6.4).

Table 6.4 THE EFFECTS OF FEEDING A COMPLETE DIET OR WHEAT ON A FREE CHOICE BASIS ON THE GIZZARD SIZE AND OOCYST OUTPUT OF BROILER CHICKENS

Replicate	*Complete diet*		*Free choice wheat*	
	Gizzard size (% body wt)	*Oocyst count*	*Gizzard size* (% body wt)	*Oocyst count*
1	1.09	180,000	2.28	<10,000
2	1.21	170,000	2.20	<10,000
3	1.55	150,000	1.74	10,000
4	1.87	140,000	2.21	<10,000
5	1.65	150,000	1.86	20,000
6	1.77	140,000	2.01	<10,000
7	1.17	170,000	2.07	<10,000
8	1.70	150,000	1.96	10,000
9	1.61	150,000	1.82	10,000
10	2.01	100,000	2.53	<10,000
Average	1.56	150,000	2.07	<11,000

(Courtesy of Cumming, 1991)

Within each group, the birds with the largest percentage gizzard tended to have the lowest oocyst output. The birds fed free choice wheat had on average 0.5% more of their body weight as gizzard. Cumming observed that bigger gizzards ground feed material, plus any ingested oocysts, more powerfully and for longer.

Legality

The question of legality arises because additives covered by product licences are used in table poultry feeds and the terms of each licence specify quite specifically how each product should be used. Also, as part of the procedures to protect both animal and human health, manufacturers of feeds containing licensed additives have to be registered.

There is not really a problem with layers' feed because normally they contain no products which require a licence. However, in the case of broiler and turkey feeds, there is at least one, and usually two or even three, products the use of which are covered by product licences included in the feed. These are basically the performance enhancer, the anti-coccidial and the anti-blackhead drug in the case of turkeys. If the compound feed is diluted with wheat after manufacture or at the farm then the inclusion levels of these products are diluted outside the terms of their product licence.

The Ministry advice is that the Medicated Feed Regulations do not apply where a standard feeding stuff containing a non-prescription medicine such as a coccidiostat is sold and labelled as a final medicated feed and used with a computer based system such as Flockman. Therefore, users of such systems will not be required to register their premises as on-farm mixers under Category B of The Medicines (Medicated Animal Feeding Stuffs) Regulations 1992.

To protect the efficacy of Prescription Only Medicines (POMs), which may be prescribed by a veterinarian under a Veterinary Written Directive (VWD), it is important that any whole cereal incorporation is suspended for the period during which any final medicated feed containing a POM product is fed. The exception to this is where the veterinarian advises an alternative strategy which allows whole cereal feeding to continue.

Therefore, if whole wheat is blended into feed, sold and labelled as a final medicated feed, on farm by equipment approved by Ministry of Agriculture, then no registration as a Category B home mixer is required. However, if feed delivered to the farm contains a prescription product such as an antibiotic only included with the authority of a veterinarian, sometimes the case with broiler starter diets, or the levels of the non POM medicines in the feed have been increased to compensate for the dilution effect of the wheat, then if whole wheat is to be added to that feed the owner of the premises on which this is to take place would have to register as a Category B mixer with the Royal Pharmaceutical Society of Great Britain.

Conclusions

The combination of a better understanding of the way that the daily intake of key nutrients, and particularly amino acids, influence the important performance objectives of the modern broiler chicken and the availability of cost effective computer based control equipment capable of operating in poultry houses has meant that it is now a commercial reality to be able to design feeding programmes, even specifically to one crop of birds, in order to achieve the desired performance objective with that crop. The sophistication of systems is such that it is even possible to modify the feeding programme for a crop of birds on a daily basis through the growing cycle if it is failing to achieve the desired objectives in terms of growth and feed conversion and to monitor the outcome of any change of programme.

By adopting such programmes in the field, it has become clear that the genetic potential of the modern broiler chicken is well in excess of that achieved under normal commercial conditions. However, with a combination of good flock health, general management and the application of a modified feeding programme, high levels of performance can be achieved in a cost effective way. One significant effect of this approach is that target body weights may be achieved much earlier than in the past. A commercial effect of this is the possibility of increasing the throughput of growing facilities with a corresponding increase in margin per unit of growing space per year. The requirement for increased growing capacity required by a company to meet increased demand may be reduced which could be important in light of planning and environmental difficulties when building new sites. Also, the number of birds that could be fed from an integrated feed mill of limited capacity can be increased if part of the final feed fed is whole cereal.

Such modified feeding programmes may involve the scientific feeding of whole wheat during the growing period. The legality and safety of this practice has been questioned. At the end of the day it seems those concerns are now unfounded. Whether whole wheat is used or simply combinations of complete feeds, the implications are that less complete feeds may have to be manufactured because final feeds, which theoretically should be changed daily during the growing period, could be blended on site.

Despite the availability of suitable equipment to undertake on-farm wheat blending and the clear financial advantages of so doing along with other possible beneficial effects resulting from it, both environmental and bird welfare, there is still reluctance to adopt it by many farmers. Clearly the broiler industry, and possibly turkeys, offer the greatest potential financial benefits because of the wide price differential between the price of the relevant compound feeds and whole wheat.

References

Anon (1992) *Poultry World*, **146**(3), 19
Cowan, P. J. and Michie, W. (1978a) *British Poultry Science*, **19**, 1–6
Cowan, P. J. and Michie, W. (1978b) *British Poultry Science*, **19**, 149–152
Cumming, R. B. (1991) In *Recent Advances in Animal Nutrition in Australia 1991*, pp 339–344
Filmer, D. (1991) *Feeds and Feeding*, **1**(2), 30–33
Rose, S. P. and Michie, W. (1982) *British Poultry Science*, **23**, 547–554
Rose, S. P. and Michie, W. (1984) *Animal Feed Science and Technology*, **11**, 221–229
Rose, S. P., Burnett, A. and Elmajeed, R. A. (1986) *British Poultry Science*, **27**, 215–224

7

REARING THE LAYER PULLET — A MULTIPHASIC APPROACH

RENE P. KWAKKEL
Department of Animal Nutrition, Agricultural University, Haagsteeg 4, 6708 PM Wageningen, The Netherlands

Introduction

Egg production is the ultimate goal for rearing a layer pullet. It is the integration of the genetic capability of the bird on the one hand and, for the main part, the result of nutritional, and other environmental, conditions during the laying period, on the other. It is well known, however, that laying performance can also be influenced by the occurrence of physiological alterations in the growing bird, due to the application of certain feeding strategies during rearing (Table 7.1). Such physiological alterations might result in distinctive changes in body characteristics during the pre-lay period or even alter processes relating to organ and tissue development.

In this chapter the importance of studying growth and development of pullets using a multiphasic approach is being addressed. Multiphasic growth functions describe growth of the body and its constituents in layer pullets. Results from multiphasic growth studies may help to explain differences in laying performance due to nutritional conditions during rearing.

The modern layer is continuously improving egg production, due to a genetic selection for optimum performance: today's young hen matures several weeks

Table 7.1 POSSIBLE MODE OF ACTION OF PULLET FEEDING STRATEGIES

Input	—	Restricted energy or protein (amino acids)
	—	Severity of restriction
	—	Period of restriction
	—	Moment of cessation of restriction
Mechanism	—	Critical periods in organ and tissue growth
	—	Alterations in nutrient partitioning
	—	Changes in multiphasic growth patterns
	—	Endocrine control of development processes
Output	—	Onset of lay / rate of lay
	—	Egg size and quality
	—	Feed intake and adult body growth

earlier than the hen of two decades ago. Onset of lay is now around 19 weeks of age, and the hen reaches peak production only a few weeks later (Summers and Leeson, 1983; Summers, Leeson and Spratt, 1987). The consequence of this improvement is that pullet feeding strategies have to be reconsidered and evaluated every few years in order to match the feeding strategy to the physiological requirements of the growing pullet (Summers, 1983; Leeson, 1986).

In the next paragraph a brief review of two decades of pullet feeding research is presented.

Research on pullet feeding strategies: 1970–1992

For 30 years, the optimization of nutritional conditions for rearing hens in order to maximize adult laying performance has been a key subject for researchers (for reviews see Lee, Gulliver and Morris, 1971; Balnave, 1973; Karunajeewa, 1987). Lee *et al.* (1971) presented a comprehensive review on the effects of pullet feeding strategies during the sixties. They concluded that, although restricted-fed pullets delayed onset of lay, egg output was increased, due to a more efficient utilization of nutrients during lay. The main advantage of restricted feeding was the economic benefit in saving rearing feed costs (Balnave, 1973).

TARGET WEIGHTS

Some 15 years ago, breeding companies emphasized the importance of achieving a so-called 'target weight' for a ready-to-lay pullet (Wells, 1980; Balnave, 1984). Physiological relationships between target weights and production features were not very clear. Body weight served as a convenient tool for judging the rearing stage under practical conditions (Singh and Nordskogg, 1982). The rearing period was considered as a 'non-profitable' stage and nutritional requirements were therefore not properly defined (Leeson and Summers, 1980). The main goal was to achieve target weights with a minimum of nutritional input. All rearing methods were based on this least-cost principle (Leeson, 1986). Most of the experiments described in the literature were conducted to evaluate each feeding strategy on the basis of this economic policy.

Hence, this kind of empirical research did not explain any of the physiological mechanisms involved in pullet growth and maturation (Kwakkel, De Koning, Verstegen and Hof, 1991).

GROWTH PATTERNS

However, Wells (1980) stated that 'body weight at 18 weeks of age is not a reliable indicator of subsequent laying performance, when considered in isolation from the pattern of growth leading to that weight'. Undoubtedly, a minimum pre-lay body weight is necessary for onset of lay (Dunnington, Siegel, Cherry and Soller, 1983; Dunnington and Siegel, 1984), but the shape of the body growth curve may give additional information on subsequent performance ability (Leeson and Summers, 1980). From the early 1980s, research on feeding strategies for layer pullets became more focused on growth pattern than on target weights (Leeson and Summers, 1980; Wells, 1980; Summers, 1986).

Pullets' feeding strategies were usually 'step-down' protein programmes, a

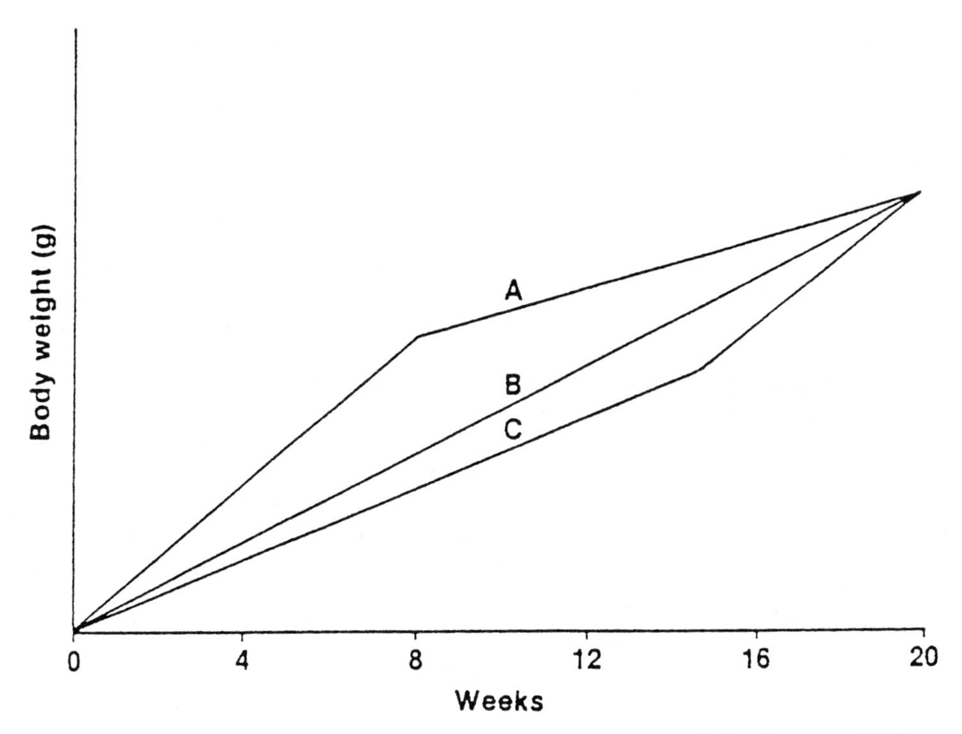

Figure 7.1 Different growth patterns for pullets as proposed in literature (After Summers, 1983)

decrease in protein content with advancing age, characterized by high growth rates in the starter period (high protein). In order not to overrule target weights, pullets were then restricted in feed intake during the grower period. As a consequence, the growth curve followed pattern A (Figure 7.1; Summers, 1983). The young hen, however, was not able to build up any body reserve for the first few laying weeks when feed intake did not yet match nutrient requirements. Post-peak production dips were a common phenomenon (Leeson and Summers, 1980).

Leeson and Summers (1979; 1980) stated on the basis of 'choice-feeding' experiments, that the relative requirements for protein increased with advancing age, and that pullets were fed 'back to front' up to that time. They postulated that the birds would be better fed with low protein starter and high protein grower diets, a 'step-up' regime, than with the conventional step-down diets. Their theory was based on a rather slow rate of muscle growth during the starter period and an increased ovary and oviduct growth during the grower period, which would increase appetite for high protein diets in that period (Leeson and Summers, 1980). The step-up method resulted in a linear growth curve B (Figure 7.1). This new method saved feed costs during early rearing and delayed sexual maturity. Unfortunately, it also decreased egg sizes and increased mortality during the rearing and laying period. These problems were attacked by higher protein levels during the first few weeks of rearing (the 'modified' step-up regimes: Bish, Beane, Ruszler and Cherry, 1984; Robinson, Beane, Bish, Ruszler and Baker, 1986) which resulted in normal egg sizes and lower mortalities.

Wells (1980) suggested a so-called 'mid-term' feed restriction, from about 7 to 15 weeks of age, to be suitable for the layer pullet. This method assumed that it would be better to restrict the bird during periods of overconsumption and prevent excessive fat deposition. However, by allowing the young hen free access to feed three weeks before onset of lay, she could compensate growth retardation and increase her appetite.

A few years later, pullet rearing specialists Summers and Leeson promoted a quick body growth during the starter period in order to enable an early maturation of the size of the body (skeleton growth; Leeson, 1986; Summers, 1986). Problems with prolapse in young hens at the start of lay was the basis for this advice (Leeson, personal communication). They suggested a diet/body weight strategy instead of one based on diet/age. This was a step forward in pullet rearing in that dietary changes should take place if body weight levels were reached, rather than if a certain chronological age had been reached.

Leeson postulated that the aim of a good rearing programme should be to produce large pullets carrying sufficient reserves of body protein and especially fat, at point of lay (Leeson, personal comm.). Energy intake as the promoter of body weight gain during the grower period had to be stimulated (Summers, 1986; Summers *et al.*, 1987). The bird should grow fast during late rearing (energy as body reserve). A growth reduction during mid-term rearing followed by high density diets during late rearing seemed to be the most appropriate method to promote feed consumption just before lay (curve C; Figure 7.1).

BODY COMPOSITION

Differences in the pattern of growth might affect body composition during and at the end of rearing. Experiments were designed to elucidate the role of body composition in order to explain differences in onset of lay and overall performance (Brody, Eitan, Soller, Nir and Nitsan, 1980; Soller, Eitan and Brody, 1984; Chi, 1985; Johnson, Cumming and Farrell, 1985; Leeson, 1986; Zelenka, Siegel, Dunnington and Cherry, 1986; Summers *et al.*, 1987). Body composition during rearing and at point-of-lay does play an important role in the maturation of the young hen. However, the exact causal pathways are not yet well understood (Dunnington and Siegel, 1984). Onset of lay may be determined by a number of interrelated factors such as age, body weight, body fat and/or fat free tissue (both in relative and absolute amounts), body size, and genetic strain (Brody *et al.*, 1980; Brody, Siegel and Cherry, 1984; Dunnington *et al.*, 1983; Dunnington and Siegel, 1984; Soller *et al.*, 1984; Summers *et al.*, 1987; Zelenka *et al.*, 1986). Age and body weight, in particular, were thought to be important as threshold factors for onset of lay (Dunnington *et al.*, 1983; Dunnington and Siegel, 1984). However, Dunnington and Siegel (1984) postulated that a limiting body weight may require a certain body composition to commence production. Work on broiler breeders showed that a specific percentage of body fat seemed to be required for onset of lay (Brody *et al.*, 1980, 1984; Zelenka *et al.*, 1986). Soller *et al.* (1984) confirmed these results but they also thought that a certain lean body mass was required.

Most scientists agree with the statement of Summers *et al.* (1987) that a profitable young hen must probably attain a minimum body weight in combination with a 'particular' body composition in order to initiate egg production (Brody *et al.*, 1984; Soller *et al.*, 1984; Zelenka *et al.*, 1986).

OTHER PROPOSED MECHANISMS

A mild stress caused by a feed restriction should stimulate the development of endocrine glands, and so induce a higher rate of lay (Hollands and Gowe, 1961). Long term metabolic alterations due to feed restrictions were suggested by Hollands and Gowe (1965).

An altered gonadotrophin output or an increased sensitivity of the ovary and oviduct to gonadotrophins should induce bigger oviducts, faster rates of follicle growth and an improved production by restricted birds (Frankham and Doornenbal, 1970; Watson, 1975).

Johnson, Choice, Farrell and Cumming (1984) concluded that the feed or energy intake after cessation of restriction could play an important role in increasing egg sizes.

Hocking, Gilbert, Walker and Waddington (1987) recently focused more on the biology of reproduction in the young layer. He examined the hierarchical structure of follicle development in ovaries of rearing hens fed different quantities of feed. Hocking, Waddington, Walker and Gilbert (1989) postulated that the incidence of shell-less and broken eggs in *ad libitum* fed birds, leading to an unsatisfactory egg production, could be explained by the large amount of 'ready' yellow follicles, inducing multiple and internal ovulation. He suggested therefore a feeding programme with a moderate restriction until point of lay.

Katanbaf, Dunnington and Siegel (1989) dissected several body structures from pullets having different feeding strategies, and found treatment differences in growth rate for some of them. Unfortunately no conclusions were drawn concerning differences in performance.

DISCREPANCIES IN THE LITERATURE

Results in the literature which describe egg performance as a result of particular feeding strategies, are hard to compare and sometimes even conflicting (Lee, 1987). These discrepancies are the result of differences in:

- body weight at point of lay (Kwakkel *et al.*, 1991),
- management: e.g., uniformity, feeding space allowance (Robinson and Sheridan, 1982),
- levels of essential amino acids in low protein diets (Balnave, 1973),
- strain (Abu-Serewa, 1979; Lee, 1987),
- the severity of the restriction (Balnave, 1984; Lee, 1987),
- start and cessation of restriction (Johnson *et al.*, 1984),
- method of restriction (Wells, 1980).

Lessons from the past and future procedure

Three basic questions on pullet feeding strategies remain:

1. At which physiological stage(s) of development do organs and tissues receive signals resulting in irreversible preparations, a kind of 'setting', for subsequent egg production?
2. What system (e.g. endocrine or nervous) controls this setting of the body?
3. What is the relationship between a certain feeding strategy and the physiological alterations related to this setting for egg production? In other words:

in what way, to what extent, and during which period are nutrient restrictions appropriate?

Answers to these questions need an interdisciplinary biological approach.

A new approach: Multiphasic growth curve analyses

MONOPHASIC VERSUS MULTIPHASIC GROWTH

Growth can be considered as a discontinuous process. The existence of several distinguishable growth 'waves', which is the basis of the multiphasic growth theory, is widely accepted and has been incorporated into human medical research.

In 1777, the Count of Montbayard was one of the first who perceived the multiphasic nature of growth. He described the pubertal growth spurt of his son by collecting his son's height data from birth till 18 years of age on a yearly basis (among others: Short, 1980; Koops, 1989).

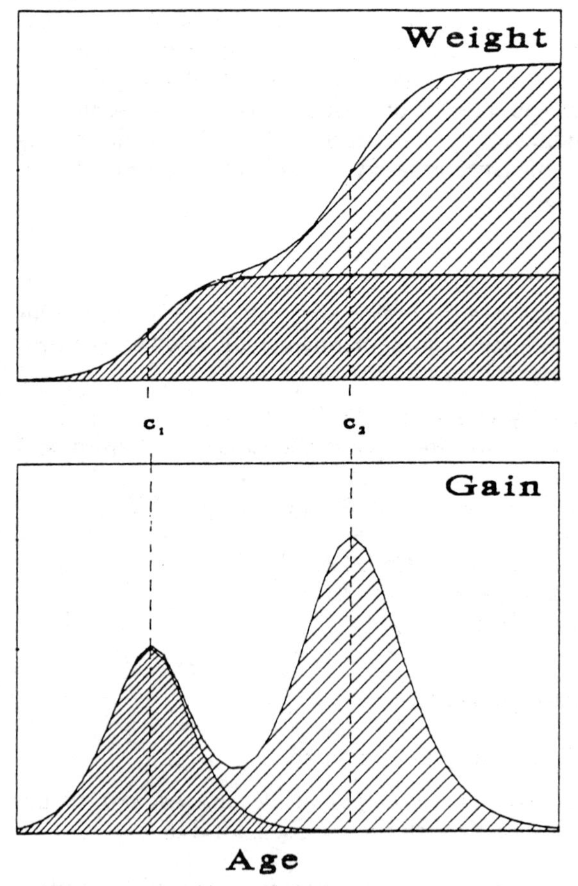

Figure 7.2 An example of a diphasic weight-age (upper) and gain-age (lower) curve. C_1 and C_2 are points of inflection (ages at maximum gain) for the first and second phase, respectively (After Koops and Grossman, 1991a)

In contrast, growth in farm animals, e.g. in poultry, has usually been described by the well-known monophasic S-curve, assuming only one inflection point, in which growth rate is at a maximum. The main reason for this was not the acceptance of discontinuous growth by animal researchers, but the lack of detailed observations over time, a prerequisite for the multiphasic growth approach (Koops, 1989). An example of a diphasic growth function is presented in Figure 7.2 (Koops and Grossman, 1991a).

CRITICAL PERIODS

If nutrient supply is not limited, each body organ or tissue will follow its own distinctive maturation curve (Ricklefs, 1975). As a consequence, there will be a variation in the nutritional demand of every organ or tissue in the course of time, related to the development of that respective body structure (Ricklefs, 1985). This means that the supply of nutrients for certain organs may be critical at particular stages of immature growth (McCance, 1977). Ignoring such 'critical periods' might influence subsequent performance negatively. An overall moderate feed restriction for layer pullets during the entire rearing period is common practice in the Netherlands. However, it does not take into account the existence of such critical periods (Kwakkel, 1992).

Nutritional programmes for pullets have to be adjusted to take into account the stages of development of important body structures (e.g., the ovary and oviduct for the layer). For example, multiple regression analysis, done by Elahi and Horst (1985), indicated that the influence of oviduct weight on egg weight is almost three times larger than the influence of body weight on egg weight. Critical periods need to be identified, and as a consequence, nutrient restrictions at these stages avoided (Kwakkel, 1992).

THE MULTIPHASIC GROWTH FUNCTION

The mathematical function presented here, enables one to study distinct stages of development of particular body parts (organs or tissues). The multiphasic growth function (MGF), developed at the Wageningen Agricultural University (Koops, 1989), is based on a summation of n sigmoidal curves. Each curve or growth cycle is described by a logistic function. The curve represents growth data as a function of age. The MGF function [1] defines number of phases, growth within phases, age at maximum gain in each growth phase, and duration of each phase. The initial parameters as described by Koops (1986), have been reparameterized for an easier biological interpretation (Kwakkel and Koops, 1991):

$$y_t = \sum_{i=1}^{n} \left(2\frac{A_i}{B_i} \{1\text{-tan}h^2 [\frac{4}{B_i}(t-C_i)]\}\right) \quad [1]$$

where y_t is the predicted value of mean weight gain in week t; n is the number of phases; and in each phase i, A_i is the asymptotic weight gained (grams), B_i is duration of the phase (weeks) and, C_i is the age at maximum gain (weeks). The hyperbolic tangent is *tan h*.

Multiphasic body growth in non-restricted pullets

Multiphasic analyses of the entire body of *ad libitum* fed pullets will give some idea about specific growth spurts during stages of development in 'normal' reared hens. Therefore, body growth data of pullets fed adequate *ad libitum* starter and grower diets (Kwakkel *et al.*, 1991) were fitted using the MGF function. A four phase growth function described the data most accurately (Figure 7.3; Kwakkel, Ducro and Koops, 1993). The first two post-hatch growth spurts accounted for 82% of mature body weight (asymptotic weight gains were 1,150 and 215 g, respectively; Table 7.2). A third growth spurt, with a sharp increase at around 19 weeks of age, the so-called 'maturity spurt', consisted of over 70% of growth of the reproductive tract (Kwakkel, Unpublished). The first two post-hatch growth spurts seem to represent some kind of general growth, e.g., the skeleton, muscles and the gut (Lilja, Sperber and Marks, 1985), before processes of sexual development start. In the fourth phase depot fat is likely to be accrued.

Grossman and Koops (1988) fitted body growth data of layers to a diphasic growth function. It is assumed, however, that their first growth spurt is a combination of our first two growth spurts (compare parameter estimates of both curves in Table 7.2). They postulated that the second growth spurt at around 27 weeks of age was associated with sexual maturity. The difference in age at maximum gain of the maturity spurt between that work and ours can partly be explained by the genetic changes for earlier sexual maturity (data of Grossman and Koops (1988) came from birds reared in 1967). Grossman and Koops (1988) also recognized a distinguishable growth spurt beyond the maturity spurt, but, because their model did not improve with the addition of an extra phase, they did not include this growth spurt in the model.

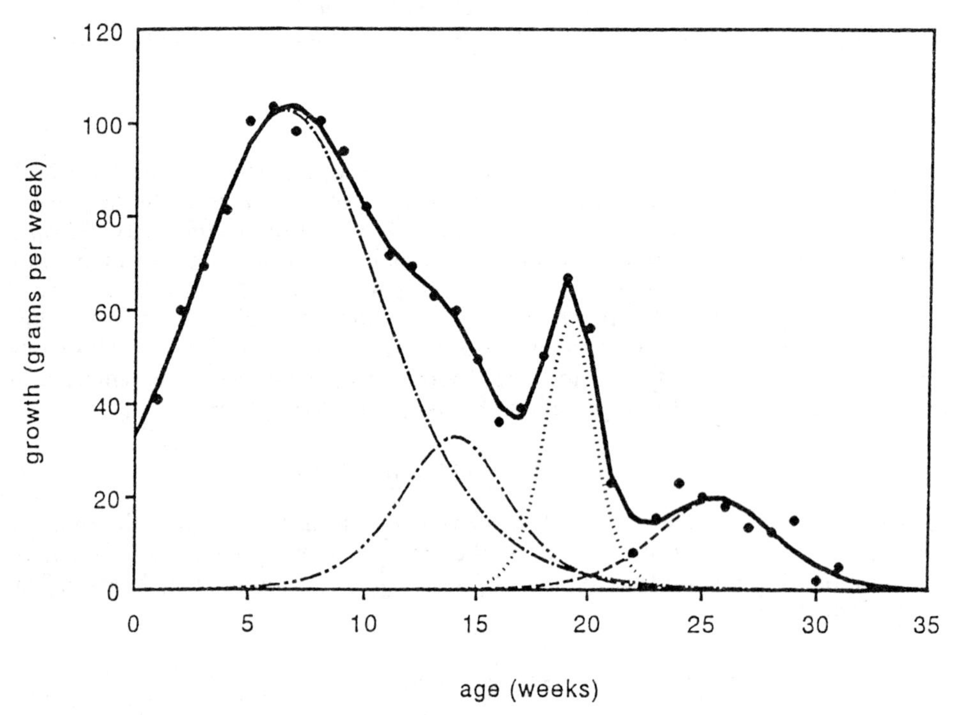

Figure 7.3 Body weight gain of *ad libitum* fed pullets from 0 to 32 weeks of age, described by a four phasic growth function (Kwakkel *et al.*, 1993)

Table 7.2 PARAMETER ESTIMATES OF A FOUR PHASIC[a] AND DIPHASIC[b] FUNCTION OF BODY GROWTH IN NON-RESTRICTED PULLETS AND HENS

		Parameter estimates			
Type of function[c]	*Phase i*	*Asymptotic weight gained*[d] (g)	(%)	*Duration*	*Age at maximum gain* (weeks)
Four phasic	1	1150	69	22.4	6.6
	2	215	82	13.1	14.0
	3	169	92	5.8	19.2
	4	135	100	13.9	25.6
Diphasic	1	1499	82	33.3	11.8
	2	334	100	11.9	27.3

[a] After Kwakkel *et al.* (1993)
[b] Calculated from Grossman and Koops (1988) by Kwakkel *et al.* (1993)
[c]

$$y_t = \sum_{i=1}^{n} (2\frac{A_i}{B_i}\{1\text{-}\tanh^2[\frac{4}{B_i}(t\text{-}C_i)]\})$$

where, in each phase i, A_i is the asymptotic weight gained (g), B_i is the duration (weeks), and C_i is the age at maximum gain (weeks)
[d] Cumulative weight gained as a percentage of mature BW

Comparing feeding strategies: an experimental design

At the Wageningen Agricultural University, a 2 × 2 factorial (plus added control) experiment had been conducted to investigate the effects of method (low lysine or a restricted amount of feed) and period of restriction (starter or grower) during rearing, on body development and egg performance in pullets and hens. Five to eight pullets were sacrificed weekly per treatment, defeathered and the gut-fill removed. Measurements on organ growth were carried out. Following a 3-week interval, emptied bodies, including all organs and viscera, were frozen, minced and freeze dried for analyses of dry matter (DM), crude fat, crude protein and ash. The design of the experiment is outlined in Table 7.3. More details on management

Table 7.3 DESIGN OF THE EXPERIMENT[a]

Treatment		*0–6 weeks of age*	*7–18 weeks of age*
C	Control	Starter[b] diet	Grower[c] diet
RLs	Lysine restricted Starter phase	4.0 g/kg digestible lysine	Grower[c] diet
RFs	Feed restricted Starter phase	Pair gained to RLs	Grower[c] diet
RLg	Lysine restricted Grower phase	Starter[b] diet	3.0 g/kg digestible lysine
RFg	Feed restricted Grower phase	Starter[b] diet	Pair gained to RLg

[a] Restricted treatments were only restricted in one of the two rearing phases; lysine-deficient diets were adequate in all other nutrients
[b] Starter diet: 8.5 g/kg digestible lysine; ad libitum feeding level
[c] Grower diet: 6.0 g/kg digestible lysine; ad libitum feeding level (From Kwakkel *et al.*, 1991)

and experimental procedures are described in Kwakkel *et al.* (1991) and Kwakkel, *et al. (1993).*

Onset of lay related to the maturity growth spurt

No differences in body weight were observed between the four restricted groups at 18 weeks of age. Onset of lay, however, was significantly ($P<0.01$) affected by the method of restriction. Lysine-restricted pullets started to lay 5 days later on average than did the 'pair gained' feed-restricted pullets, irrespective of the period of restriction ('pair gained' means that two groups were fed such that the gain in liveweight was similar between groups; Table 7.4). A difference in onset of lay between pullets restricted in lysine or feed was already observed by Gous (1978) and Wells (1980). The differences in onset of lay in our experiment were clearly reflected by differences in age at maximum gain of the maturity growth spurt. The interval between the maturity spurt and onset of lay (in terms of 50% production) was about 15 days. If this interval, with the prerequisite of a given management and a non-restricted feeding level, is quite fixed, our next question is: What triggers the onset of the maturity growth spurt? or in other words 'What kind of body conformation is needed to trigger physiological (endocrine) systems to start the development of the reproductive tract?' Another question follows directly from this 'What was the body composition at the beginning of the maturity growth spurt?'

Body composition at the beginning of sexual development

The start of the maturity growth spurt was estimated by subtracting half the duration (see Table 7.2) from the age at maximum gain of the maturity spurt.

Table 7.4 INTERVAL (IN DAYS) BETWEEN THE MATURITY PEAK AND ONSET OF LAY, AS WELL AS BODY COMPOSITION (IN ABSOLUTE g) AT START OF SEXUAL DEVELOPMENT OF PULLETS ON DIFFERENT FEEDING STRATEGIES

	Treatments[a]				
	C	*RLs*	*RFs*	*RLg*	*RFg*
Maturity peak[b]	134	142	139	141	136
Onset of lay[c]	149	157	153	156	150
Interval	15	15	14	15	14
Start of sexual development[d]	113	121	118	120	115
Crude fat[e]	178.1	126.0	153.6	175.9	111.7
Crude protein[e]	201.3	176.9	187.5	187.3	191.8
Ash[e]	41.0	37.7	38.2	38.1	38.5

[a] Treatments: see Table 7.3
[b] Maturity peak is defined as the age of maximum gain of the maturity growth spurt (in days)
[c] Onset of lay is defined as the age (in days) at which 50% rate of lay is reached
[d] 'Start of sexual development' has been estimated by subtracting half of the duration (see Table 7.2) from the age at maximum gain of the maturity peak
[e] Body composition at 'start of sexual development' was predicted from observations on chemical composition (Kwakkel, *et al.* 1993)

All treatments started the maturity phase between 113 and 121 days of age (Table 7.4). For all treatments, amounts of crude fat, crude protein and ash at the estimated start of the maturity spurt were predicted from observations on chemical body composition at regular intervals in all treatments (Kwakkel *et al.*, 1993). It seems that a particular amount of body protein determines the moment of sexual development: all restricted groups had about 187 (177 to 201) g protein within the body, whereas the fat content fluctuated between 112 and 178 g at that point. The relatively large amount of protein in the control group might be the result of reaching the 180 g protein level before another threshold level (age ?) had passed. It seems that some general protein growth must be complete before sexual development can start. Thus, it seems that body composition, especially protein, is critical in determining the onset of lay (Johnson *et al.*, 1984; 1985; Summers *et al.*, 1987), even 5 weeks before the first eggs are being laid. Fat content seems to be of minor importance as a 'setting' factor for onset of lay.

Multiphasic growth of body components in non-restricted pullets

The shape of a growth curve is determined by the accretion of protein, fat and ash in the body. The moment of deposition of these components in individual body organs determines the birds' physiological age and stage of maturity (Ricklefs, 1985).

The accretion of DM, crude protein, crude fat and ash in *ad libitum* fed pullets showed a diphasic growth pattern using the MGF (Figure 7.4; Kwakkel and Koops, 1991). The sum of estimated weight gain in crude protein, crude fat and ash in the first phase (212 ± 23, 123 ± 31, and 42 ± 3 g, respectively) and in the second phase (20 ± 16, 137 ± 29, and 9 ± 8 g, respectively) was equal to the total DM accretion in each phase (380 ± 94 and 170 ± 82 g, respectively), which confirms the accuracy of the fits. In the first phase of crude protein, crude fat and ash, similar durations (25.2 ± 3.3, 22.6 ± 4.8, and 20.6 ± 2.0 weeks, respectively) and approximately same ages at maximum gain (7.9 ± 0.5, 8.5 ± 1.1, and 6.8 ± 0.3 weeks, respectively) suggest a functional relationship between these components within that time interval (Kwakkel *et al.*, 1993). It is postulated that growth in crude protein and crude fat in the first phase consists mainly of muscle growth and related intramuscular fat deposition (Walstra, 1980; Koops and Grossman, 1991a). Crude protein growth in the second phase is related to growth of the oviduct, whereas growth of crude fat in this phase is related to abdominal fat and other fat storage depots (Kwakkel, 1992). The diphasic nature of fat growth has also been recognized in pigs (Walstra, 1980).

The well-known growth order of bones, muscles and lipids in men and animals, as proposed by Hammond (1932) in his classical theory, is confirmed by these data: the diphasic growth function showed that of total DM growth in the first phase, 56% was crude protein growth and only 33% was crude fat growth. On the contrary, 12% of DM growth in the second phase was crude protein growth and almost 83% was crude fat growth. About 90% of total crude protein growth is accrued in the first phase of rearing. This is in total disagreement with the assumptions made for introducing the 'step up' rearing method by Leeson and Summers (1980).

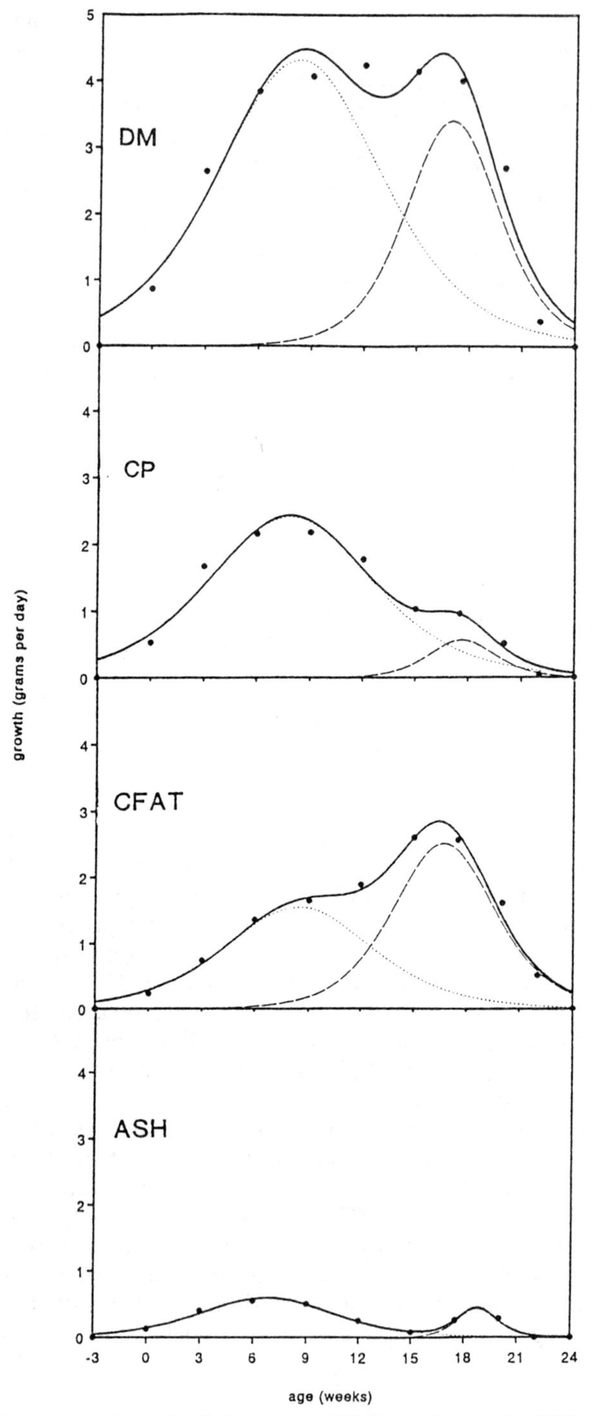

Figure 7.4 Growth of dry matter (DM), crude protein (CP), crude fat (CFAT), and ash in non-restricted pullets, described by a diphasic growth function (Kwakkel *et al.*, 1993)

THE MULTIPHASIC ALLOMETRIC FUNCTION

Another multiphasic model, the multiphasic allometric function (MAF) was applied to compare relative accretion of chemical body components in pullets which had been reared on the different feeding strategies as mentioned in Table 7.3.

The MAF function, also developed by Koops and Grossman (1993), is an extension of the simple allometric model [2]. This model is often used in the literature (Ricklefs, 1975; Lilja *et al.*, 1985) to describe the relationship between two body constituents (Koops and Grossman, 1991b):

$$y_x = \alpha x^{\beta}$$

or

$$\ln(y_x) = \ln(\alpha) + \beta \ln(x) \qquad [2]$$

where, y_x is the weight (or length) of one body constituent, x is the weight (or length) of the other constituent, α is the scale parameter ($\ln(\alpha)$ is intercept), and β is the allometric growth coefficient (slope).

This simple allometric model has become popular because it incorporates the allometric growth coefficient, a parameter which is easy to compare between treatments. However, a change in the relationship between the two body structures, which in the light of the multiphasic growth theory is quite conceivable, is not possible with this simple model. Researchers tried to overcome this problem by fitting different linear log-log relationships for different data areas 'by eye' (Ricklefs, 1975; Lilja *et al.*, 1985). The MAF function [3] allows a smooth transition from one allometric level to another:

$$\ln(y_x) = \ln(\alpha_{i-1}) + \beta_i \ln(x) - \sum_{i=1}^{n-1} \{(\beta_i - \beta_{i+1})\ r_i\ \ln[1+\left(\frac{X}{C_i}\right)^{\frac{1}{r_i}}]\} \qquad [3]$$

where y_x, x, and α are as in Equation [2]; n is the number of phases; and in each phase i, β_i is the allometric growth coefficient, c_i is the estimated breakpoint between phase i and $i + 1$, and r_i is a smoothness parameter.

Multiphasic allometry between body components in restricted fed pullets

In all treatment groups, simple allometric functions of crude protein, crude fat, and ash, each as a function of DM, did not fit the data very well. Figure 7.5 illustrates this for the control group. Monophasic fits showed periodic deviations. In all cases the simple allometric fit was significantly improved by including a second phase in the model (Kwakkel *et al.*, 1993).

No differences between treatments could be found in the relationship between live body weight (LW) and empty body weight (EBM; on average, EBM = 0.85 × LW) as well as in the allometric relationship between DM and EBM (Kwakkel, In preparation). Parameter estimates of the mono- and diphasic allometric relationships between crude protein and DM, crude fat and DM, and ash and DM for *ad libitum* fed pullets (C) and lysine-restricted pullets during the starter phase (RLs) are presented in Table 7.5 (preliminary results).

The suggested biological relationship in *ad libitum* fed pullets between crude protein, crude fat, and ash in the first phase of the diphasic growth function (MGF; Figure 7.4) is strengthened by the observation of similar allometric growth

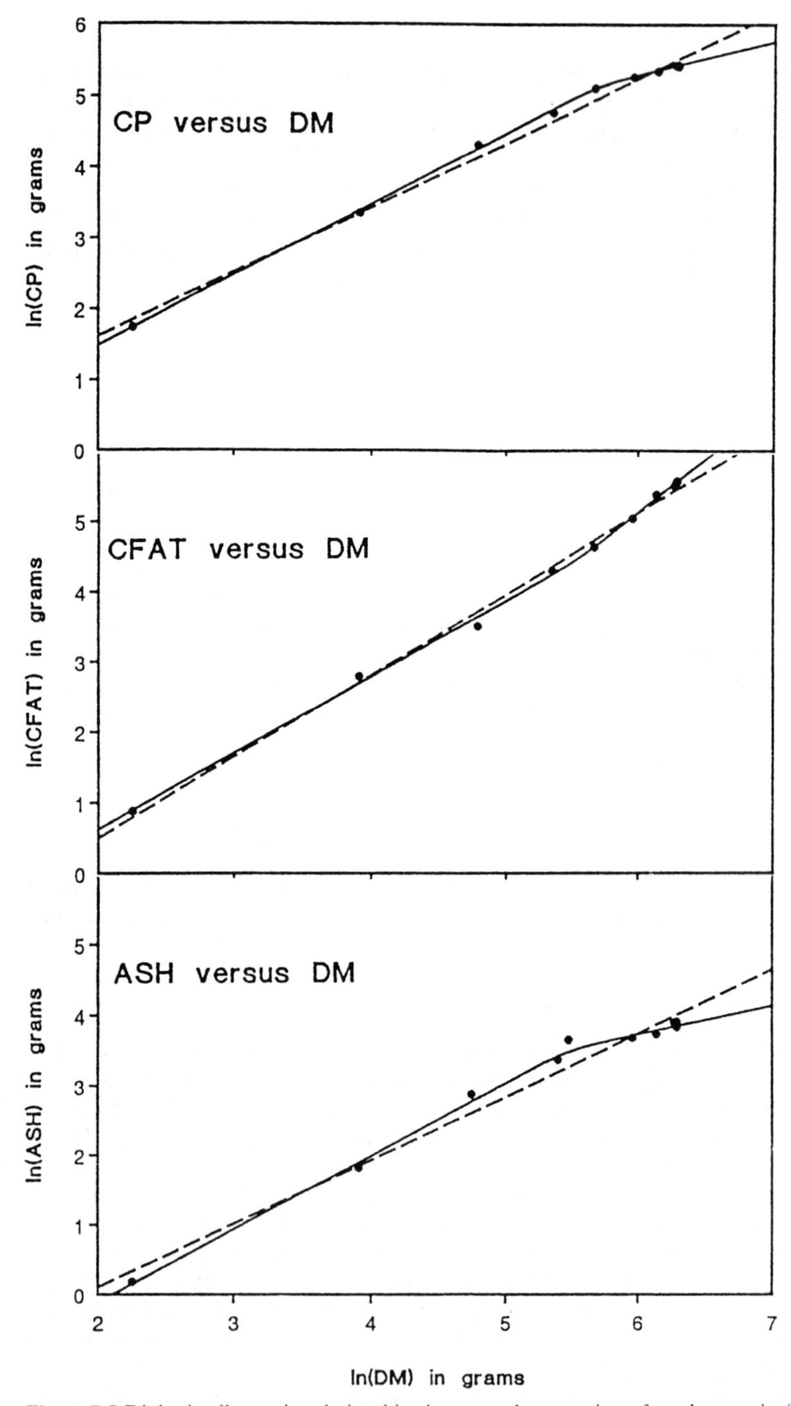

Figure 7.5 Diphasic allometric relationships between the accretion of crude protein (CP) and dry matter (DM), crude fat (CFAT) and dry matter, and ash and dry matter in non-restricted pullets (Kwakkel *et al.*, 1993)

Table 7.5 PARAMETER ESTIMATES OF THE MONO- AND DIPHASIC ALLOMETRIC RELATIONSHIPS[a] BETWEEN CRUDE PROTEIN AND DRY MATTER, CRUDE FAT AND DRY MATTER, AND ASH AND DRY MATTER IN PULLETS FED TWO DIFFERENT FEEDING STRATEGIES[b]

	C		*RLs*	
	Monophasic	*Diphasic*	*Monophasic*	*Diphasic*
		Crude Protein / Dry Matter		
$\ln(\alpha)$	−0.20	−0.51	−0.30	−0.61
$ß_1$	0.91	1.00	0.93	1.03
$ß_2$		0.46		0.58
$\ln(c)$		5.66		5.41
		Crude Fat / Dry Matter		
$\ln(\alpha)$	−1.79	−1.53	−1.63	−1.33
$ß_1$	1.15	1.08	1.11	1.02
$ß_2$		1.51	1.66	
$\ln(c)$	5.61	5.64		
		Ash / Dry Matter		
$\ln(\alpha)$	−1.71	−2.21	−1.71	−2.39
$ß_1$	0.91	1.05	0.92	1.13
$ß_2$		0.40		0.65
$\ln(c)$		5.46		4.69

(Kwakkel, Unpublished)

[a] $$\ln(y_x) = \ln(\alpha) + ß\ln(x) \quad \text{(monophasic)}$$

$$\ln(y_x) = \ln(\alpha) + ß_1\ln(x) - \{0.1 \times (ß_1 - ß_2) \times \ln[1 + e^{(\ln(x)-\ln(c))/0.1}]\} \quad \text{(diphasic)}$$

where y_x = crude protein, crude fat or ash, and x = dry matter

All diphasic models improved the fit significantly compared to the monophasic models. The parameter r was fixed at a value of 0.1 (Kwakkel *et al.*, 1993)

[b] Treatments: see Table 7.3

coefficients, close to unity, which means isometry in growth, in the first phase of the diphasic allometric function for crude protein/DM, crude fat/DM, and ash/DM (Table 7.5).

A remarkable phenomenon becomes evident when examining Figure 7.6: it is postulated that the lysine-restricted pullets in the starter phase did not change their body composition during the restrictive period as compared with the *ad libitum* fed group on a relative basis (compare also $ß_1$'s of the diphasic model in Table 7.5). Early fat growth in the lysine-restricted birds seems to be closely related to protein growth. These pullets were thought to become fatter than the control pullets because of their imbalanced diet. However, they did not. On a body weight basis they even consumed less feed than did the control pullets.

A graphical representation of ln(ash) as a function of ln(crude protein) for all treatments (Figure 7.7; preliminary results) illustrates the fixed allometric relationship between protein and ash growth in the body, irrespective of the applied feed restriction.

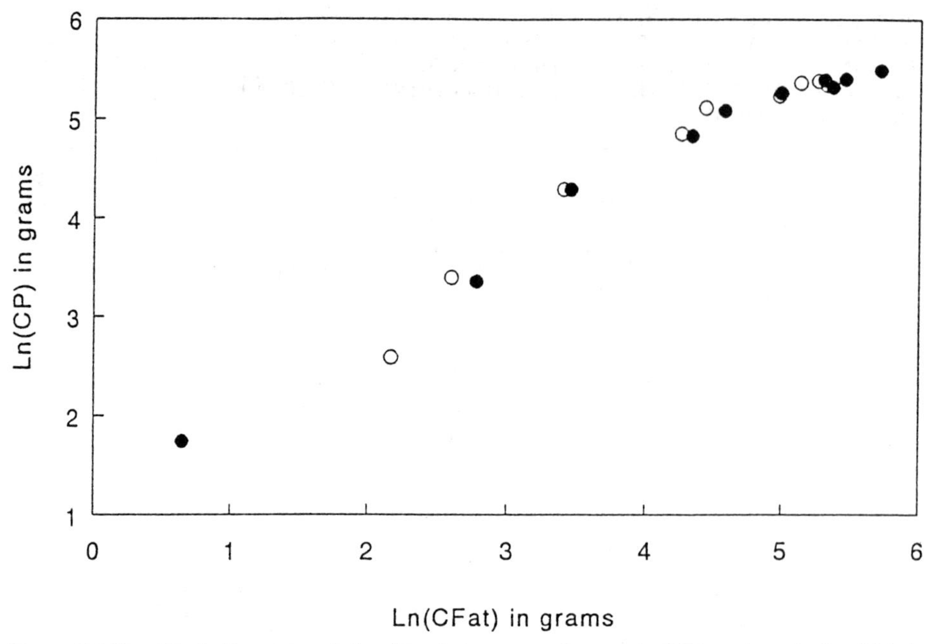

Figure 7.6 Non-fitted allometric relationships between crude protein (CP) and crude fat (CFat) growth in *ad libitum* fed (C: ●) and starter lysine-restricted (RLs: ○) pullets

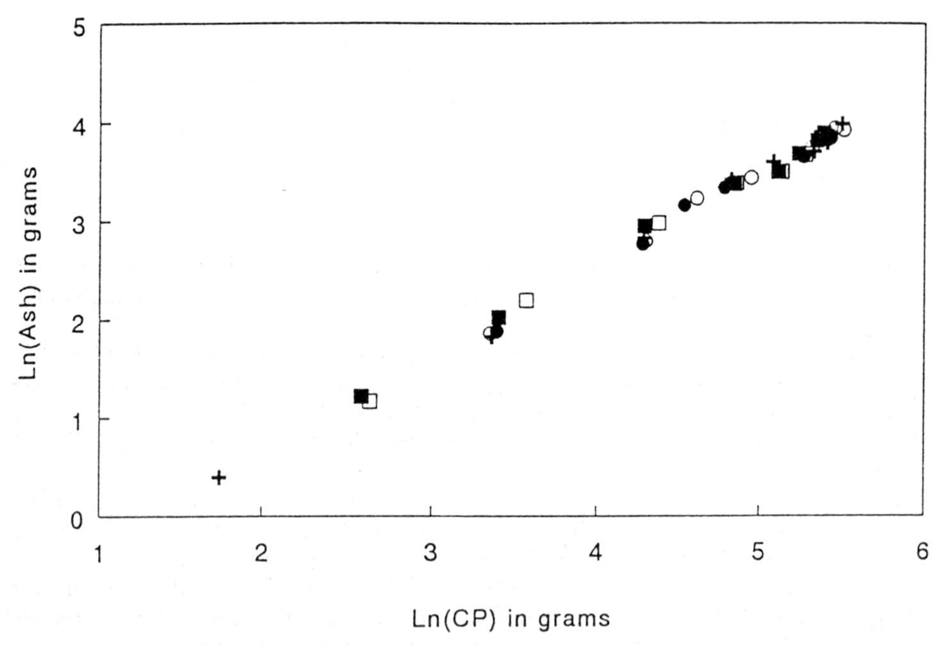

Figure 7.7 Non-fitted allometric relationships between ash and crude protein (CP) growth in pullets on different feeding strategies (C: +; RLs: ■; RFs: □; RLg: ●; RFg: ○)

Is mature gut weight responsible for heavier eggs?

The applied feeding strategies in Table 7.3 yielded another interesting contrast: pullets restricted during the grower period laid, on average, a 1.7 g heavier egg ($P<0.01$) than those restricted during the starter period. The method of restriction did not affect egg weight (Kwakkel *et al.*, 1991).

In our study, egg size was related neither to chronological age, as postulated by Williams and Sharp (1978) in Johnson *et al.* (1984), nor to body weight, as suggested by Summers (1983) and Summers and Leeson (1983) (Kwakkel *et al.*, 1991).

The mature asymptote of tibia weight (as a marker for frame size), a possible explanation for egg size differences, seems to be the same for all treatments (Figure 7.8). However, a heavier mature gut weight tends to accompany the heavier eggs in grower restricted birds (Figure 7.9). In our experiment, daily feed intake during lay was comparable for all treatments. A heavier gut may be related to an increased absorptive capacity in the grower restricted pullets which might have affected egg sizes. Mature sizes of oviduct and uterus have not been examined as yet.

CONCLUDING REMARKS ON THE USE OF MULTIPHASIC FUNCTIONS

The main advantage of these multiphasic functions is that the parameters are biologically interpretable. That means that the fit is not only a way to describe some related data, but it is also suitable to compare treatments, on the basis of changes in clearly defined parameters which represent biological processes.

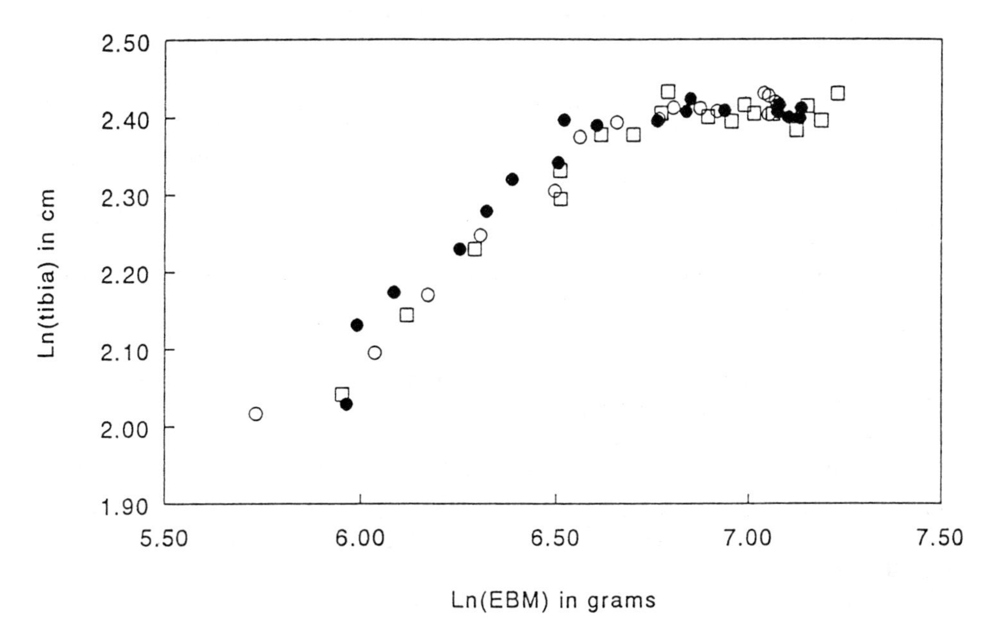

Figure 7.8 Non-fitted diphasic allometric relationships between length of the tibia and empty body mass (EBM) in *ad libitum* fed (C:□), starter (Rs: ○) and grower (Rg: ●) restricted pullets

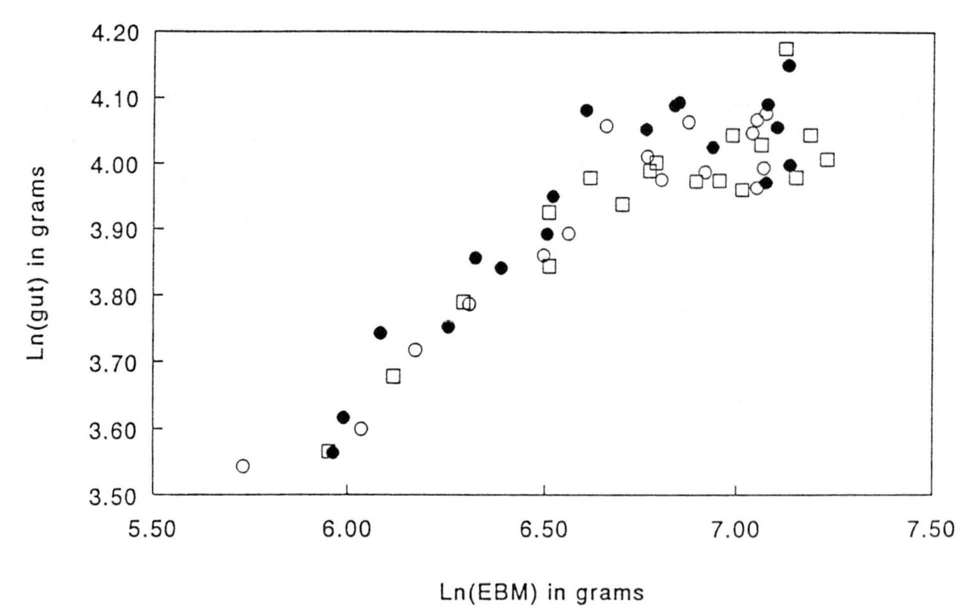

Figure 7.9 Non-fitted diphasic allometric relationships between weights of the gut and empty body mass in *ad libitum* fed (C: □), starter (Rs: ○) and grower (Rg: ●) restricted pullets

The *multiphasic growth function* enables us to study distinguishable stages of growth of the body or other compound structures in non-restricted pullets and hens. It is also useful in determining phases of development, growth spurts, of particular body organs, in order to discover possible 'critical periods' for the respective body parts.

The *multiphasic allometric function* is particularly useful in relative growth studies where feeding strategies are being compared. Growth of a body constituent y is dependent on constituent x (e.g. x is body weight, thus 'weight dependent'), if the allometric growth coefficients (slopes) are the same. Growth of a body constituent y is more 'age dependent' if the allometric coefficients are different between treatments.

Both multiphasic functions were fitted using non-linear regression procedures of the SAS statistical package (SAS Institute, 1991). Each fit was checked by four 'goodness-of-fit' criteria:

1 Asymptotic standard error for each parameter,
2 Residual standard deviation,
3 Coefficient of determination, and
4 Durbin-Watson statistic (a test for autocorrelation).

In this chapter a few interpretations of observed differences in performance by use of the multiphasic functions have been postulated. In the near future data sets from literature will be used to verify certain proposed relationships.

Fits on data of restricted fed birds have to be considered critically. Computer fit procedures may lead to some problems if changes in the feed allowance interfere with 'natural' growth spurts. In that case, it might be difficult to fit the data to the logistic, the 'bell-shape', or allometric curve. In general, Ricklefs stated that growth curves of birds under conditions of severe starvation are often so distorted that they can no longer be meaningfully fitted by a mathematical function (Ricklefs, 1968).

Prospects and recommendations for a feeding strategy for layer pullets

In order to give answers to the proposed questions in the former paragraph, one has to realize that research into multiphasic growth processes in layer pullets has only just started. It is our aim to include endocrine mechanisms in this field of research, whereas, since in our opinion, no repartitioning of nutrients in periods of suboptimal nutrition can occur without the control of the hormonal system (Short, 1980; Decuypere and Siau, 1989). Related to the process of maturation, changing sex hormone profiles, due to feeding strategies, might be of great importance in young hens.

After a thorough 'mapping' of the interactions between physiological and nutritional factors involving the maturing pullet, decisions about the use of specific feeding strategies during rearing can be made (Kwakkel, Corÿn and Bruining, 1988). It is now desirable to enhance our knowledge on the relationships between multiphasic growth of body constituents and ultimate laying performance.

Some recommendations, based on our preliminary results, are:

1. An early lysine (or protein) restriction affects adult performance negatively (delay in onset and low egg weights; Kwakkel *et al.*, 1991); also mentioned in earlier reports (Leeson and Summers, 1980 (step up); Chi, 1985). It may prevent 'normal' muscle and skeletal development, illustrated by the strong protein/ash relationship (Figure 7.7). In *ad libitum* situations, metatarsus (shank), tibia and keel mature before 13 weeks of age (Kwakkel *et al.*, in preparation), and restrictions may cause problems at onset of lay while birds are still 'catching up' their body frame size. However, a permanently smaller body frame of a restricted reared pullet is unlikely to occur (Leeson and Summers, 1984; Figure 7.8; Kwakkel, in preparation).
2. The 'set-point' for onset of lay is already at around 16 weeks of age (Table 7.4). Body protein seems to be a major determinant in initiating processes of puberty. High protein grower diets up to that age seem appropriate.
3. A mid-term feed restriction is still advisable. A gradual cessation of the restriction from 14 weeks of age onward might increase daily feed intake slowly and will affect performance positively (Johnson *et al.*, 1985; Bowmaker and Gous, 1989; Kwakkel *et al.*, 1991). In contrast to a direct switch to *ad libitum* feeding, this gradual transition might prevent too many irregular eggs at the onset of lay (Hocking *et al.*, 1987).

Multiphasic growth analysis seems to be an indispensable tool in studies in which the effects of feeding strategies on body development need to be assessed, to explain differences in adult performances.

Acknowledgements

I would like to thank Dr G. Hof, Dr S. Tamminga, Dr M.W.A. Verstegen and Dr B.A. Williams for their valuable comments on the several steps in preparing this chapter. I am most grateful to the undergraduate students who assisted during the feeding trials and subsequent dissections.

References

Abu-Serewa, S. (1979) *Australian Journal of Experimental Agriculture and Animal Husbandry*, **19**, 547–553

Balnave, D. (1973) *World's Poultry Science Journal*, **29**, 354–362

Balnave, D. (1984) *Australian Journal of Agricultural Research*, **35**, 845–849

Bish, C.L., Beane, W.L., Ruszler, P.L. and Cherry, J.A. (1984) *Poultry Science*, **63**, 2450–2457

Bowmaker, J.E. and Gous, R.M. (1989) *British Poultry Science*, **30**, 663–675

Brody, T., Eitan, Y., Soller, M., Nir, I. and Nitsan, Z. (1980) *British Poultry Science*, **21**, 437–446

Brody, T.B., Siegel, P.B. and Cherry, J.A. (1984) *British Poultry Science*, **25**, 245–252

Chi, M.S. (1985) *British Poultry Science*, **26**, 433–440

Decuypere, E. and Siau, O. (1989) In *Trends and Developments in the Feed Industry*. pp. 77–85, Athens, Greece

Dunnington, E.A., Siegel, P.B., Cherry, J.A. and Soller, M. (1983) *Archiv Geflugelkunde*, **47**, 87–89

Dunnington, E.A. and Siegel, P.B. (1984) *Poultry Science*, **63**, 828–830

Elahi, F. and Horst, P. (1985) *Archiv Geflugelkunde*, **49**, 16–22

Frankham, R. and Doornenbal, H. (1970) *Poultry Science*, **49**, 1619–1621

Gous, R.M. (1978) *British Poultry Science*, **19**, 441–448

Grossman, M. and Koops, W.J. (1988) *Poultry Science*, **67**, 33–42

Hammond, J. (1932) *Journal of the Royal Agricultural Society*, **93**, 131–145

Hocking, P.M., Gilbert, A.B., Walker, M.A. and Waddington, D. (1987) *British Poultry Science*, **28**, 495–506

Hocking, P.M., Waddington, D., Walker, M.A. and Gilbert, A.B. (1989) *British Poultry Science*, **30**, 161–174

Hollands, K.G. and Gowe, R.S. (1961) *Poultry Science*, **40**, 574–583

Hollands, K.G. and Gowe, R.S. (1965) *British Poultry Science*, **6**, 287–295

Johnson, R.J., Choice, A., Farrell, D.J. and Cumming, R.B. (1984) *British Poultry Science*, **25**, 369–387

Johnson, R.J., Cumming, R.B. and Farrell, D.J. (1985) *British Poultry Science*, **26**, 335–348

Karunajeewa, H. (1987) *World's Poultry Science Journal*, **43**, 20–32

Katanbaf, M.N., Dunnington, E.A. and Siegel, P.B. (1989) *Poultry Science*, **68**, 359–368

Koops, W.J. (1986) *Growth*, **50**, 169–177

Koops, W.J. (1989) *Multiphasic Analysis of Growth*. Ph.D. Thesis, Agricultural University, Wageningen, The Netherlands

Koops, W.J. and Grossman, M. (1991a) *Journal of Animal Science*, **69**, 3265–3273

Koops, W.J. and Grossman, M. (1991b) *Growth, Development and Aging*, **55**, 203–212

Koops, W.J. and Grossman, M. (1993) *Growth, Development and Aging*, **57** in press

Kwakkel, R.P. (1992) In *Proceedings of the 19th World's Poultry Congress*. pp. 480–484, Amsterdam, The Netherlands: WPSA Dutch branch
Kwakkel, R.P., Corijn, P.C.M.Z. and Bruining, M. (1988) *Netherlands Journal of Agricultural Science*, **36**, 187–190
Kwakkel, R.P., De Koning, F.L.S.M., Verstegen, M.W.A. and Hof, G. (1991) *British Poultry Science*, **32**, 747–761
Kwakkel, R.P. and Koops, W.J. (1991) In *Proceedings of the 8th European Symposium on Poultry Nutrition*. pp. 368–371, WPSA Italian branch, Venice, Italy
Kwakkel, R.P., Ducro, B.J. and Koops, W.J. (1993) *Poultry Science* **72** in press
Lee, K. (1987) *Poultry Science*, **66**, 694–699
Lee, P.J.W., Gulliver, A.L. and Morris, T.R. (1971) *British Poultry Science*, **12**, 413–437
Leeson, S. (1986) In *Proceedings of the 47th Minnesota Nutrition Conference and Monsanto Technical Symposium*. pp. 227–234, Minnesota Agric. Extension Service, Minnesota, USA
Leeson, S. and Summers, J.D. (1979) *Poultry Science*, **58**, 681–686
Leeson, S. and Summers, J.D. (1980) In *Recent Advances in Animal Nutrition* — 1980 pp. 202–213. Edited by W. Haresign. London: Butterworths
Leeson, S. and Summers, J.D. (1984) *Poultry Science*, **63**, 1222–1228
Lilja, C., Sperber, I. and Marks, H.L. (1985) *Growth*, **49**, 51–62.
McCance, R.A. (1977) In *Growth and Poultry Meat Production*, pp. 3–11. Edited by K.N. Boorman and B.J. Wilson. Edinburgh: British Poultry Science Ltd.
Ricklefs, R.E. (1968) *The Ibis*, **110 (4)**, 419–451
Ricklefs, R.E. (1975) *The Condor*, **77 (1)**, 34–45
Ricklefs, R.E. (1985) *Poultry Science*, **64**, 1563–1576
Robinson, D. and Sheridan, A.K. (1982) *British Poultry Science*, **23**, 199–214
Robinson, F.E., Beane, W.L., Bish, C.L., Ruszler, P.L. and Baker, J.L. (1986) *Poultry Science*, **65**, 122–129
SAS Institute. (1991) In *SAS System for Regression, 2nd Edition*, pp. 169–188, Cary, New York, USA: SAS Institute Inc.
Short, R.V. (1980) In *Growth in Animals*, pp. 25–45. Edited by T.L.J. Lawrence. London: Butterworths
Singh, H. and Nordskogg, A.W. (1982) *Poultry Science*, **61**, 1933–1938
Soller, M., Eitan, Y. and Brody, T. (1984) *Poultry Science*, **63**, 1255–1261
Summers, J.D. (1983) In *Proceedings of the Maryland Nutrition Conference for Feed Manufacturers*, pp. 12–18 Belville, USA.
Summers, J.D. (1986) In *Proceedings of the Maryland Nutrition Conference for Feed Manufacturers*, pp. 21–26 Belville, USA.
Summers, J.D. and Leeson, S. (1983) *Poultry Science*, **62**, 1155–1159
Summers, J.D., Leeson, S. and Spratt, D. (1987) *Poultry Science*, **66**, 1750–1757
Walstra, P. (1980) *Growth and Carcass Composition from Birth to Maturity in Relation to Feeding Level and Sex in Dutch Landrace Pigs*. Ph.D. Thesis, Agricultural University, Wageningen, The Netherlands
Watson, N.A. (1975) *British Poultry Science*, **16**, 259–262
Wells, R.G. (1980) In *Recent Advances in Animal Nutrition* — 1980, pp. 185–202. Edited by W. Haresign. London: Butterworths
Zelenka, D.J., Siegel, P.B., Dunnington, E.A. and Cherry, J.A. (1986) *Poultry Science*, **65**, 233–240

8

DIET AND LEG WEAKNESS IN POULTRY

B.A. WATKINS
Purdue University, West Lafayette, Indiana, USA
(Journal paper 13667 of the Purdue University Agricultural Experiment Station)

Introduction

Long bone growth and modelling in poultry are regulated by complex interactions between the animal's genetic potential, environmental influences and nutrition. These interactions produce a bone architecture that balances functionally appropriate morphology with the skeleton's involvement in calcium and phosphorus homeostasis. In growing poultry the long bones increase in length and diameter by the process called modelling. Bone modelling represents an adaptive process that is distinct from bone remodelling, which is the term used to describe the resorption and formation of mineralized tissue which maintains skeletal mass and morphology in adult poultry. As many of the skeletal lesions which afflict poultry are the consequence of abnormalities in bone modelling, not bone remodelling, an appreciation of the differences between these two contrasting processes is a prerequisite for understanding the pathogenesis of skeletal lesions in poultry.

Numerous growth regulatory factors are present in bone tissues. In general, the prostaglandins and cytokines affecting the skeletal system are produced locally by chondrocytes, osteoblasts, monocytes/macrophages and lymphocytes found in or associated with bone. These compounds are biosynthesized and secreted by the aforementioned cells either from induction by systemic endocrine hormones such as parathyroid hormone, oestrogen and vitamin D_3, or by autocrine or paracrine signalling agents within bone.

Several nutrients influence the growth and development of long bones in poultry. The effects of calcium, phosphorus and $1,25(OH)_2$vitamin D_3 on bone growth are well known. Low calcium intakes result in reduced serum calcium, osteoporosis or low calcium rickets, thin eggshells and reduced egg production. A severe phosphorus deficiency can cause rickets, but serum phosphorus levels are usually maintained during deficiency. Vitamin D_3, the antirachitic vitamin, affects several aspects of bone metabolism. Since the discovery of vitamin D_3 (cholecalciferol) and its chemical synthesis, poultry diets are easily supplemented with this vitamin to facilitate total confinement rearing. The active metabolite of vitamin D_3 actions is 1,25-dihydroxyvitamin D_3 ($1,25(OH)_2D_3$). The ($1,25(OH)_2D_3$) elevates calcium and phosphorus levels in plasma. In many respects, the response of $1,25(OH)_2D_3$ on target tissues is similar to that of a classical steroid hormone; however, new evidence suggests that $1,25(OH)_2D_3$ elicits biological responses via a nongenomic pathway.

The goal of this chapter is to explain the process of bone modelling and remodelling in poultry; discuss the roles of prostaglandins, cytokines and growth factors involved in the local regulation of bone metabolism; and explain the vitamin D_3 endocrine system and discuss the actions of $1,25(OH)_2D_3$ on bone resorption.

Bone cells and bone metabolism

Bone is a metabolically active, multifunctional tissue comprising populations of chondrocytes, osteoblasts, osteoclasts, osteocytes, endothelial cells, monocytes, macrophages and lymphocytes. The complex milieu of cells found in bone tissue produces a variety of biological regulators that control bone metabolism. Endocrine hormones (parathyroid hormone (PTH), oestrogens and $1,25(OH)_2D_3$) and autocrine and paracrine factors (prostaglandins, cytokines and insulin-like growth factors) co-ordinate the principal activities of bone metabolism to increase the length and diameter of long bones as poultry grow. The activities controlling bone growth are bone matrix formation, mineralization of new bone matrix and resorption of bone apatite. Osteoblasts participate in bone formation and osteoclasts are responsible for resorption of bone.

The prostaglandins (PG) and cytokines affecting the skeletal system are produced locally by chondrocytes, osteoblasts, monocytes/macrophages and lymphocytes found in or associated with bone. These compounds are biosynthesized and secreted by bone cells or cells associated with the skeleton either from induction by systemic endocrine hormones or by autocrine or paracrine signalling agents within bone. Most PGs and cytokines influence metabolic processes in bone to stimulate or inhibit matrix formation, mineralization and resorption as well as induce mitogenic effects on bone cells.

Bone remodelling

The skeletal morphology of adult poultry represents a sophisticated compromise between structural obligation and metabolic responsibility, serving the animal in support and locomotion while actively participating in the regulation of calcium homeostasis (Bain and Watkins, 1993). This compromise is accomplished through the animal's genetic potential for growth and intricate interactions between nutrition, metabolism and endocrine factors. Hormones and certain nutrients modulate the autocrine and paracrine cellular relationships (actions of PG and cytokines) responsible for the maintenance of bone mass and architecture. In the adult skeleton, the coordination of bone resorbing and bone forming activities is termed the 'bone remodelling cycle'.

The regulation of bone remodelling and its corresponding role in the maintenance of adult bone mass (as in the laying hen or breeding stock) is distinctly different from the processes which control skeletal growth and bone modelling in young poultry (for example rapidly growing meat-type poultry). As the name implies, modelling is responsible for altering bone shape. Modelling of bone is an adaptive process, providing order and specificity to the more generalized increases in bone mass which accompany tissue growth (Bain and Watkins, 1993).

Bone remodelling is responsible for the maintenance of tissue mass and architecture in the adult skeleton (Frost, 1973). Groups of cells participate in co-ordinated activities to function as units to remove and replace bone mineral at discrete skeletal sites. These organized groups of cells are called 'basic multicellular units', or BMU (Frost, 1963). The most important members of the BMU are osteoblasts and osteoclasts.

Parfitt (1990) has divided the cellular interactions associated with a remodelling cycle into four main events; activation, resorption, reversal and formation. A remodelling cycle begins when a non-remodelling bone surface becomes 'activated'. The signals of activation are not fully understood but it is believed that the bone lining cells covering inactive surfaces may initiate this event.

Osteoclasts attach to the bone surface and resorb bone in discrete units or packets of mineral during the resorption phase (Parfitt, 1979). The osteoclast becomes attached to the bone surface in a membrane bound microenvironment which can be optimized for the enzyme actions and cell activities associated with dissolution of the mineralized matrix and release of bone Ca^{2+}. As the period of bone resorption subsides, the reversal phase occurs in preparation for osteoblast recruitment and deposition of new bone matrix.

The formation phase is initiated by groups of osteoblasts being recruited to the site. The osteoblasts synthesize and deposit new bone matrix (osteoid) into the excavated cavity. The osteoid becomes the site for mineralization. The bone remodelling cycle is complete after osteoblasts refill the cavity left during resorption of bone. The events of resorption and formation are believed to be 'coupled' in such a way that resorption is always followed with new bone formation. There is a hypothesis that a reservoir of growth factors and cytokines reside in bone to maintain the bone remodelling cycle (Canalis, 1988). According to this hypothesis, osteoclastic bone resorption releases regulatory molecules into the local microenvironment, where they in turn produce the autocrine and paracrine interactions that are associated with the recruitment of the osteoblast to the remodelling site (Farley, Tarbaux, Murphy, Masuda and Baylink, 1987).

Bone modelling

In contrast to bone remodelling, bone modelling lacks local coupling of resorption with bone formation on the modelling bone surface. The resorption and formation in bone modelling occurs on separate surfaces; therefore, surface activation in modelling bone may be followed by either resorption or formation (Frost, 1973; Burr and Martin, 1989). The timing and sequence of the cellular events and the extent of their activities in bone remodelling and modelling processes are fundamentally different (Table 8.1).

To summarize the activities of bone cells during modelling, osteoclasts resorb bone on the inner, endosteal surface (marrow cavity) while osteoblasts add matrix on the outer, periosteal surface. As the bone grows the osteoblastic and osteoclastic activities will lead to increases in bone size and changes in longitudinal and cross-sectional geometry which are characteristic of individual skeletal components. Modelling drifts reflect the bone's ability to sculpt its morphology in response to functional demands. For example, increasing muscle mass and physical activity of the bird will influence how the bone is modelled. Figure 8.1 illustrates the results of these activities controlling bone modelling when influenced by diet.

In rapidly growing meat-type poultry, the functional demands on the bone necessitate that the diaphysis must grow via an advancing front of radial lamellae (Riddell, 1981). These lamellae are assembled by groups of osteoblasts continually

Table 8.1. COMPARISON OF BONE MODELLING AND REMODELLING ACTIVITIES

	Bone Remodelling	*Bone Modelling*
Local coupling	Formation and resorption are coupled	Formation and resorption are **not** coupled
Timing and sequence of activity	Cyclical: A[a]-RS[b]-RV[c]-F[d]; formation always follows resorption	F and RS are continuous and occur on separate surfaces
Extent of surface activity	20% of surfaces are active	100% of surfaces are active
Anatomical objectives	Skeletal maintenance	Gain in skeletal mass and changes in skeletal form

[a] A = activation; [b] RS = resorption; [c] RV = reversal; [d] F = formation

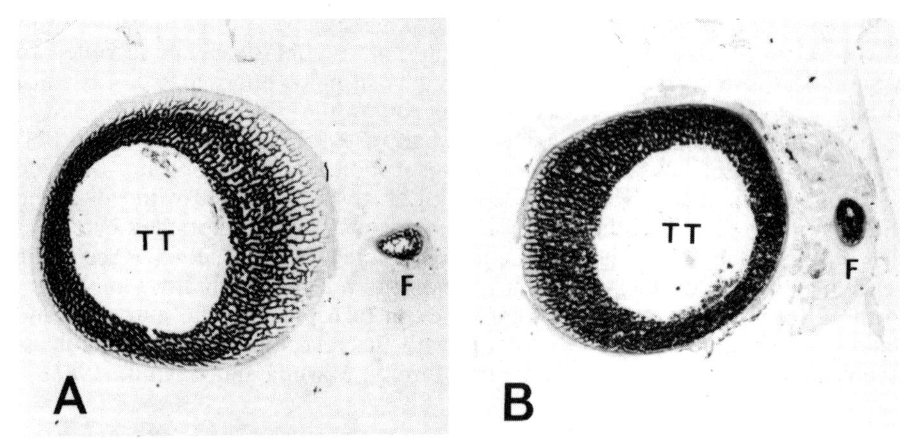

Figure 8.1. Photomicrographs of cortical bone modelling patterns at the mid-diaphysis of the tibiotarsus (TT) in three-week-old broiler chicks fed biotin-adequate (A) and biotin-deficient (B) diets. Note the difference in the pattern of formation drifts and cortical thickness relative to the fibula (F).

recruited from precursor cells in the thick periosteum (membrane covering the outer bone surface). As poultry mature, the scaffold of primary osteons will be consolidated (Riddell, 1981). With periosteal expansion in this system capable of exceeding 100 μm/day (Bain, Newbrey and Watkins, 1988; Watkins, Bain and Newbrey, 1989), poultry may well represent the ultimate bone modelling system with growth rates and modelling drifts that far exceed most mammalian farm animals.

Bone mineralization

MATRIX VESICLES AND BIOACTIVE PHOSPHOLIPIDS

The calcifying cartilage in long bones of growing chickens contains chondrocytes which elaborate matrix vesicles that initiate mineralization (Wuthier, 1988). Matrix vesicles are also present in developing long bones of the embryonic chick. Wuthier (1988) describes the matrix vesicle structure as a lipid enclosed microenvironment with acidic phospholipids (such as phosphatidylserine) that exhibit high-affinity binding for Ca^{2+}. A major proportion of matrix vesicle phosphatidylserine is complexed with Ca^{2+} and P_i (Wuthier, 1988). Matrix vesicles contain ion-transport proteins for P_i and Ca^{2+} and possess several active phosphatases, especially alkaline phosphatase. The matrix vesicle rapidly accumulates P_i and Ca^{2+} to yield octacalcium phosphate which forms apatite (Wuthier, 1988). The developing mineralized crystals eventually rupture the matrix vesicle membrane.

Factors involved in the local regulation of bone metabolism

SYSTEMIC HORMONES

Parathyroid hormone is a stimulator of osteoclastic bone resorption in poultry but the effect may be mediated through another cell type by a paracrine interaction. Likewise, $1,25(OH)_2D_3$ stimulates bone resorption but cytosolic receptors for this form of vitamin D_3 have been found in the osteoblast and not in the osteoclast (Norman, 1985; Suda, Takahashi and Abe, 1992). As with PTH, $1,25(OH)_2D_3$ may activate bone resorption via the osteoblast or by a localized compound that regulates bone cell function.

PROSTAGLANDINS

The essential fatty acid linoleic is metabolically converted to polyunsaturated fatty acids (PUFA) in the liver of the chicken (Watkins, 1991). Linoleic acid 18:2n6 is referred to as an n-6 PUFA because the terminal double bond nearest the methyl end of the molecule in its carbon chain is located at the sixth position. Enzymatic desaturation (addition of double bonds) and chain-elongation (by 2-carbon units) of 18:2n6 leads to the formation of arachidonic acid an n-6 PUFA (Figure 8.2). The Δ-6 desaturase is probably the major rate-regulating step in PUFA synthesis for poultry. Hormonal and nutritional regulation of the Δ-6 and Δ-5 desaturases controls the rates of conversion of 18:2n6 to its respective long-chain n-6 PUFA (Watkins, 1991).

Specific PUFA serve as substrates for the biosynthesis of a variety of oxygenated compounds called eicosanoids (prostacyclins, prostaglandins, thromboxanes, H(P)ETEs (hydroperoxy and hydroxy acids), lipoxins and leukotrienes). For example, 20:3n6, 20:4n6 and 20:5n3 are substrates for the 1, 2 and 3 series prostaglandins, respectively. The synthesis of eicosanoids is ubiquitous in poultry and these oxygenated C_{20} carboxylic acids affect nearly all physiological systems. Some of their established biological actions in chickens include stimulation of myoblast, chondrocyte and bone cell differentiation, and mediation of oviposition and bone resorption (Watkins, 1991).

Prior to the synthesis of PG, substrate must be made available. Activation of phospholipases cleaves substrate from membrane phospholipids (Figure 8.2). The liberated fatty acid is available to undergo oxidative transformation via the cyclooxygenase or lipoxygenase pathways. Various stimuli (physical, hormonal, chemical and toxic) cause the release of prostaglandins in tissues. Prostaglandins, prostacyclins and thromboxanes are the major products evolved from the cyclooxygenase pathway while H(P)ETEs, leukotrienes and lipoxins are those emanating from the lipoxygenase pathway (Figure 8.2). Cyclooxygenase is inhibited by

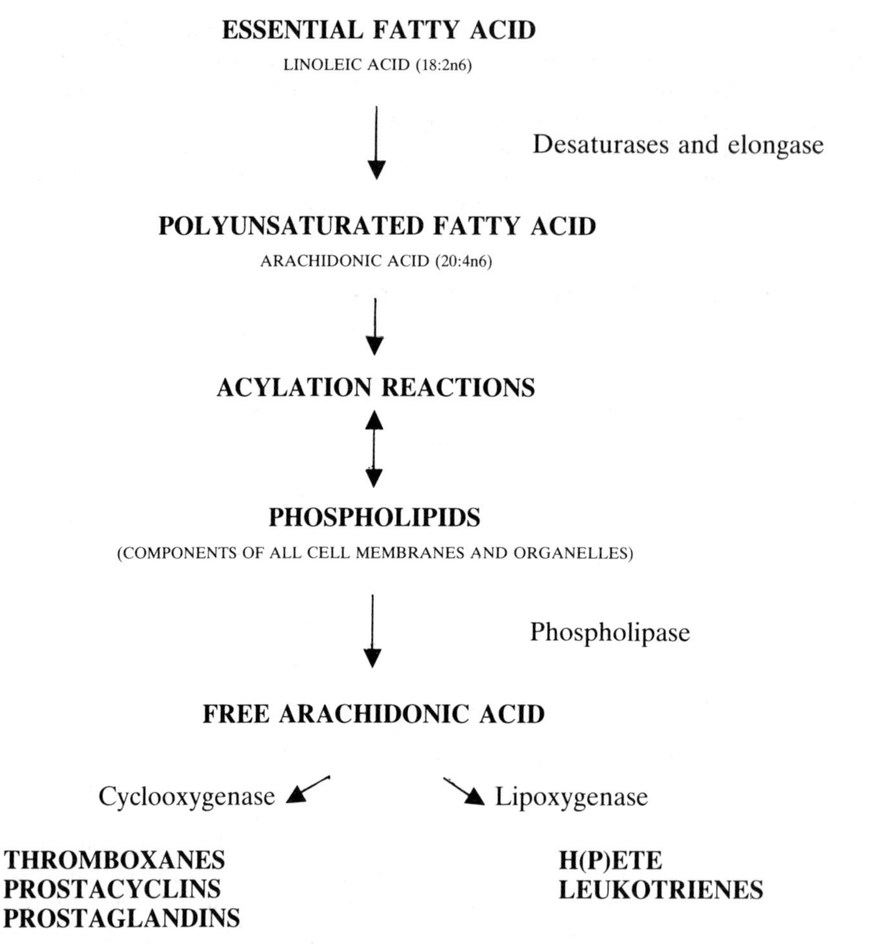

Figure 8.2. Conversion of the essential fatty acid linoleic acid to arachidonic acid, its incorporation into phospholipids and subsequent biosynthesis into eicosanoids

aspirin and indomethacin. Arachidonic acid is the primary substrate for most of the eicosanoids produced. Once formed, the eicosanoids exert localized, often autocrine or paracrine effects on individual cells. The eicosanoids are short-lived biological regulators which produce immediate responses before being rapidly degraded by enzymes in tissues where they are synthesized.

In 1970, prostaglandin E_2 (PGE_2) was observed to cause calcium release from bone tissue indicating effects on bone resorption (Raisz and Martin, 1983). Production of PG has been measured in chick cartilage tissue (Chepenik, Ho, Waite and Parker, 1984; Gay and Kosher, 1985), bone organ culture (Raisz and Martin, 1983) and osteoblasts (Feyen, van der Wilt, Moonen, Bon and Nijweide, 1984). Physical stress (Somjen, Binderman, Berger and Harell, 1980) and systemic and local bone regulatory factors (PTH, epidermal growth factor (EGF), platelet-derived growth factors (PDGF), transforming growth factors (TGF) and interleukin-1 (IL-1)) stimulate PG synthesis and release in osteoblast or bone organ cultures (Krane, Goldring and Goldring, 1988; Yang, Gonnerman and Polger, 1989).

In 1972, Blumenkrantz and Sondergaard reported that PGE_2 and prostaglandin $F_{2\alpha}$ ($PGF_{2\alpha}$) stimulated collagen synthesis in embryonic chick tibia. Besides regulating chondrogenesis, PGE_2 stimulates bone formation in chick calvaria bone (Chyun and Raisz, 1984). The current evidence suggests that PGEs stimulate formation and resorption of bone but an overproduction of PGE is associated with pathology, as in osteomyelitis and avian osteopetrosis.

Prostaglandin E_2 (10^{-8} M) was a potent stimulator of bone resorption when tested on embryonic chick long bone *in vitro* (Satterlee, Amborski, McIntyre, Parker and Jacobs-Perry, 1984). Yang *et al.* (1989) also reported that PGE_2 induced bone resorption as measured by ^{45}Ca release from prelabelled 17-day-old embryonic chick calvaria and PTH enhanced PGE_2 production. Shaw and Dacke (1989) reported that intravenous injection of chicks with PTH and 16,16-dimethyl PGE_2 caused an inhibition of ^{45}Ca uptake into femur and calvarium. When osteoclast-enriched cultures isolated from two-week-old White Leghorn chicks were treated with PGE_2 (10^{-6} M) resorption was stimulated but below the basal level (de Vernejoul, Horowitz, Demignon, Neff and Baron, 1988).

CYTOKINES

The cytokines are extracellular signalling proteins secreted by effector cells which act on nearby target cells. Cytokines exert their effects at low concentrations in autocrine or paracrine cell-to-cell communications. Cytokines stimulate anabolic processes in cells but also may inhibit cell activity; hence, they could be called biological modifiers. Some cytokines influence cell behaviour through endocrine hormones or PG. For example, the synthesis and release of PGE_2 is associated with the response produced by the action of cytokines (Krane *et al.*, 1988).

The cytokines involved in bone modelling and remodelling include EGF, fibroblast growth factor (FGF), interferon-γ (IFN-γ), interleukins (IL-1, IL-6), PDGF, transforming growth factors (TGF-α, TGF-β), tumour necrosis factor-α (TNF-α) and insulin-like growth factors (IGF-I, IGF-II) (Canalis, McCarthy and Centrella, 1988; 1991; Krane *et al.*, 1988). Table 8.2 provides a brief summary of effects produced by eicosanoids, cytokines and regulatory peptides on bone cells and tissues. Whereas, most cytokines are potent stimulators of bone resorption, few in fact enhance bone formation (Table 8.2). Basic FGF and TGF-β stimulate proliferation and differentiation of collagen synthesizing cells.

Transforming growth factor-β regulates the proliferation of chick growth plate chondrocytes isolated from the growth plate of three- to five-week-old chicks. In embryonic chick limb bud mesenchymal cells TGF-β stimulated chondrogenesis and the cultures seem to endogenously produce TGF-β based on immunofluorescent staining of polyclonal antibody against TGF-β (Leonard, Fuld, Frenz, Downie, Massague and Newman, 1991). PTH, IL-1 and $1,25(OH)_2D_3$ stimulate bone resorption by increasing TGF-β activity in bone organ culture, and depending on the cell type, TGF-β induce increases, decreases or biphasic effects on cell replication and matrix synthesis (Centrella, McCarthy and Canalis, 1988).

Many of the cytokines exert a positive effect on bone resorption. EGF, IL, PDGF, TGF-α and TNF-α all stimulate bone resorption *in vitro* (Table 8.2). IFN-γ inhibits *in vitro* bone resorption in the presence of cytokines (IL-1, TNF-α and TNF-β) and calcium-regulating hormones (PTH and $1,25(OH)_2D_3$).

Table 8.2. REPORTED RESPONSES OF AUTOCRINE AND PARACRINE FACTORS IN BONE[a]

Responses observed in bone	*Cytokine, eicosanoid or peptide growth factor*[b]
Bone formation or matrix production	FGF, IGF, PGE_2, TGF-β
Bone resorption	EGF, IL, LT, PDGF, TGF-α, TNF-α
Collagen synthesis	FGF, IGF, TGF-β

[a] Adapted from: Canalis, McCarthy and Centrella, 1991; Norrdin, Jee and High, 1990; Spencer, 1991. [b] Epidermal growth factor = EGF; Fibroblast growth factor = FGF; Interleukin = IL; Insulin-like growth factor = IGF; Leukotriene = LT; Platelet-derived growth factor = PDGF; Prostaglandin E_2 = PGE_2; Transforming growth factor = TGF-α, TGF-β; Tumour necrosis factor = TNF-α

The mechanism by which IFN-γ inhibits bone resorption is controversial, but recent investigations suggest that this cytokine interferes with osteoclast formation (Canalis *et al.*, 1991).

INSULIN-LIKE GROWTH FACTORS

The insulin-like growth factors (IGF) which are also called somatomedins, are described as paracrine or autocrine regulatory polypeptides of cells. These compounds stimulate growth and the synthesis of DNA, RNA and proteins in cells. IGF are mitogenic and stimulate differentiation in a variety of cell types (D'Ercole, Stiles and Underwood, 1984). Pituitary growth hormone (GH) controls the tissue biosynthesis and secretion of IGF-I (insulin-like growth factor I or somatomedin C) postnatally (Spencer, 1991; Clemmons and Underwood, 1991). Serum concentrations of IGF-I are maintained by liver synthesis under the influence of GH. Much of the circulating IGF is bound to plasma IGF binding proteins (IGFBP) (Spencer, 1991). IGF-I mediates the effects of GH. IGF-II is produced in bone tissues and it is stored at higher concentrations than IGF-I in chicken skeletal tissues (Bautista, Mohan and Baylink, 1990). The liver is the principle source of IGF-I posthatch, but extrahepatic tissues contribute much more during prehatch in the chicken (Serrano, Shuldiner, Roberts, LeRoith and de Pablo, 1990). Interestingly, receptors for IGF-I are present very early (three to six days) and dominate compared with the insulin receptors in chick embryonic head and brain (Bassas, Girbau, Lesniak, Roth and de Pablo, 1989).

The plasma IGF concentrations in the growing chicken increase progressively from zero to three weeks of age but plateau from three to seven weeks (McGuinness and Cogburn, 1990). When chickens were fasted for 24 hrs the concentrations of IGF decreased (Ballard, Johnson, Owens, Francis, Upton, McMurtry and Wallace, 1990). Receptors for IGF in chicken liver also peak at three weeks and then drop from three to 10 weeks of age (Duclos and Goddard, 1990).

Throughout embryonic development, IGF receptor numbers and distribution are regulated and IGF-I receptors are present at most stages of embryogenesis in chick growing limbs (Bassas and Girbau, 1990). *In vitro* studies with embryonic chick sternal chondrocytes showed a rapid stimulation of RNA and proteoglycan synthesis but delayed glycosaminoglycan and DNA synthesis with IGF-I treatment

(Kemp, Kearns, Smith and Elders, 1988). Stimulation of glycosaminoglycan synthesis seems to be an early action of IGF-I. Recent indirect evidence would suggest that chick embryonic pelvic cartilage contains somatomedin-C based on *in vitro* tissue proliferation and peptide analyses (Burch, Weir and van Wyk, 1986).

Vitamin D_3

VITAMIN D_3 FORMATION, ABSORPTION AND METABOLISM

Vitamin D exists in two forms ergocalciferol (vitamin D_2) and cholecalciferol (vitamin D_3). Ergocalciferol and its provitamin (ergosterol) are present in plants, but cholecalciferol and its precursor 7-dehydrocholesterol are most prevalent in animals. The provitamin D sterol 7-dehydrocholesterol is a precursor of vitamin D_3 as well as a product of cholesterol metabolism. In poultry, previtamin D_3 is formed in the skin from 7-dehydrocholesterol by action of ultraviolet light. The reaction results in the opening of the B ring of the sterol nucleus. The previtamin D_3 undergoes isomerization to form vitamin D_3. The newly formed vitamin D_3 in the skin is transported by the blood bound to α-globulin.

Dietary sources of vitamin D_3 are absorbed from the small intestine of poultry. Efficiency of vitamin D_3 absorption is dependent upon adequate fat digestion and absorption. Bile salts facilitate vitamin D_3 absorption into the gut mucosa. Once absorbed, vitamin D_3 is transported with neutral lipids as portomicrons via the portal blood to the liver.

In liver, vitamin D_3 undergoes hydroxylation at carbon 25 on the side chain to form 25-OH vitamin D_3. The vitamin D-25-hydroxylase in liver is a microsomal enzyme requiring cytochrome P_{450}. A second important hydroxylation takes place at carbon 1 of the A ring of $25(OH)D_3$ in the kidney by action of a mitochondrial 25-OH-vitamin 1-hydroxylase. The fully hydroxylated vitamin ($1,25(OH)_2D_3$) has 500 times more biological activity than $25(OH)D_3$; however, $25(OH)D_3$ is the main circulating form of the vitamin. Most of the vitamin D_3 metabolites are transported in blood by vitamin D-binding proteins. Other dihydroxylated metabolites of $25(OH)D_3$ are produced by the kidney ($24,25(OH)_2D_3$) but their importance is not fully understood.

The activity of the kidney 1-hydroxylase is regulated by the concentrations of circulating Ca^{2+} and phosphorus, and by PTH, calcitonin and $1,25(OH)_2D_3$. PTH, low plasma calcium and calcitonin elevate the activity of kidney 1-hydroxylase while $1,25(OH)_2D_3$ results in feedback inhibition of the enzyme. Besides the hydroxylase activity in kidney, extrarenal 1-hydroxylase has been reported in other tissues (cells) of animals.

PHYSIOLOGICAL FUNCTIONS OF VITAMIN D_3 METABOLITES

The best known physiological role of vitamin D_3 is the maintenance of calcium and phosphorus homeostasis (Norman, 1979). A lack of vitamin D_3 results in rickets or osteomalacia in poultry. The active form of vitamin D, $1,25(OH)_2D_3$, acts very much like a steroid hormone. The binding of $1,25(OH)_2D_3$ to cell receptors of target tissues and the relocation of the receptor-ligand complex to the cell nucleus induces the synthesis of mRNA for Ca-binding proteins (calbindins). Cell receptors for $1,25(OH)_2D_3$ have been characterized biochemically and are found in a variety of tissues in poultry (Pike, 1991; Minghetti and Norman, 1988). The calbindins are widespread in poultry with the most notable being the intestinal calbindin which is responsible for calcium absorption of which the bulk is vitamin D_3-dependent. Calbindins are present in avian uterus, kidney, brain, bone and skin. $1,25(OH)_2D_3$ also stimulates intestinal absorption of phosphorus. Both calcium and phosphorus resorption in kidney is enhanced by $1,25(OH)_2D_3$.

Bone resorption is stimulated by vitamin D_3, the effect of calcium release is primarily a response due to $1,25(OH)_2D_3$. The resorption of bone is a result of osteoclastic activity, yet the osteoclast unlike many cells lacks receptors for vitamin D_3 (Suda *et al.*, 1992). The osteoblast does contain receptors for $1,25(OH)_2D_3$ and is probably responsible for mediating vitamin D_3 and PTH effects on bone by affecting osteoclastic activity.

New evidence is emerging to suggest roles for $25(OH)D_3$ and $24,25(OH)_2D_3$ in normal bone biology. Both forms of vitamin D_3 are contained in bone tissue but their precise function is not known.

Vitamin D_3 also plays a role in bone formation although the relationships of its action with PTH on bone and its function in bone growth are not well defined. Some studies suggest that vitamin D_3 metabolites enhance the uptake of Ca^{2+} in bone and stimulate the synthesis of growth factor receptors or production of anabolic cytokines and growth factors. All of these effects would contribute to *in vivo* bone formation. Not all of vitamin D_3 effects can be explained by genomic mechanisms. One nongenomic action of $1,25(OH)_2D_3$ is the rapid stimulation of calcium transport from the brush border to the basal lateral membrane of the chick intestinal enterocyte (Zhou, Nemere and Norman, 1992). This process has been termed 'transcaltachia'.

VITAMIN D_3 EFFECTS ON BONE RESORPTION

The major target cell for $1,25(OH)_2D_3$ in bone is the osteoblast (Suda *et al.*, 1992). Although osteoclastic activity is responsible for bone resorption induced by $1,25(OH)_2D_3$, the osteoblast appears to mediate the effect of vitamin D_3. Osteoblasts release soluble factors (proteins) to stimulate differentiation of osteoclast progenitors. Some of the proteins released by osteoblasts are bone and matrix Gla proteins (vitamin K-dependent calcium-binding proteins) and osteocalcin (another calcium-binding protein). The production of proteins that stimulate differentiation of osteoclast precursors by the osteoblasts is believed to contribute to the bone resorbing response of vitamin D_3.

Conclusions

This review provides a description of bone modelling and remodelling processes in poultry. Participation of osteoblasts and osteoclasts in bone formation and bone resorption activities determine bone architecture in growing poultry. The regulation of bone cell activity is controlled by systemic and localized hormones. PTH, $1,25(OH)_2D_3$ and PGE_2 are activators of bone resorption. On the other hand PG and $1,25(OH)_2D_3$ both can stimulate cartilage and bone formation. The cytokines induce multiple effects on bone cells influencing resorptive and osteogenic activities. Most cytokines stimulate resorption, and synergistic effects have been observed on bone resorption with IL-1 and TNF-α. The IGFs also participate in the local regulation of bone growth in poultry. IGFs function as anabolic agents in bone during bone modelling in fast-growing poultry but may contribute to remodelling of bone since they can be found in mature bone tissue.

A better understanding of bone metabolism in poultry will occur as the interactions between diet and localized factors regulating bone biology are described. As more information becomes available on how vitamin D metabolites stimulate the production of soluble proteins in osteoblasts to affect osteoclastic bone resorption, nutritionists will be better equipped to provide the best nutrition during bone modelling and remodelling in poultry.

References

Bain, S.D., Newbrey, J.W. and Watkins, B.A. (1988) *Poultry Science*, **67**, 590-595

Bain, S.D. and Watkins, B.A. (1993) *The Journal of Nutrition*, **123**, 317–322

Ballard, F.J., Johnson, R.J., Owens, P.C., Francis, G.L., Upton, F.M., McMurtry, J.P. and Wallace, J.C. (1990) *General and Comparative Endocrinology*, **79**, 459–468

Bassas, L., Girbau, M., Lesniak, M.A., Roth, J. and de Pablo, F. (1989) *Endocrinology*, **125**, 2320–2327

Bassas, L. and Girbau, M. (1990) In *Progress in Comparative Endocrinology*, pp. 99–104. Edited by Bathazart. New York, USA: Wiley-Liss

Bautista, C.M., Mohan, S. and Baylink, D.J. (1990) *Metabolism*, **39**, 96–100

Blumenkrantz, N. and Sondergaard, J. (1972) *Nature New Biology*, **239**, 246

Burch, W.M., Weir, S. and van Wyk, J.J. (1986) *Endocrinology*, **119**, 1370–1376

Burr, D.B. and Martin, R.B. (1989) *American Journal of Anatomy*, **186**, 186-216

Canalis, E. (1988) *Triangle*, **27**, 11-19

Canalis, E., McCarthy, T. and Centrella, M. (1988) *Journal of Clinical Investigation*, **81**, 277–281

Canalis, E., McCarthy, T.L. and Centrella, M. (1991) *Annual Review of Medicine*, **42**, 17–24

Centrella, M., McCarthy, T.L. and Canalis, E. (1988) *FASEB Journal*, **2**, 3066–3073

Chepenik, K.P., Ho, W.C., Waite, B.M. and Parker, C.L. (1984) *Calcified Tissue International*, **36**, 175–181

Chyun, Y.S. and Raisz, L.G. (1984) *Prostaglandins*, **27**, 97–103

Clemmons, D.R. and Underwood, L.E. (1991) *Annual Review of Nutrition*, **11**, 393–412

D'Ercole, A.J., Stiles, A.D. and Underwood, L.E. (1984) *Proceedings of the National Academy of Sciences in the United States of America*, **81**, 935–939

De Vernejoul, M., Horowitz, M., Demignon, J., Neff, L. and Baron, R. (1988) *Journal of Bone and Mineral Research*, **3**, 69–80

Duclos, M.J. and Goddard, C. (1990) *Journal of Endocrinology*, **125**, 199–206

Farley J.R., Tarbaux N., Murphy, L.A., Masuda, T. and Baylink, D.J. (1987) *Metabolism*, **36**, 314-321

Feyen, J.H.M., van der Wilt, G., Moonen, P., Bon, A.D. and Nijweide, P.J. (1984) *Prostaglandins*, **28**, 769–781

Frost, H.M. (1963) In *Bone Remodelling Dynamics*. Springfield, IL, USA: Charles C. Thomas

Frost, H.M. (1973) In *Bone Remodelling and its Relationship to Metabolic Bone Disease*, pp. 28-53. Springfield, IL, USA: Charles C. Thomas
Gay, S.W. and Kosher, R.A. (1985) *Journal of Embryology and Experimental Morphology*, **89**, 367–382
Kemp, S.F., Kearns, G.L., Smith, W.G. and Elders, M.J. (1988) *Acta Endocrinologica*, **119**, 245–250
Krane, S.M., Goldring, M.B. and Goldring, S.R. (1988) In *Cell and Molecular Biology of Vertebrate Hard Tissue*, pp. 239–256. Edited by D. Evered and S. Harnett. Chichester, U.K.: John Wiley and Sons
Leonard, C.M., Fuld, H.M., Frenz, D.A., Downie, S.A., Massague, J. and Newman, S.A. (1991) *Developmental Biology*, **145**, 99–109
McGuinness, M.C. and Cogburn, L.A. (1990) *General and Comparative Endocrinology*, **79**, 446–458
Minghetti, P.P. and Norman, A.W. (1988) *FASEB Journal*, **2**, 3043–3053
Norman, A.W. (1979) *Vitamin D: The Calcium Homeostatic Steroid Hormone*, **490**, New York: Academic Press
Norman, A.W. (1985) *Physiologist*, **28**, 219–231
Norrdin, R.W., Jee, W.S.S. and High, W.B. (1990) *Prostaglandins Leukotrienes and Essential Fatty Acids*, **41**, 139–149
Parfitt, A.M. (1979) *Calcified Tissue International*, **28**, 1–5
Parfitt, A.M. (1990) In *Progress in Basic and Clinical Pharmacology*, pp. 1–27. Edited by J.A. Kanis. Basel Switzerland: S. Karger AG
Pike, J.W. (1991) *Annual Review of Nutrition*, **11**, 189–216
Raisz, L.G. and Martin, T.J. (1983) In *Bone and Mineral Research Annual*, pp. 286–310. Edited by W.A. Peck. New York, USA: Elsevier Science B.V.
Riddell, C. (1981) In *Advances in Veterinary Science and Comparative Medicine*, pp. 277–310. Edited by C.E. Cornelius and C.F. Simpson. New York, USA: Academic Press
Satterlee, D.G., Amborski, G.F., McIntyre, M.D., Parker, M.S. and Jacobs-Perry, L.A. (1984) *Poultry Science*, **63**, 633–638
Serrano, J., Shuldiner, A.R., Roberts, C.T., LeRoith, D. and de Pablo, F. (1990) *Endocrinology*, **127**, 1547–1549
Shaw, A.J. and Dacke, C.G. (1989) *Calcified Tissue International*, **44**, 209–213
Somjen, D., Binderman, I., Berger, E. and Harell, A. (1980) *Biochimica Biophysica Acta*, **627**, 91-100
Spencer, E.M. (1991) In *Modern Concepts of Insulin-like Growth Factors*. New York, USA: Elsevier Science Publishing Company
Suda, T., Takahashi, N. and Abe, E. (1992) *Journal of Cellular Biochemistry*, **49**, 53–58
Watkins, B.A., Bain, S.D. and Newbrey, J.W. (1989) *Calcified Tissue International*, **45**, 41-46
Watkins, B.A. (1991) *The Journal of Nutrition*, **121**, 1475–1485
Wuthier, R.E. (1988) *ISI Atlas Science Biochemistry*, **1**, 231–241
Yang, C.Y., Gonnerman, W.A. and Polgar, P.R. (1989) In *Advances in Prostaglandin, Thromboxane and Leukotriene Research*, pp. 435–438. Edited by B. Sammuelsson, P.Y.K. Wong and Sun, F.F. New York, USA: Raven Press Ltd
Zhou, L., Nemere, I. and Norman, A.W. (1992) *Journal of Bone and Mineral Research*, **7**, 457–463

III

Legislation

9

LEGISLATION AND ITS EFFECTS ON THE FEED COMPOUNDER

D.R. WILLIAMS
BOCM Pauls Ltd, Selby, North Yorkshire, YO8 7DT, UK

Introduction

New legislation covering animal feeding stuffs will be introduced in 1993. The European Commission is the main source of these changes and it is likely that the stream of controls will continue, particularly in the general areas of public health, the environment and health and safety at work. The main changes expected in 1993 are highlighted.

Raw materials and straights

The main purpose is to amend the Straight Feeds Directive (77/101/EEC) to make it apply to all feed raw materials, not only straights. A new definition is proposed which retains the concept of straights but broadens the scope as follows:

> 'Feed materials means various products of vegetable or animal origin, in their natural state, fresh or preserved, and products derived from the industrial processing thereof, and organic or inorganic substances whether or not containing additives, which are intended for use for oral animal feeding, whether direct as such, in a processed form, in the preparation of compound feeding stuffs or as carriers of premixtures'.

The requirements for compulsory and optional declarations of analytical constituents are still to be agreed. The labelling of materials may take three forms namely 'feed material' or 'feed material for industrial use' depending on the levels of undesirable substances present, or 'single feed material' if sold for direct oral feeding.

There are suggestions that the non exclusive list of feed materials in this draft directive should also serve as the reference list of materials for declaration of feed ingredients which was published in 1992.

Marketing of compounds

The principle change to be introduced during the latter part of 1993 is the introduction of the reference list of ingredients for declaration purposes. This list was published in the EEC Official Journal in October 1992. There are 12 chapters which contain a non exclusive list of the main ingredients used in the manufacture of feeds. The published names and limits concerning botanical purity and permitted impurities define the materials, and are the names which must be used when making the statutory declaration of ingredients. Compounders who declare ingredients by category are already required to use the categories listed in current feed regulations. There are no compositional limits for the list of ingredients whose sole purpose is for declaration of ingredients unlike the list of raw materials/straights which lists compositional limits (and compulsory and optional declarations to be used when materials are sold). A list of abbreviations which may be used in place of the full names is under discussion with the Ministry of Agriculture, Fisheries and Food. An abbreviated list is not permitted in the EEC legislation and there may be difficulties in permitting such abbreviations! The new lists and conditions will appear in the UK Feeding Stuffs Regulations during 1993.

Undesirable substances

The new Directive introducing new controls was published in October 1992 for implementation by 31 December 1993.

- Regulations will be extended to cover animals living freely in the wild where they are nourished with feeding stuffs.
- The maximum levels of undesirable substances or products set apply in general from the date on which the raw materials and feeding stuffs are put into circulation, including all stages of marketing and in particular from the date of their importation.
- It is forbidden to mix severely contaminated consignments above maximum permitted levels with other consignments of raw materials to comply with the set maxima. These restrictions apply to aflatoxin levels above 0.2 mg/kg in groundnut, copra, palm kernel, cotton, babassu, maize and their by-products and to cadmium and arsenic in phosphates above 10 and 20 mg/kg respectively.
- Maxima set for undesirable substances will apply to feeding stuffs exported to third countries.
- Importers, producers, or a person who by virtue of professional activities, possesses, or has possessed, or has had direct contact with a consignment of raw material or of feeding stuffs, and has knowledge to the effect that because of contamination with undesirable substances or product covered by the regulations that they do not meet the regulations and consequently pose a serious risk for animal and public health shall immediately inform the official authorities even if destruction of the consignment is envisaged.
- Authorities shall ensure that in the case of such contaminated consignments they are not used in animal nutrition and disposed of in a safe manner.
- Provisions for exchanging information between enforcement authorities about highly contaminated consignments of raw materials and feeding stuffs have been reinforced.

The Trade Associations UKASTA and GAFTA are reviewing raw material purchasing contract terms in the light of these changes, and the legal liabilities will also be considered with regard to ownership of consignments.

Control of additives

The fourth amending directive is still under discussion and major changes are likely, namely the extension of controls to 'additives when used in animal nutrition' as distinct from the current application which defines additives as '. . . when incorporated in feeds'. Water borne additives, forage additives, and boluses would be covered.

Two new categories would also be introduced namely enzymes and probiotics (preparations containing micro organisms). Manufacturers or suppliers of these substances must send lists of their products to the Commission without delay. Fees will be payable on submission of the list which would be offset against the full registration costs incurred when the full dossier is submitted within 2–3 years. Data on safety, quality and efficacy will be required and it is anticipated that it will take 5–6 years to clear all dossiers.

Dietetic feeds

This directive is making slow but steady progress with the main thrust coming from Germany. These feeds will be designed to redress nutritional imbalances or other physiological changes for which medicated feeds or those subject to control by existing directives are not appropriate.

Manufacturers have been requested to submit details of their dietetic feeds with the claims being made for their application. The potential advantages of such feeds would emerge if they provided a means to deliver higher levels of trace elements and vitamins for example than permitted under additives directive controls. The Commission are planning to have agreement on the directive in 1993.

EEC packaging and packaging waste directive

The proposal for a directive was published in the Official Journal in October 1992. All packaging waste, e.g. paper and plastic bags, is covered and it is likely that the feed industry will have to comply. The main requirements at this stage are that member states plan certain key actions. These cover provision of information on packaging and waste for each industry, and preparation of plans for management systems for return, recovery or re-use systems whilst maintaining health and safety requirements. Detailed statistics on the volume, characteristics and economics of packaging and waste are required. In addition, specific targets are set for example, priorities, in order of importance, are: prevention, recovery, disposal.

Within 10 years, 90% of waste packaging should be removed for recovery. Within 10 years, 60% of the waste to be recycled. National interim targets of 60% and 40% respectively are proposed. The estimated costs to the UK feed manufacturers have been based on current estimates that from a total of 10 million tonnes of feed, 2.5 million tonnes are sold in 25 kg bags, which is equivalent to 100 million bags. The overall estimated costs of recovery of bags is about £10m which equates to £1/t on all feeds or £4/t on bagged feeds!

Health and safety: manual handling operations

These Regulations replace earlier requirements concerning maximum weights for lifting but now go much further to cover all aspects of handling articles manually (i.e. transporting or supporting them). This is one of the seven main injury priority areas in the food industry.

The Regulations require a logical approach:

Judge all manual handling activities as to whether they might be hazardous. Past in-company or industrial sector experience of where manual handling injuries occur will provide guidance as to what is likely to be hazardous and establish relative priorities. The Health and Safety Executive (HSE) guidance indicates the range of weights and movements which are unlikely to be hazardous. The greater the departure from these ranges, the greater the likely need to carry out an assessment of that particular manual handling operation.

Avoidance of the manual handling operation which might involve a risk of injury (as determined from the above judgement) is required so far as is reasonably practicable. This will have a major application in the food industry as the onus will be on the employer to prove that eliminating the risk by instituting mechanical handling or by work practice changes are not reasonably practicable. Many applications of packing and handling sacks and work in progress could probably justify these changes.

For those work activities where it was judged previously that there might be a risk of injury, and where that risk cannot be avoided, then an assessment of the risk must be made and usually recorded. The assessment should highlight the contributing factors: the nature of the task, the load (including any person and any animal), the environment and the individual. The Regulations require employees to report to the employer if he is unlikely to be able to handle the loads which might be expected of his or her type for physical reasons. This implies that the employer need only consider the broad range of capabilities that might reasonably be expected for each age or sex of employee unless otherwise advised by such employees.

These contributing factors must be removed or reduced to the lowest level reasonably practicable. These measures will include information and training of employees (including specifically, where reasonably practical, the weight and centre of mass of the article to be lifted).

Reducing the number of assessments required

A reduction in the number of assessments required can be achieved by:

- instituting mechanical handling instead;
- grouping similar manual handling operations together under one assessment. This could even be done at sector level. 'Manual handling operations' means any transporting or supporting of a load, including the lifting, putting down, pushing, pulling, carrying or moving thereof by hand or by bodily force;
- setting safe operating limits which, provided they were not exceeded, would

carry no risk and hence require no further assessment. This approach might be particularly useful for deliveries etc. It could be supplemented by measures to be taken for any manual handling operation outside the range.

Any duty imposed by these Regulations on an employer in respect of his employees shall also be imposed on a self-employed person in respect of himself.

Duties of employers

Each employer shall so far as is reasonably practicable, avoid the need for his employees to undertake any manual handling operations which involve a risk of their being injured, or where it is not reasonably practicable to avoid the need for his employees to undertake any manual handling operations which involve a risk of their being injured:

- make a suitable and sufficient assessment of all such manual handling operations to be undertaken by them, having regard to the factors which are specified;
- take appropriate steps to reduce the risk of injury to those employees arising out of their undertaking any such manual handling operations to the lowest level reasonably practicable; and
- take appropriate steps to provide any of those employees who are undertaking any such manual handling operations with general indications and, where it is reasonably practicable to do so, precise information on the weight of each load, and the heaviest side of any load whose centre of gravity is not positioned centrally.

Any assessment shall be reviewed by the employer who made it if there is reason to suspect that it is no longer valid, or there has been a significant change in the manual handling operations to which it relates.

Where, as a result of any such review, changes to an assessment are required, the relevant employer shall make them.

Duty of employees

Each employee shall make full and proper use of any system or work provided for his use by his employer.

Health and safety — general

Personal protective equipment (PPE) at work regulations came into force on 1st January 1993. They require employers to provide PPE free of charge to protect against risks not controlled by other means e.g. under Control of Substances Hazardous to Health (COSHH) regulations. Before providing PPE, risks to health and safety which have not been otherwise controlled must be assessed and records kept. The PPE must be kept in clean and efficient working order, and the employer must provide information, instructions and training. Employees have a duty to use PPE and report any defects or loss.

The display screen equipment regulations also came into force on 1st January 1993. Since most feed manufacturers make widespread use of VDUs in their computing operations, they are covered by these regulations. The employer has a duty to:

- analyse work stations (guidelines are published) to assess health and safety risks;
- ensure work stations meet requirements (guidelines published);
- plan daily work to include breaks or changes of activity;
- train staff and provide information on all aspects of health and safety;
- provide, free of charge, appropriate eye and eyesight testing.

New systems have to comply now but existing work stations have to meet the standards by 31st December 1996.

Conclusion

The impact of new legislation continues to play a significant role in feed manufacturing. All the legislation discussed in this paper represents increased costs which must be recovered in the price paid by the customer. There is an increasing burden of legislation that is applicable to industry in general, not only feed manufacturing, and this is deemed to be for the protection of health and safety of employees and consumers, and the environment. This is a commendable principle but it requires the active participation of representatives of the industry, through trade associations in particular, to ensure that sensible and practicable legislation emerges from the consultations between industry and government departments.

10

THE REPORT OF THE EXPERT GROUP ON ANIMAL FEEDINGSTUFFS

G.E. LAMMING
University of Nottingham, Sutton Bonington Campus, Sutton Bonington, Loughborough, Leics, LE12 5RD, UK

Background to the report

The review arose following the House of Commons Agriculture Committee's fifth report on Bovine Spongiform Encephalopathy (1989/90) which recommended that 'the Government establish an expert committee to examine the whole range of animal feeds and advise on how the industries which produce them should be regulated'. The Government accepted this recommendation and in late 1990 appointed the following to serve:

Dr E.M. Cooke, Deputy Director, Public Health Laboratory Service;
Mr C. Maclean, Technical Director, Meat and Livestock Commission;
Professor P.C. Thomas, Principal & Chief Executive, Scottish Agricultural College;
Professor G.E. Lamming, Professor of Animal Physiology, University of Nottingham. (Chairman); and
Mrs Elizabeth Owen, Ministry of Agriculture, Fisheries and Food (MAFF), an expert on UK and EC feedingstuffs regulations as secretary, assisted by Miss Jill Powis.

The terms of reference were:

> 'To review the existing framework covering the animal feed industry in the United Kingdom. To advise on whether any improvements are required in the mechanisms by which the responsible Departments take account of food safety requirements in regulating the industry and to report to ministers by the end of 1991.'

The committee first met in February 1991 and submitted its report to Ministers in June 1992 (Her Majesty's Stationary Office, 1992a). It consulted widely with staff of Departments, Industry; with other relevant committees and EC authorities. It also received advice from independent experts.

The Government speedily examined the report and published its response to each recommendation on July 16th, 1992 (Her Majesty's Stationery Office, 1992b). It also noted the various comments on feedingstuffs issues made by the Group.

It will not be possible to cover all the recommendations contained in the 61 paragraphs of the report's summary. In the space available important aspects of the report will be selected for discussion. In the Government's response there is the stated intention to accept many of the recommendations, including the recommendation to establish an Animal Feedingstuffs Advisory Committee.

Main recommendations

THE ANIMAL FEEDINGSTUFFS ADVISORY COMMITTEE

Members of the Expert Group were aware of public concern for the safety of food sources, especially following the increased incidence of *Salmonella* infections in man, particularly *Salmonella enteritidis*, and also because of the increased number of cases of spongiform encephalopathy which were occurring in cattle in the UK. The Expert Group, having reviewed all the available information, concluded that animal feedingstuffs form an integral part of the feed/farm animal/human food chain. For the purposes of ensuring food safety, the Group accepted that there were major problems to be addressed, but were convinced that suitable strategies could be developed. This would include surveillance and control of all animal feedingstuffs from both home production and imported materials.

The group concluded that the feed industry is regulated by extensive controls which, in general, are effective. However, there are some gaps in legislation and in its enforcement. These areas could be the subject of new legislation. In addition, there were clearly some gaps in knowledge available to the Group which, while allowing problem areas to be generally identified, prevented clear recommendations from being made to ministers by the Expert Group at this stage. For these, and additional reasons, the Expert Group recommended that an independent Animal Feedingstuffs Advisory Committee (AFAC) be established to take an overview of all feedingstuffs issues in relation to human and animal health. Recommendations were made concerning its remit, scope of work, membership and formal links with other committees, relationship with surveillance authorities and its method of reporting. The recommended interface of the proposed committee with other established committees is illustrated in Figure 10.1 which depicts the two way interchange envisaged between five existing advisory committees whose work impinges on feedingstuffs issues; the Veterinary Products Committee, the Spongiform Encephalopathy Advisory Committee, The Advisory Committee on Pesticides, the Steering Group on the Microbiological Safety of Food and the Steering Group on Chemical Aspects of Food Surveillance.

The Group felt there is a need for a single committee to be aware of and consider all technical developments in raw materials manufacturing processes and feed treatments. It should coordinate information received from other committees concerning feedingstuffs issues and advise on policy and legislative measures and on EC proposals. It should also keep under review and advise on the nutritional characteristics and the safety of feed ingredients, the safety, quality and efficacy of feed additives, undesirable substances in feedingstuffs and the recycling of own-species material. These are issues of vital importance to all, but particularly for those at managerial level in control of the animal feedingstuffs industries of Europe and indeed, the whole world. The feed compound trade and suppliers of raw materials and straights have a special role in terms of quality control of animal feedingstuffs because of the extensive nature of the clientele they serve. Equally, it is important that all producers of animals, supplying

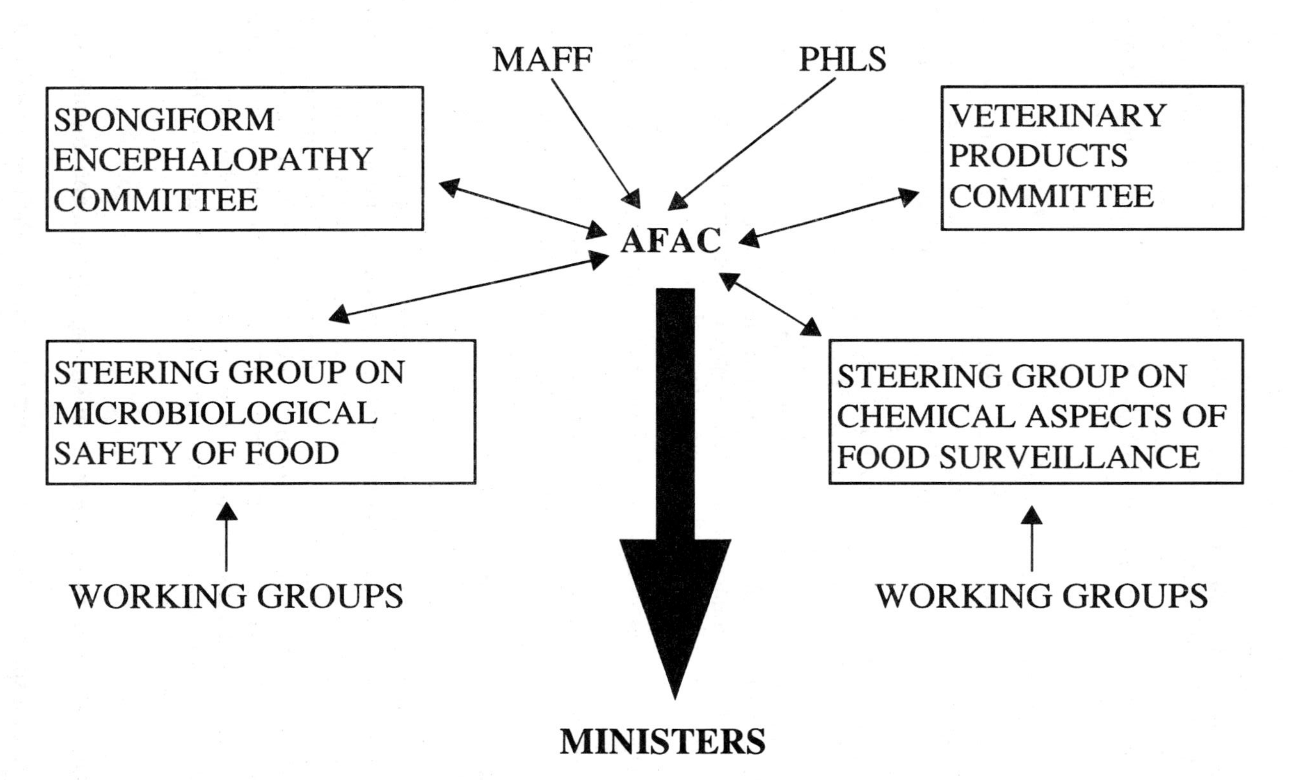

Figure 10.1 Structure of Animal Feedingstuffs Advisory Committee

products for human consumption, are brought within a common legislative framework for both surveillance and legislative control. Ministers have accepted this recommendation and it is understood that arrangements are in hand to establish the AFAC. The Expert Group divided their work into four areas: the Spongiform Encephalopathies, *Salmonella* and other Pathogens, Medicated Feedingstuffs, and Feed Composition and Marketing. Detailed reviews and recommendations for each area are made in separate chapters of the report. A brief review of the major recommendations for these areas is given here. In order to avoid misinterpretation of the Expert Group's views, in some cases the recommendations are quoted directly from the Report with appropriate annotation.

THE SPONGIFORM ENCEPHALOPATHIES

The disease Bovine Spongiform Encephalopathy (BSE) has probably provided UK veterinary epidemiologists with the most challenging problem they have faced during the twentieth century. Epidemiological studies indicate that BSE was probably caused initially by feeding cattle rations containing meat and bone meal prepared from animal material which included tissues infected with scrapie, a disease of sheep endemic in the UK and also in several other countries. Subsequently, the epidemic was amplified in cattle by feeding meat and bone meal containing material from BSE infected cattle. Since, in the UK, material from cattle provides about 45% of the total material processed into meat and bone meal used in animal rations, within species amplification of the disease is now thought to have played a significant role in the increasing incidence of BSE. It is now recognised that the outbreak probably was related, among other things, to changes in the early 1980's in the rendering processes (including the discontinuation of solvent use to extract the tallow which had involved an additional heat process to remove the solvents). These changes allowed a sufficiently large infective dose of the scrapie agent to survive the rendering process, and then not only to cross the species barrier to infect the cattle, but also to cause clinical disease within their natural commercial life span. Further references to the discussions and conclusions of the relevant expert committees (Working Party on Bovine Spongiform Encephalopathy [Southwood Committee]; Consultative Committee on Research into Spongiform Encephalopathy [1st Tyrrell Committee] and Spongiform Encephalopathy Advisory Committee [2nd Tyrrell Committee]) are given in the main report which also contains full details of the epidemiological data for the BSE outbreak up to June 1992, immediately prior to the publication of the report. To that date there had been 58,880 cases in 18,547 herds 92.5% from dairy cattle herds and 7.5% from beef herds. The more important additional details are that to date (January 1st, 1993) there has been 79,057 confirmed cases on 22,510 farms (Her Majesty's Stationery Office, Press Release, December 1992).

Current and future plans for control of spongiform encephalopathies

The absence of a specific diagnostic test for infectivity in the live animal makes it necessary for extensive control measures to be in place. The three most important current control measures are:

1 Destruction of carcases of cattle infected with BSE.
2 Ban on feeding ruminant protein to ruminants. (Ruminant Protein Ban, RPB).
3 Ban on the inclusion of specified bovine offal in all animal rations. (Specified Bovine Offals Ban, SBOB)

The destruction of carcasses from BSE suspects removes this material from the feed chain and, furthermore, the SBOB removes any currently suspect tissue from possibly infected animals which are not exhibiting clinical signs of the disease. The SBOB includes brain, spinal cord, thymus, tonsils, spleen and intestines from bovine animals over 6 months of age and since, so far, the BSE agent has not been identified in any tissue except brain, these measures reinforce the view that the SBOB removes any suspect tissue from the feed of both animals and birds. In the case of the RPB, a direct quote from the Report is given to summarize exactly the importance the Expert Group allocated to this measure.

> 'This measure, if properly enforced, will be effective in preventing the exposure of cattle to BSE infection via feed and in protecting all other ruminant species from exposure to food-borne sources of ruminant SE agents. The inclusion of both ovine and bovine sources of protein in the ban will prevent the infection of cattle with scrapie from sheep, and will stop the recycling via feedingstuffs of SEs in cattle and sheep. It will also protect exotic *Bovidae*, some of which have succumbed to SEs in British zoos, and deer which are also susceptible to an SE (chronic wasting disease) though no such disease has occurred in the UK. Although the ban came into force in July 1988 it was expected that it could require up to six months for all ruminant rations, manufactured before the ban, and containing ruminant protein, to be cleared from the feed chain. Current epidemiological evidence suggests that the effective lag in clearing these feed sources has been about three months.'

In fact, although the RPB was instituted in July 1988, it now appears that rations infected with the BSE agent were in use throughout 1988 and the current high incidence of cases indicates both that contaminated rations were being fed well after July 1988 and further that the degree of infectivity of the BSE agent must have been high in rations containing meat and bone meal during this final period. Up to the November 20th 1992, 599 cases of BSE had been confirmed in animals born after the July 1988 feed ban. All but two of these had access to feed containing ruminant protein and investigation of the feeding of these two was still in progress. By that date, studies of the age specific incidence showed the anticipated decline in the percentage of the two youngest groups (2–3 and 3–4 years old) for the first 9 months of 1992 compared to the years 1990 and 1991 (Table 10.1), continuing the expected trend first reported earlier in 1992 by Wilesmith (Wilesmith, Ryan, Hueston and Hoinville 1992a; Wilesmith, Hoinville, Ryan and Sayers 1992b; Wilesmith and Ryan, 1992). All the evidence continues to point to contaminated ruminant protein as the origin of the epidemic and to justify confidence in the feed bans.

The House of Commons Agriculture Committee recommended that the Expert Group investigate the feasibility of developing a rendering process capable of destroying the agent. The Government's response to the Agriculture Committee report was to confirm that a large volume of research was already in progress,

Table 10.1 AGE SPECIFIC INCIDENCE OF BOVINE SPONGIFORM ENCEPHALOPATHY TO 30 SEPTEMBER 1992

Age (years)		*Incidence* (% of herds affected)	
	1990	*1991*	*1992*
2–3	0.09	0.03	0.003
3–4	1.33	1.86	0.34
4–5	3.81	5.34	5.09
5–6	3.29	3.65	3.31
6–7	1.91	1.77	1.40
7–8	0.73	0.86	0.58
8–9	0.18	0.33	0.34
9–10	0.09	0.17	0.19
10–11	0.07	0.07	0.06

sponsored by several authorities. The final results of these studies using BSE-infected material will not be available for some time and further work with material from scrapie infected sheep has now commenced. A final view on these issues must await these data. However, sufficient interim evidence is already available to indicate that the infective agent is extremely resistant to the normal time/temperature heat treatments used in processing meat and bone meal. The Expert Group felt that there is considerable doubt that rendering plants could be expected to consistently meet the demanding criteria which are likely to be required to guarantee destruction of the infective agent. For example, it is contemplated that because of the nature of the agent, a very high margin for safety would be required and there is a real prospect that the resulting protein material would be denatured, thus making the process nutritionally damaging. The Group was also aware of the substantial problems of disposing the vast quantity of animal material (some 1.75 million tons/annum) if it is not made acceptable as animal feed. In view of the evidence cited above, the Expert Group recommend that

> 'the feed bans be retained even after the results of the inactivation study become available, unless the results provide unequivocal information on the inactivation of the scrapie/BSE agents, and the necessary conditions can be consistently achieved by the rendering industry. In our view, the feed bans are, at the present state of our knowledge, the only certain safeguard in preventing disease transmission via feed in animals.'

The Group spent some time considering the possibility that UK livestock could be infected by other SE agents. Scrapie was probably the original source of BSE so the primary incident required the species barrier from sheep to cattle to be breached. Thereafter, infected cattle material was recycled from meat and bone meal fed to cattle producing several serial passages and selection of the cattle adapted strains. This emphasizes the potential dangers, dependent on the species, of intra species recycling. The transmission between species is dependent on the dose given, the route of administration (i.e. orally or parenterally) and the species barrier effect. Pigs and marmosets have been experimentally infected by administering a high dose of the infective agent intracerebrally. So far, no experimental pigs have succumbed to feeding infected material.

The experts who advised the Group recommended that the greatest risk of another BSE-like epidemic in domestic species would require inter species exposure and then intra species recycling. Intra species recycling is not confined to domestic animals; it has been seriously implicated recently in France (Anon, 1992) in a higher incidence of Creutzfeldt-Jacob disease (CJD) in human patients treated parenterally with extracts of human pituitaries to provide Growth Hormones (GH) activity. It is postulated that immediately prior to the cessation of use of human pituitary extracts in France there must have been a high level of infectivity in the pituitary extracts used which, of course, were administered by injection and not orally as with BSE. A similar situation occurred in the disease Kuru as a result of cannibalistic rites practised in New Guinea resulting in the recycling within the species. Informed opinion suggests that Kuru may have started with a case of CJD. Thus, no species barrier was involved in initiating nor sustaining the disease. Since the cessation of cannibalism, Kuru has become a rare disease.

There are currently two known reservoirs of SE agents in livestock in the UK: Scrapie in sheep and BSE in cattle. The extent of current exposure to these known reservoirs is given in Figure 10.2 together with the areas where preventative measures apply. It immediately becomes obvious that mainly the pig, and to some extent poultry, are now the 'sink' for the disposal of meat and bone meal made from ruminant tissues. No case of SE has occurred in pigs naturally nor, so far, in pigs experimentally given infected material orally. The Group recommended a high level of surveillance of possible SEs in other livestock should they occur. While the Group recommended that further research is required, especially with respect to the nature of the species barrier as it relates to spongiform encephalopathy and specifically on the nature of the prion protein (PrP) gene in the pig, it felt that the current controls were adequate.

The Group also urged that when information is available, as a result of the compulsory notification of scrapie, the appropriate committees should be invited to examine methods designed to reduce (and hopefully eventually eliminate) the reservoir of scrapie. Considering the importance of the spongiform encephalopathies, it is this author's opinion (a view believed to be shared by the Group as a whole)

Known reservoir	Preventative measures	Species exposed via feed
SHEEP	------ / RPB / ------➤	Cattle & other ruminants
SHEEP	-----------------➤	Pigs
SHEEP	-----------------➤	Poultry
SHEEP	-----------------➤	Fish
CATTLE	------ / RPB / ------➤	Sheep & other ruminants
CATTLE	------/ SBOB / -----➤	Pigs
CATTLE	------/ SBOB / -----➤	Poultry
CATTLE	-----------------➤	Fish

RPB — ruminant protein ban SBOB — specified bovine offals ban

Figure 10.2 Possibilities of infection of UK livestock by spongiform encephalopathies

that the RPB should be retained for the foreseeable future. In addition, the whole question of the safety of species recycling should be researched and critically examined by the AFAC. We must be convinced that intra species recycling in pigs and poultry does not contain hidden dangers of other spongiform encephalopathies or any condition to create a zoonosis.

SALMONELLA AND OTHER PATHOGENS

The Report considers human diseases potentially transmissible via animal feeding-stuffs (anthrax, *Salmonella*, *Listeria*, *Yersinia*, *E coli*. 0157 and *Campylobacter*) but the greatest emphasis in this section is related to the salmonellas and particularly *Salmonella enteritidis* Phage Type 4 which in recent years has been responsible for a dramatic rise in *Salmonella* infections in man in the UK (see Figure 10.3). There

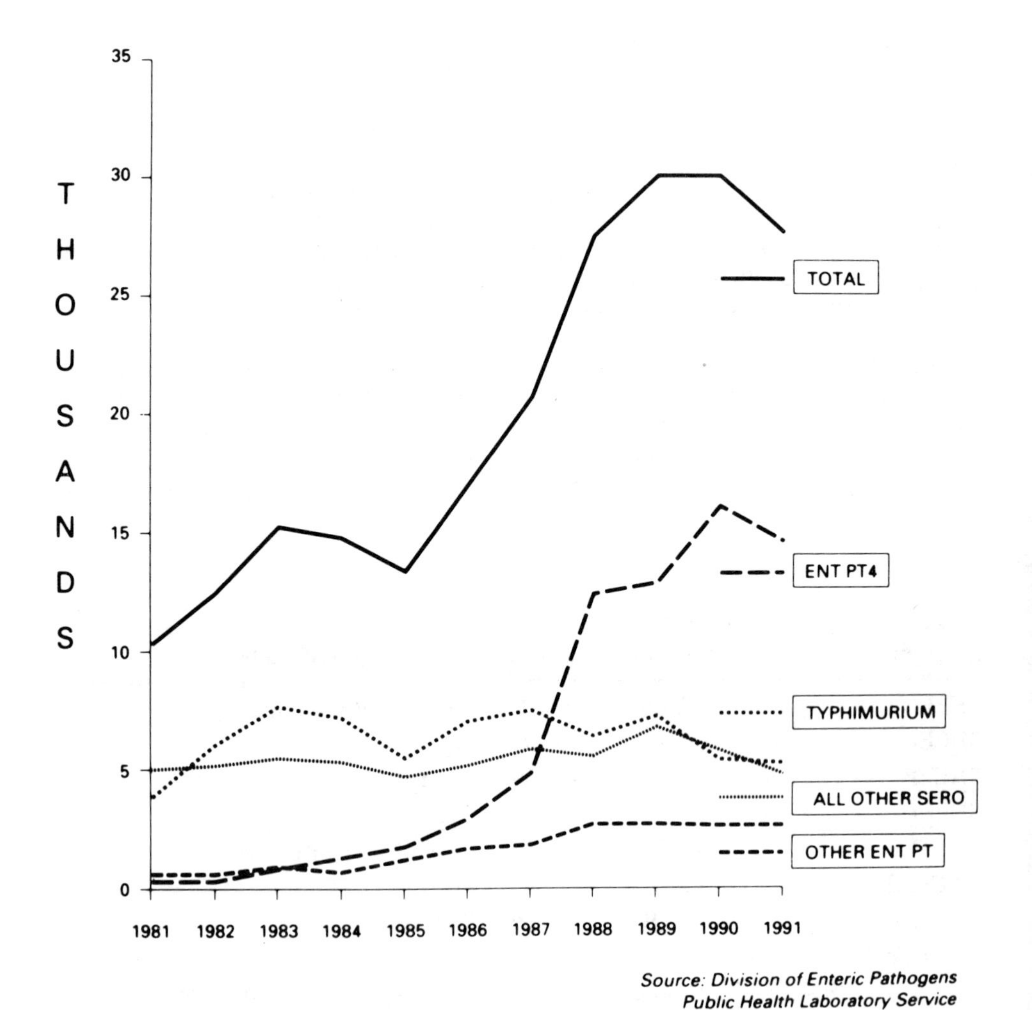

Figure 10.3 Current isolations in humans, animals, poultry and processed animal protein

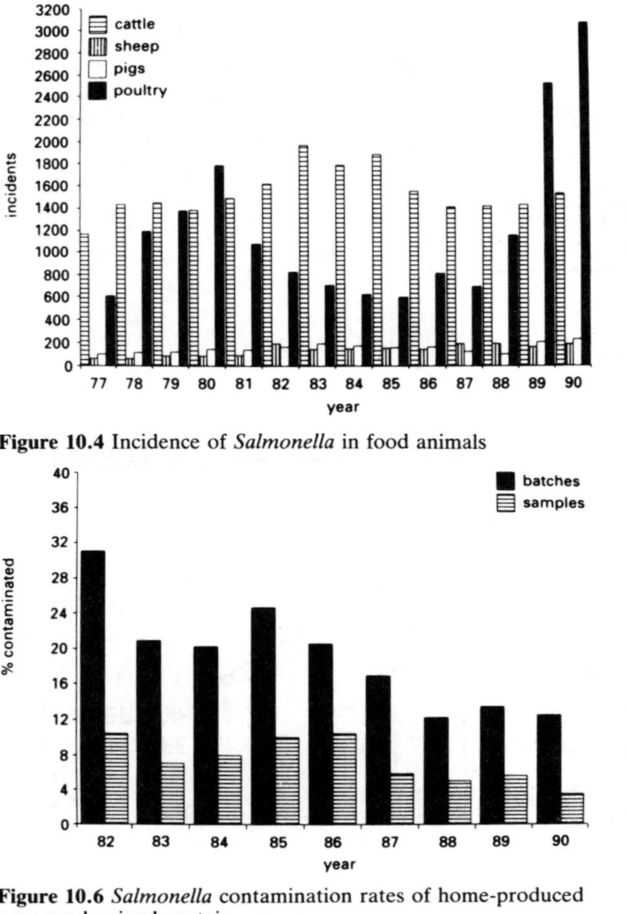

Figure 10.4 Incidence of *Salmonella* in food animals

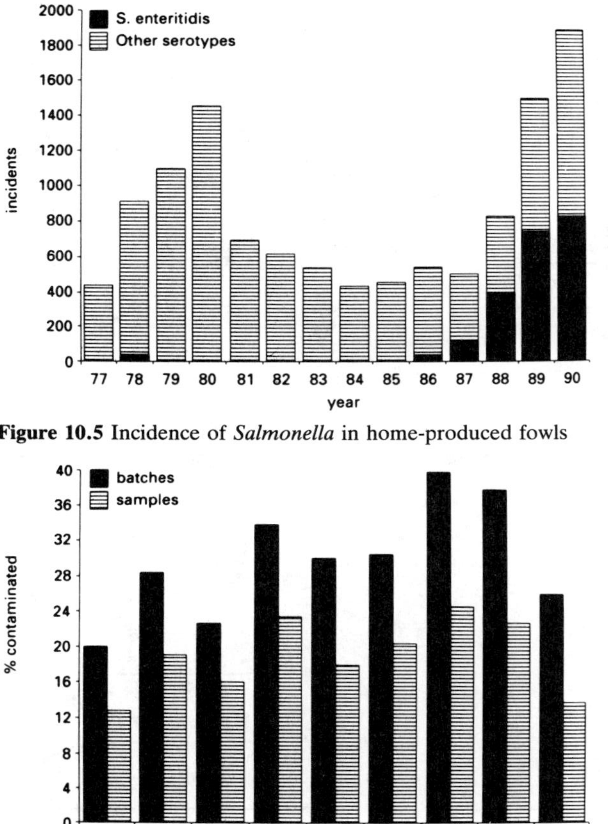

Figure 10.5 Incidence of *Salmonella* in home-produced fowls

Figure 10.6 *Salmonella* contamination rates of home-produced processed animal protein

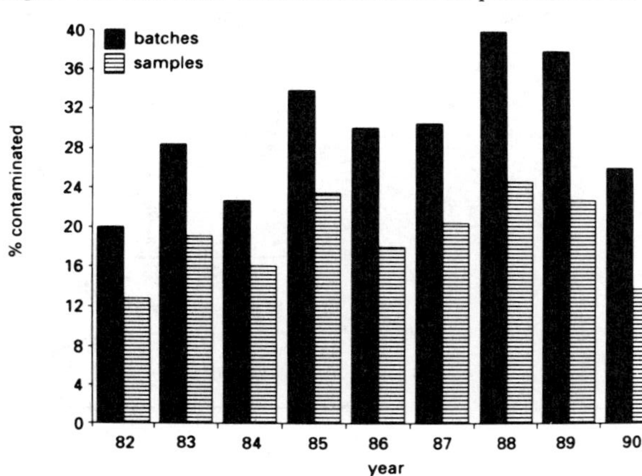

Figure 10.7 *Salmonella* contamination rates of imported processed animal protein

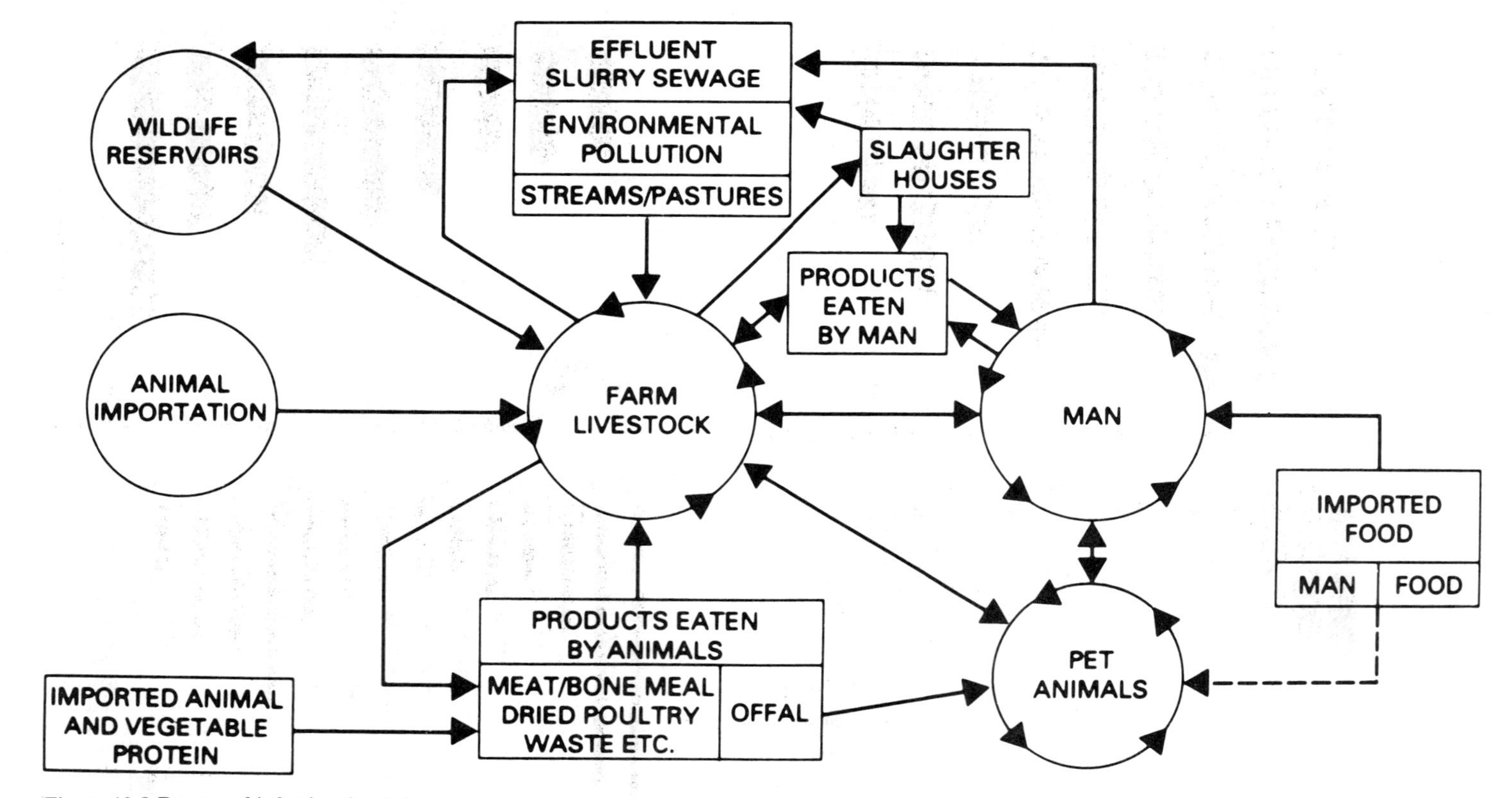

Figure 10.8 Routes of infection for *Salmonella*

are also data on trends in incidents in food animals (Figure 10.4), home-produced fowl (Figure 10.5), home-produced processed animal protein (Figure 10.6) and imported processed animal protein (Figure 10.7). There are several potential routes of infection of *Salmonella* for both man and farm livestock (Figure 10.8). A number of committees (e.g. The Richmond Committee) and expert groups have reported on these issues, leading to the establishment of the Steering Group on the Microbiological Safety of Food to coordinate work on microbiological food safety.

There is insufficient time here to discuss the full details of evidence submitted to the Group in this area of its work and the main report must be examined for a full review. Of particular importance in the present context are the results of the voluntary and mandatory tests for *Salmonella* in feedingstuffs carried out by MAFF and feed processors and by laboratories reporting positive test results from samples submitted voluntarily by suppliers and manufacturers in the feed chain.

In recent years, the incidence of *Salmonella* in home-produced animal protein has shown an encouraging decline (Figure 10.6), imported animal protein sources (Figure 10.7) generally showing higher, but also declining, contamination rates, with some imported fish meal supplies of particular concern. Some concern is expressed that some vegetable protein sources, including home produced rape, may be a risk as sources of infection. These sources are not currently subjected to mandatory testing and we recommended they should be tested. The Group also recommended that further surveys be undertaken to monitor feedingstuffs and vegetable materials and that all results (not just positive results) are reported and made available to the AFAC. It further recommended that various feed treatments (heat treatments including pelleting, irradiation and acid treatments) designed for control of infection in feedstuffs be examined and that the AFAC should examine this area in detail.

Considering the importance of maintaining supplies of uncontaminated feed, particularly in poultry flocks, a direct quote from the main report is appropriate:

> 'The consensus view presented to us is that feed is important in introducing contaminations into flocks and that every effort should be made to reduce contaminations from this source. Not least is the arguments that all efforts to disinfect houses and maintain clean flocks are negated if infection is reintroduced by feedingstuffs. This argument, sustained by all the experts we met is supported by the examples cited in the main report. We have been told that the impact of contaminated feed will be greater if it enters breeding flocks and that concentrated efforts to reduce contamination at that point would have maximum effect. Nevertheless the removal of feed contamination alone at this stage of the epidemic would not necessarily have a significant effect on the current *S. enteritides* disease in the human population.' . . . 'We recommend that the Animal Feedingstuffs Advisory Committee should consider all available information on *Salmonella* and other pathogens in feedingstuffs with particular reference to the possibility that serotypes found in feed may subsequently become important for animals, and man, and advise on any measures which need to be taken. Further we recommend that authorities charged with monitoring *Salmonella* and other organisms in animal feed, animals, food, and humans be represented on this Committee.'

MEDICATED FEEDINGSTUFFS

This section examined the mechanisms for licensing veterinary medicines and their incorporation into compound feed by compounders and farmers. In as much as the Medicines Act 1968 provides effective legislation for the control of manufacture, sale and supply of veterinary medicinal products in the UK, this section of the present paper can be correspondingly brief. The work of the Veterinary Medicines Directorate (VMD) and the contribution of the Veterinary Products Committee (VPC) in advising ministers on the safety, quality and efficacy of veterinary medicines is well known and the public should be generally reassured on these issues. Equally, the EC authorities are moving towards harmonization of licensing procedures including the development of standard maximum residue limits (MRLs). Of major importance to feed manufacturers is the Expert Group's recommendation that fish farmers who use medicated feeds should be registered with the Royal Pharmaceutical Society of Great Britain or the Department of Agriculture in Northern Ireland and that on-farm mixers and integrated units using medicinal additives and intermediate medicated feeds should also be registered and therefore subjected to surveillance and control of feedingstuffs manufacture.

In respect to veterinary drug residues, the Group was satisfied with the progress made as evidenced by recent reports of the MAFF Working Party on Veterinary Residues (WPVR) delivered through the Steering Group on the Chemical Aspects of Food Surveillance as well as the results of statutory chemical analyses by the VMD under the National Surveillance Scheme for residues in meat, which is now published annually. These reports present the survey data on residues of antimicrobials, antihelmentics and coccidiostats, residues of synthetic and natural hormones, nonhormonal growth promoters and tranquilizers in meat. The general decline in the number of samples showing positive residues is very encouraging and only in respect of residues of sulphonamides did the group express some concern, although it noted a steady decline in the incidence of sulphonamide residues above the MRLs between 1986 and 1991. Nevertheless, it recommended that careful consideration be given to licensing of drugs for in-feed use which require long withdrawal periods.

The Expert Group states:

> 'These findings which are published in the WPVR Report, have been assessed by the various Committees which have responsibilities on these matters, namely the Committee on Toxicity of Chemicals in Food, Consumer Products and the Environment, the Food Advisory Committee and the Veterinary Products Committee'.

It concluded that the reports

> 'provide reassurance that, in general, foods of animal origin contain little or no residue of veterinary products and that where present they pose no significant risk to human health. They recommended that surveillance should be continued and should be increased in certain areas (poultry, eggs, fish)'.

Finally, in the area of medicinal substances in feeds, the report draws attention for VPC consideration to some evidence of the increasing multiple antibiotic resistance of salmonellas and the question of apramycin/gentamycin resistance. Generally, the view is developing that the prophylactic use of antibiotics with cross resistance to those used in human medicine should be discouraged. This is an important area of interface between the AFAC, the VPC and the Steering Group on the Microbiological Safety of Food and the contributions of the MAFF investigation groups and the Public Health Laboratory Service to future research and the debate are of paramount importance.

COMPOSITION AND MARKETING

A number of recommendations and observations in this area are made and of particular note in this context are the recommendations for identification, sampling and inspection of on-farm mixers to ensure more complete coverage of the feedingstuffs industry and for clear cut guidance and a code of practise aimed primarily at on-farm mixers and integrated units.

Summary

The feed industry must be prepared to accept stringent controls to maintain the safety of feed sources in the animal feed/farm animal/human food chain, but the controls must be applied to all sources and all segments of the feedingstuffs industry. The necessary structures in the relevant ministries, government agencies and independent committees to ensure that the safety of feed ingredients and manufactured feedingstuffs can be maintained, thus helping to ensure food safety. The Group recommended that the interrelationship between the relevant authorities and committees should be formalized under a new *Animal Feedingstuffs Advisory Committee*. This should be an independent committee empowered to suggest new areas for research and surveillance. It should report direct to ministers.

Acknowledgements

Figures 10.2 to 10.8 were taken from the Expert Group Report. The source of Figure 10.3 was the Division of Enteric Pathogens, Public Health Laboratory Service while that for Figures 10.4 to 10.7 was MAFF, ADAS Animal Salmonellosis 1990.

References

Anon (1992) French officials panic over rare brain disease outbreak. *Science*, **258 December 4**, 1571

Her Majesty's Stationery Office (1992a) *The Report of the Expert Group on Animal Feedingstuffs to the Minister of Agriculture, Fisheries and Food, the Secretary of State for Health and the Secretaries of State for Wales, Scotland and Northern Ireland*. London: HMSO

Her Majesty's Stationery Office (1992b) *Report of the Expert Group on Animal Feedingstuffs. Recommendation and Government's Response*. London: HMSO

Wilesmith, J.W., Ryan, J.B.M., Hueston, W.D. and Hoinville, L.J. (1992a) Bovine spongiform encephalopathy — epidemiologic features 1985 to 1990. *Veterinary Record*, **130 (5)**, 90–94

Wilesmith, J.W., Hoinville, L.J., Ryan, J.B.M. and Sayers, A.R. (1992b) Bovine spongiform encephalopathy — aspects of the clinical picture and analyses of possible changes 1986 — 1990. *Veterinary Record*, **130 (10)**, 197–201

Wilesmith, J.W. and Ryan, J.B.M. (1992) Bovine spongiform encephalopathy — recent observations on the age-specific incidences. *Veterinary Record*, **130 (22)** 491-492

11

MAXIMUM RESIDUE LIMITS — THE IMPACT OF UK AND EC LEGISLATION

K.N. WOODWARD
Veterinary Medicines Directorate, Woodham Lane, New Haw, Addlestone, Surrey, KT15 3NB, UK

Historical perspective

The safety evaluation of various types of chemicals has long depended on the concept of the maximum residue limit or MRL for the protection of consumers. For residues of pesticides, the Joint FAO/WHO Meeting on Pesticides Residues (JMPR) has been establishing MRLs for over 25 years (Anon, 1990a) while the Joint FAO/WHO Expert Committee on Food Additives (JECFA) elaborated MRLs for a number of antibiotics used in veterinary medicine as long ago as 1969 (Anon, 1969). However, it was not until 1984, at the international level, that veterinary drug residues came of age, when a Joint FAO/WHO Expert Consultation held in Rome made a number of recommendations which eventually led to the establishment of regular meetings of JECFA to consider MRLs for veterinary drugs (Anon, 1985). The first of these JECFA meetings dedicated to the establishment of MRLs for veterinary drugs was held in Rome in 1987 (Anon, 1988) and since then there have been regular meetings of the Committee.

In the European Community (EC), veterinary medicinal products are dealt with in two ways. Those intended for therapeutic use or for the alteration of physiological function are covered by the so-called Veterinary Medicines Directives, 81/851/EEC and 81/852/EEC, and their amending Directives. On the other hand, substances used for prophylaxis and growth promotion and which are added to feed, are dealt with by the directive known widely as the feed additives directive, 70/524/EEC (Woodward, 1991a). To emphasise the separation, the former are dealt with by one Directorate General within the European Commission (DG III) while the latter are dealt with by another (DG VI). At the present time, only the former group of substances, what might be regarded as the conventional veterinary medicines, are formally subject to MRLs within the EC. The latter group, the medicinal feed additives await this type of treatment.

The Working Group on the Safety of Residues (WGSR) of the Committee for Veterinary Medicinal Products (CVMP) had been examining the toxicology of veterinary medicines and establishing MRLs on an ad hoc basis for several years when, in 1990, an EC regulation, Council Regulation 2377/90 was introduced (Anon, 1990b). This legislation has two main consequences for the veterinary drugs covered by the veterinary medicines directives:

- from January 1st 1992, no new active ingredient or pharmacologically active excipients may be introduced on to the markets of Member States unless there is a Community MRL
- existing agents are subject to a systematic call-up so that MRLs can be established for these by 1997.

Rather obviously, the Regulation does not apply to products used in companion animals. After 1 January 1997, the administration to food producing animals of medicines containing active ingredients which are not in one of Annexes I to III of 2377/90 is prohibited. All substances considered under 2377/90 will enter into one of four Annexes the specifications of which are as follows:

Annex I – full MRL
Annex II – no MRL required (eg for naturally occurring compounds such as some hormonal products)
Annex III – provisional MRL (pending further data and subject to an expiry date)
Annex IV – use in food-producing animals is unsafe and no MRL can be established (effectively prohibited in the EC for use in food-producing species).

When established, Community MRLs for veterinary drugs serve four main and inter-related purposes:

1. they effectively specify safe concentrations in foodstuffs;
2. they form the basis for the establishment of withdrawal periods;
3. they act as standards for residues surveillance;
4. they remove (or should remove) barriers to trade in sales of produce of animal origin, within the Community.

The task of establishing MRLs is still carried out by the WGSR with final ratification by the CVMP.

The scientific basis of MRLs

Two important concepts lie behind the elaboration of MRLs for veterinary drugs — the no-observed effect level in toxicological (or microbiological) studies and the acceptable daily intake.

The no-observed effect level (NOEL) is generally taken as the dose in toxicological or microbiological studies at which and below which adverse effects do not occur (Anon, 1987; Anon, 1990a). It is further defined by stipulating that these must be the most sensitive effects in the most sensitive species studied. The acceptable daily intake (ADI) has been used by JECFA for a number of years. It is derived from the NOEL by dividing this by a suitable safety factor. This safety factor is usually 100 (10 for animal-human variation × 10 for human – human variation) but its value can be greater depending on the nature of the effect and the quality of the data package (Bigwood, 1973; Perez, 1977; Vettorazi and Radaelli-Benvenuti, 1982; Woodward, 1991b) or smaller, e.g. if it is based on human data. It has recently been suggested that a more scientific approach to the safety factor might involve studies of trans-species factor such as pharmacokinetic parameters (Renwick, 1991) but this would involve a substantial increase in costs arising from the extra research involved and the author himself concludes that the current approach is pragmatic if not totally scientific. The ADI is usually cited in mg of the drug/kg body weight/day:

$$ADI = \frac{NOEL}{SF\ (100)}$$

ADI = Acceptable daily intake (mg/kg body weight/day); *NOEL* = No-observed effect level; *SF* = Safety factor

or, if the standard human body weight of 60 kg adopted by JECFA is used in the calculation, in mg of drugs per person per day:

$$ADI = \frac{NOEL \times 60\ kg}{SF\ (100)}$$

ADI = Acceptable daily intake/person (mg/person/day)

Establishment of the MRL once the ADI has been calculated is less straight forward, and there is no simple equation or procedure available. In the United States, the concept of the safe concentration has evolved:

$$Safe\ concentration = \frac{ADI \times 600\ kg}{500\ g \times food\ consumption\ factor}$$

This calculation assumes a daily intake of 1500 g food per person of which 500 g is of animal origin. The food consumption factors employed in the USA are 3 for milk, 1 for muscle, 0.5 for liver, 0.3 for kidney and 0.25 for fat (Teske, 1992).

The safe concentration can be seen as the MRL in its very simplest of forms. Its use in this way must ensure that overall, the ADI is not exceeded and moreover, it must take account of the concentrations of the drug which will be achieved in practice as a result of metabolic processes in the animals treated with the drug.

As a consequence, the elaboration of an MRL usually involves a much more iterative approach which takes into account not only the toxicological considerations in the form of the ADI, but also the residues depletion profile of both the drug and its metabolites, in the treated animals (Anon, 1990c). Moreover, any MRL set must be 'measurable' analytically. It is futile to set a value for which there is no analytical capability — unless of course this value is determined on toxicological grounds when either a suitable method of analysis must be developed or, failing that, the drug must be prohibited. Any method of analysis should have a limit of detection well below the MRL for screening purposes and up to one order of magnitude below for regulatory purposes (Crosby, 1991).

JECFA has recommended a decision tree approach to establishing MRLs for veterinary drug residues in food of animal origin, which includes milk, eggs and honey. This considers elaboration of a MRL based upon the ADI but takes into account the residues depletion profile seen in practice and the capabilities of the analytical methods available (Anon, 1990c). The over-riding consideration remains as to whether, when the total feed intake package of muscle, fat, eggs, milk etc. is taken into account, the ADI is exceeded. If it is, the calculation must be re-analyzed.

Data requirements

For applications for MRLs under Council Regulation 2377/90, the data requirements are set out in the Annex to one of the Veterinary Medicines Directives, 81/852/EEC (Anon, 1981) as amended by Commission Directive 92/18/EEC (Anon, 1992) which basically updates and consolidates the existing requirements. The general requirements for toxicity testing are:

single dose (two species)
repeated dose — 90 days (two species)
reproductive effects
embryotoxicity/foetoxicity (teratology) (two species)
mutagenicity (point mutations/clastogenic effects)
carcinogenicity
immunotoxicity
pharmacological effects
observations in humans (toxicity and pharmacological effects)

Tests for carcinogenicity, arguably the most expensive of the studies, are generally required for drugs with structural similarities to known carcinogens, for compounds which have produced positive effects in mutagenicity studies, and for drugs which have produced suspect signs in other toxicity tests. All the studies must be conducted, where possible, in accordance with approved guidelines (e.g. OECD) and in conformity with the principles of good laboratory practice (GLP).

Directive 92/18/EEC also sets out the requirements for data relating to residues studies and these include data on pharmacokinetics and pharmacodynamics in laboratory and target species as well as data on residues depletion *per se*. It is necessary to identify the target organ(s) for residues (where residues are highest or where they accumulate) and a marker metabolite (or metabolites) where depletion behaviour is representative of residues depletion of the drug in both qualitative and quantitative respects. One of the important requirements of Council Regulation (EEC) 2377/90 is that for the development of a routine method of residues surveillance. Applicants need to specify the accuracy, specificity, limits of detection and quantification and, importantly, applicability and practicability of the method under normal laboratory conditions.

One of the major uses of the MRL is in establishing the withdrawal period although of course, many other factors also need to be considered. The withdrawal period is usually set at the time when the marker metabolite depletes to below the MRL in the target tissue in all of the animals in a group in a serial slaughter residues study. There is an obvious need even then to determine that using this method, the ADI will not be exceeded in all the animal products normally considered (liver, kidney, muscle, fat and milk). JECFA on the other hand has adopted a somewhat different approach and recommends the 99th percentile with a 95% confidence limit as a realistic representation of the withdrawal period. One of the other major purposes of establishing MRLs is surveillance — surveillance for both violations of MRLs and for the use of illegal substances such as certain anabolic hormones and thyrostatic drugs (Anon, 1991a). This is conducted in the Community under the auspices of the so-called residues directive, Directive 86/469/EEC (Anon, 1986). At the present time, Member States must annually provide a National Plan to the European Commission setting out it surveillance programme. The UK currently examines some 40,000 tissue samples for violation of MRLs or for illegal residues.

Progress with MRLs in the EC

A number of MRLs have been established in the EC since Regulation 2377/90 took effect. Some of these are Annex III entries and so are provisional pending further data from the drug sponsors. Only a few are final (Annex I) while there is only one firm candidate for Annex II (hydrogen peroxide as an ectoparasiticide for salmon) at the time of writing.

One of the major reasons for Annex III entry is the additional requirement for data on the effects of antibiotic and antimicrobial substances on the gut flora in humans. The evidence to support the possible effects of such compounds on the gastrointestinal flora is somewhat thin. Nevertheless, JECFA used the approach for establishing MRLs for spiramycin and oxytetracycline and it is a requirement of 2377/90 through Directive 92/18/EEC.

The drugs for which MRLs have been established are set out below:

benzyl penicillin
ampicillin
amoxicillin
cloxacillin
dicloxacillin
ivermectin
trimethoprim[1]
nitrofurans[1]
dapsone[1]
tetracyclines[1]
chloramphenicol[1]
tylosin[1]
amitraz
febantel[1]
fenbendazole[1]
oxfendazole[1]
albendazole[1]
thiabendazole[1]
flubendazole[1]
levamisole[1]
carazolol[1]
sulphonamides[1]
spiramycin[1]

Although the MRLs are established in the Community by way of a Regulation which does not require national implementation (unlike Directives), the UK has introduced its own legislation on MRLs for veterinary drugs (Anon, 1991b). This has been done under the Food Safety Act 1990 and goes somewhat further than the existing EC legislation. It establishes MRLs which are largely based on EC values except where the Community has yet to establish MRLs eg for the β-agonist clenbuterol. However, in addition, it introduces a number of new offences including failure to observe a withdrawal period resulting in MRL violations and it imposes penalties including the condemnation of carcasses in some circumstances, e.g. the use of illegal drugs.

Extension of MRL concept to feed additives

A recent meeting held in Asti in Italy considered the very question of the use of MRLs in relation to compounds covered by Directive 70/524/EEC and considered as feed additives in the Community. Indeed, the meeting was a joint meeting of

[1] Provisional maximum residue limits

DGIII and DGVI held under the auspices of the Scientific Committee on Animal Nutrition (SCAN). SCAN is the independent advisory body which comments on the safety of feed additives and it has seen certain benefits in following the MRL route in assuring consumer safety. Not least among these are the use of the MRL for establishing withdrawal periods for 70/524/EEC candidate compounds and for residues surveillance.

This is important now as the JECFA system makes no distinction between veterinary medicines and medicinal feed additives and it recently assessed two drugs, carbadox and olaquindox, currently regulated in the Community under 70/524/EEC and so not subject to MRLs under 2377/90. In doing so it established an MRL for carbadox and identified further work on olaquindox. Hence, there is an urgent need for the EC to look at the MRL route for feed additive compounds before the JECFA system anticipates values with little corporate European input. This is important as the JECFA values are eventually introduced into the Codex Alimentarius system for international adoption and harmonisation (Woodward, 1991c).

Conclusions

The EC and the international community, as well as national authorities, are engaged in the establishment of MRLs for veterinary drugs in food of animal origin. The basic inputs into these processes are the toxicology and residues data generated in support of the MRLs. In viewing MRLs within the Community, it seems only sensible to view medicines as one distinct group — rather than to see them, as is currently the case, largely as therapeutics and feed additives — and to establish MRLs for all. This would introduce some degree of harmonisation on this front with the JECFA/Codex Alimentarius system.

There is at this point room for consideration of one other topic. This is related to this very issue of harmonisation. At the current time the data requirements under 81/852/EEC and 2377/90 are different from those of 70/524/EEC and in turn, these are different from those of other countries. This problem has been recognised in the area of human medicines and a number of efforts have been made to introduce some degree of harmonisation (D'Arcy and Harron, 1992). It has been recognised too for veterinary medicines and discussed at a meeting of the Toxicology Forum and at the VIth International Technical Consultation on Veterinary Drug Registration held in Buenos Aires in 1992. It is a topic which needs urgent attention if MRLs are to be established with any degree of consistency whether for veterinary medicines or for feed additives regardless of whether this is at a national or international level.

The European legislation will, in effect, remove those compounds from the market which do not fulfil the current criteria of consumer safety as laid out in EC law while only allowing new compounds on to the market which to satisfy these requirements. The UK's own legislation will serve only to reinforce these EC requirements.

References

Anon. (1969) Specifications for the identity and purity of food additives and their toxicological evaluation: Some antibiotics. *Twelfth Report of the Joint FAO/WHO Expert Committee on Food Additives. Technical Report Series 430*, Geneva

Anon. (1981) Council Directive of 28 September 1981 on the approximation of the laws of Member States relating to analytical, pharmacotoxicological and clinical standards and protocols in respect of the testing of veterinary medicinal products (81/85/EEC). *Official Journal, No L317*, 16–28

Anon. (1985) Residues of Veterinary Drugs in Foods. Report of a Joint FAO/WHO Expert Consultation. *FAO Food and Nutrition Paper No 32*. Rome

Anon. (1986) Council Directive of 16 September 1986 concerning the examination of animals and fresh meat for the presence of residues. *Official Journal, No L275*, 36–45

Anon. (1987) International Programme on Chemical Safety. Principles for the toxicological assessment of food additives and contaminants in food. *Environmental Health Criteria 70*. Geneva: WHO

Anon. (1988) Evaluation of certain veterinary drug residues in food. Thirty-second Report of the Joint FAO/WHO Expert Committee on Food Additives. *Technical Report Series 763*, Geneva

Anon. (1990a) International Programme on Chemical Safety. Principles for toxicological evaluation of pesticide residues in food. *Environmental Health Criteria 104*. Geneva: WHO

Anon. (1990b) Council Regulation (EEC) No.2377/90 of 26 June 1990 laying down a Community procedure for the establishment of maximum residue limits of veterinary medicinal products in foodstuffs of animal origin. *Official Journal No. L224*, 1–8

Anon. (1990c) Evaluation of certain veterinary drug residues in Food. Thirty sixth report of the Joint FAO/WHO Expert Committee on Food Additives. *Technical Report Series 799*. Geneva: WHO

Anon. (1991a) Evaluation of certain veterinary drug residues in food. Thirty-eighth report of the Joint FAO/WHO Expert Committee on Food Additives. *Technical Report Series 815*. Geneva: WHO

Anon. (1991b) The animals, meat and meat products (examination for residues and maximum residue limits) regulations 1991. *Statutory Instruments 1991 No 2843*. London: Her Majesty's Stationery Office

Anon. (1992) Commission Directive 92/18.EEC of 20 March 1992 modifying the Annex to Council Directive 81/852/EEC on the approximation of the laws of Member States relating to analytical, pharmacotoxicological and clinical standards in respect of the testing of veterinary medicinal products. *Official Journal, No L97*, 1–23

Bigwood, E.J. (1973) The acceptable daily intake of food additives. *Critical Reviews in Toxicology*, **2**, 41–93

Crosby, N.T. (1991) *Determination of Veterinary Drug Residues in Food*, pp. 37–65. London: Ellis Horwood.

D'Arcy, P.F. and Harron, D.W.G. (1992) *Proceedings of the First International Conference on Harmonisation, Brussels 1991*. Belfast, Queens University of Belfast

Perez, M.K. (1977) Human safety data collection and evaluation for the approval of new animal drugs. *Journal of Toxicology and Environmental Health*, **3**, 837–857

Renwick, A.G. (1991) Safety factors and establishment of acceptable daily intakes. *Food Additives and Contamination*, **8**, 135–150

Teske, R.H. (1992) Chemical residues in food. *Journal of the American Veterinary Medicine Association*, **201**, 253–256

Vettorazi, G. and Radaelli-Benvenuti, B. (1982) *International Regulatory Aspects for Pesticide Chemicals, Volume II, Tables and Bibliography*, pp. 1–9. Boca Raton, Florida: CRC Press

Woodward, K.N.(1991a) The licensing of veterinary medicines in the United Kingdom — the work of the Veterinary Medicines Directorate. *Biologist*, **38**, 105–108

Woodward, K.N. (1991b) Acceptable daily intakes for veterinary drugs. *Regulatory Affairs Journal*, **2**, 787–790

Woodward, K.N. (1991c) Use and regulatory control of veterinary drugs in food production. In *Food Contaminants: Sources and Surveillance*, pp. 99–108. Edited by C.S. Creaser and R. Purchase. London: Royal Society of Chemistry

IV

Pig Nutrition

12

OUTDOOR PIGS — THEIR NUTRIENT REQUIREMENTS, APPETITE AND ENVIRONMENTAL RESPONSES

W.H. CLOSE and P.K. POORNAN
Close Consultancy, Wokingham, Berks **and** *Lys Mill, Watlington, Oxon*

Introduction

The keeping of pigs out of doors is not a new phenomenon and there are many reports in the literature of keeping 'open-air pigs'. In 1923, Bonnett wrote 'many people are under the impression that this method of pig keeping is something entirely new. As a matter of fact it is nothing of the sort, for the open-air pig of Great Britain is one of the oldest national institutions'.

During the past 5 to 10 years there has been a considerable revival in outdoor pig keeping and it is currently estimated that some 15 to 20% of the national sow herd are kept in such systems. It is believed that the number of outdoor sows will be further expanded, at the expense of the more intensive indoor systems of production. This can be attributed to a number of factors:

- the lower capital development and reduced running costs compared with the more intensive, indoor systems of production;
- the significant technological developments which have improved the husbandry, management and productivity of outdoor sows;
- the demand for wholesome meat; and
- animal welfare, statutory regulations and environmental concerns, including the disposal of slurry.

There are considerable constraints to the further development and expansion of outdoor pig production since it demands the right type of soil, a mild climate, good stockmanship and innovative management. However, providing all the limitations to the system are known and overcome, then the productivity of the outdoor sow is comparable with that of the indoor animal (Table 12.1). A major component of the system will be the provision of adequate nutrition to ensure optimal productivity.

Nutrition influences the productivity of the sow at all stages of reproduction (Aherne and Kirkwood, 1985; Cole, 1990). Therefore, the establishment of a successful feeding strategy must be based on sound knowledge of the requirements and responses of the animal at all stages of its reproductive development. Failure to provide this will prevent the sow from achieving its true reproductive potential. The purpose of this paper is therefore to review the nutritional requirements and responses of the outdoor sow, especially the consequences of the environment, and to make recommendations on appropriate feeding strategies to ensure optimal reproductive performance. It is known that the feed requirement of the outdoor

Table 12.1 COMPARISON OF RESULTS FOR OUTDOOR AND INDOOR BREEDING HERDS (Year ended September 1992)

	Outdoor breeding herds	*Indoor breeding herds*
Average number of sows and gilts	425	217
Litters per sow per year *	2.19	2.26
Pigs born alive per litter	10.82	10.73
Mortality of pigs born alive (%)	10.6	12.1
Pigs reared per sow per year *	21.2	21.3
Average weaning age (days)	23	25
Sow feed per sow per year (t)	1.395	1.230
Sow feed cost per tonne (£)	143.71	141.18
Sow feed cost per piglet reared (£)	9.46	8.15

* Per sow figures exclude unserved gilts (MLC, 1992)

sow is higher than of sows kept indoors (Meat and Livestock Commission, 1992), and since the cost of feed represents some 75% of the total cost of production, it is necessary to ensure optimum use of feed.

The establishment of requirements

There are a number of features of the outdoor sow which differ from its contemporary kept indoors and which influence its nutritional requirements. The gilt is normally older, larger and is recommended to have a greater depth of backfat at first mating. Its mature body size is greater and this influences its maintenance requirements. Similarly, it is exposed to a wider range of environmental conditions, which on the one hand, may significantly increase requirements during winter, but on the other, may dramatically reduce intake in summer. The group size is usually larger and this can lead to considerable competition between sows, especially when restrictedly fed during pregnancy. In this respect, it is interesting to note that the average herd size outdoors is double that indoors (Table 12.1). Exercise is increased and there are likely to be significantly higher levels of feed wastage. The establishment of the correct feeding regime must therefore take each of these into account.

Requirements during pregnancy

ENERGY

The requirement for energy is most conveniently calculated according to the factorial procedure which traditionally partitions total needs into that for maintenance, for maternal body gain and for conceptus development. For the outdoor animal, consideration must also be given to variations in the environment and to exercise.

Under thermoneutral conditions, the maintenance energy need has been estimated by the Agricultural and Food Research Council (1990) as 440 kJ ME/kg body weight$^{0.75}$/day. Thus, for animals between 120 and 360 kg body weight, the maintenance energy requirement will increase from 16.0 to 36.4 MJ ME/day (16.8 to 38.2 MJ DE/day), respectively.

The rate of net maternal gain allowable during pregnancy will depend upon the body weight of the sow, its condition and parity. In the current exercise a weight gain of 50 kg has been allowed during its first parity, when mated at 120 kg body weight, decreasing to zero when it has attained its mature body weight, which has been assumed to be 360 kg. The rate of maternal gain therefore, decreases from 0.44 kg/day in parity 1 to 0 kg/day at maturity. National Research Council (1988) have estimated the requirement for maternal gain as 19.8 MJ ME/kg (20.8 MJ DE/kg) and this is based on the contribution of protein and fat in the gain of sows. From the information presented by the Agricultural and Food Research Council (1990), a value of 21.0 MJ ME/kg (22.1 MJ DE/kg) may be calculated between 120 and 200 kg body weight. The mean of the two values is 20.4 MJ ME/kg (21.4 MJ DE/kg) and this value has therefore been taken as the energy requirement for maternal tissue deposition during pregnancy.

For the products of conception, the weight gain during pregnancy has been taken as 20 kg, and since the energy requirement is 4.2 MJ ME/kg (4.4 MJ DE/kg) (Verstegen, Verhagen and den Hartog, 1987), the daily requirement is calculated as 0.74 MJ ME/day (0.78 MJ DE/day).

From this information it may be calculated that the energy requirements of indoor sows under thermoneutral conditions or outdoor sows during summer, will increase from 28.9 to 37.9 MJ ME/day (30.3 to 39.8 MJ DE/day) as body weight increases from 120 to 360 kg, respectively (Figure 12.1). If a wastage factor of 5% is allowed then the energy allowances will increase from 30.3 to 39.8 MJ ME/day (31.8 to 41.8 MJ DE/day), respectively. From the beginning to the end of pregnancy, the requirements will increase by about 3.5 MJ ME/day and most of this increase is associated with the increased maintenance energy requirements as the body weight of the sow increases during pregnancy.

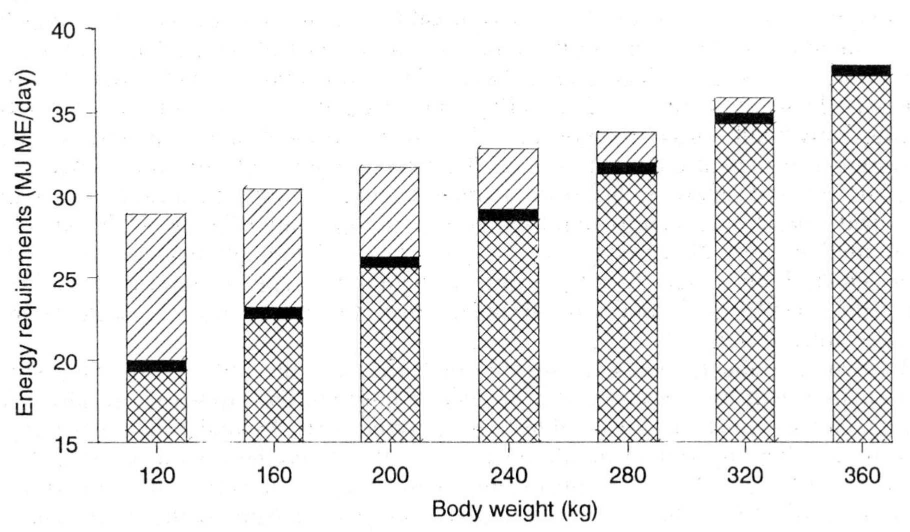

Figure 12.1 Energy requirements of the pregnant sow in thermoneutral conditions and partitioned into components of maintenance (▩), conceptus tissue (■) and maternal gain (▨)

EFFECTS OF THE ENVIRONMENT

In the outdoor situation, consideration must be given to the variable environmental conditions to which the animals may be exposed. This is especially important during pregnancy since the sow is generally fed restrictedly and has little opportunity to modify its intake in response to a changing environment. The sow's environment comprises not just temperature, but other climatic variables such as wind speed, radiation, humidity, rainfall, snow, the absence or presence of bedding and litter mates, shelter, shade and the opportunity to wallow. It is obviously difficult to directly compare the relative thermal effects of each of these predisposing factors upon the animal. The most convenient way to assess their respective responses is to calculate the equivalent environmental demands and hence the effective critical temperature of the animal.

Compared with the growing pig, there are few directly determined estimates of the lower critical temperature (LCT) of the sow under variable environmental conditions (Close, 1987). LCT represents the least temperature at which heat loss is minimal, under any given set of environmental circumstances. Estimates of LCT and of the increase in heat production at temperatures below LCT have been collated from a number of experiments and are presented in Table 12.2. This shows that for individual animals weighing between 140 and 225 kg, and depending upon feeding level, LCT varies between 12 and 23°C. For the range of intakes likely to be provided in practice, the preferred value for an individual animal living indoors in an optimum environment, but without bedding, is about 20°C. The lower critical temperature will be decreased by 5°C if the animals are group-housed (Holmes and Close, 1977; Geuyen, Verhagen and Verstegen, 1984) and may increase by 4°C if the animals are in poor body condition (Holmes and McLean, 1974; Hovell, Gordon and MacPherson, 1977). Increasing the feeding level reduces LCT and the results in Table 12.2 suggest that each 1°C decrease in LCT is associated with an increase in ME of about 40 kJ/kg body weight$^{0.75}$/day. For a 200 kg animal this represents an increased energy intake of 2.1 MJ ME/day.

Since the maintenance energy requirements of the animals have been calculated at temperatures at or below LCT, it is possible to determine the extent to which they increase as the temperature falls below the critical level. The estimates in Table 12.2 suggest increases between 7.5 kJ/kg body weight$^{0.75}$/day/°C for animals living in groups to 23.3 kJ/kg body weight$^{0.75}$/day/°C for thin animals living individually in cold conditions. The mean estimate of the increase in the maintenance heat production is 15 kJ/kg body weight$^{0.75}$/day/°C, and since the net efficiency of energy utilisation for maintenance is 0.8 (Agricultural Research Council, 1981), the energy requirements at temperatures below LCT will be increased by 18.8 kJ ME/kg body weight$^{0.75}$/day/°C. For animals between 120 and 360 kg body weight this represents an increased energy requirement of 0.68 and 1.55 MJ ME (0.71 and 1.63 MJ DE), respectively, for each 1°C decrease in temperature below LCT.

It is more difficult to assess the thermal effects of the other components of the environment on sows, because of lack of information. However, an idea of the extent to which the LCT values vary may be extrapolated from knowledge on the growing pig. For example, each 0.1 to 0.2 m/sec increase in wind speed increases LCT by about 1°C and is equivalent to a 1°C decrease in temperature (Close, 1987; 1989). Similarly, a 1 to 2°C change in radiant temperature has a similar effect to a 1°C change in air temperature (Mount, 1968; Holmes and

Table 12.2 ESTIMATES OF THE MAINTENANCE ENERGY REQUIREMENT (M: kJ/kg$^{0.75}$/DAY), LOWER CRITICAL TEMPERATURE (LCT:°C) AND INCREASE IN HEAT OUTPUT (H: kJ/kg$^{0.75}$/DAY) AT TEMPERATURES BELOW LCT IN SOWS FROM VARIOUS SOURCES

Bodyweight (kg)	*Stage of pregnancy* (days)	*ME intake*	*Condition*	*LCT* (°C)	*M*	*H*	*Source*
118–230	20–100	532–682	Individual	17 (thin) 12 (fat)	385 (23°C) 444 (18°C)	11.8	(1)
140	0–112	420	Individual	19–23	420	13.7	(2)
88–102	0	422–824	Thin	21	476 (20°C)	23.3	(3)
158–176	0	396–768	Standard	17	753 (5°C)	19.8	
164–192	46–97	535	Individual	20	—	13.7	(4)
168–186	46–97	535	Group	14	431	7.5	
162–185	43–79	487	Individual	21	420	17.5	(5)
162–185	44–80	594	Individual	18	420	13.5	
161–189	37–107	488	Individual	21	—	17.1	(6)
160–197	38–112	588	Individual	18	—	13.4	
210	39–74	487	Individual	23	430	14.2	(7)
210	39–74	626	Individual	20	430	10.9	

(1) Holmes and McLean (1974); (2) Holmes and Close (1977); (3) Hovell *et al.* (1977); (4) Geuyen *et al.* (1984); (5) Verhagen *et al.* (1986); (6) Kemp *et al.* (1987); (7) Noblet *et al.* (1989)

McLean, 1977). Holmes and Close (1977) further calculated that an 18% increase in relative humidity was equivalent to a 1°C increase in air temperature at 30°C. The presence or absence of bedding can markedly alter the animal's microclimate and a good deep bed of straw has been shown to reduce the effective critical temperature of groups of 40 kg pigs by about 4°C, compared with an insulated wooden floor (Verstegen and van der Hel, 1974). Some indication of the extent to which different environmental conditions influence the effective environmental temperature of sows in an outdoor situation is illustrated in Figure 12.2.

In terms of assessing the effects of the environment on the outdoor sow during pregnancy, it is necessary to have some idea of the meteorological conditions to which the animals are exposed. This allows the calculation of the effective environmental temperature and hence estimation of the number of days the sow is below its LCT value. If the effective temperature falls below this critical value, then additional feed must be provided to compensate for the extra thermoregulatory demand. In the present exercise it has been assumed that the lower critical temperature of a pregnant outdoor sow is 15°C since they are generally fatter than contemporary indoor animals, have access to shelter and straw bedding and can huddle in groups, if necessary. Meteorological records indicate that there are about 1600 day equivalents when the temperature will fall below 15°C. This represents the product of the number of days and the difference in temperature between the actual and the desired temperature of 15°C. If the LCT value was only 10°C, then the number of degree day equivalents would fall to about 800 days. The relationship between the effective environmental temperature and the number of degree days below LCT for a 200 kg sow is presented in Figure 12.3.

In general there will be about 300 days during the year when the animal is pregnant. Thus, the 1600 degree days below an LCT of 15°C suggests a daily

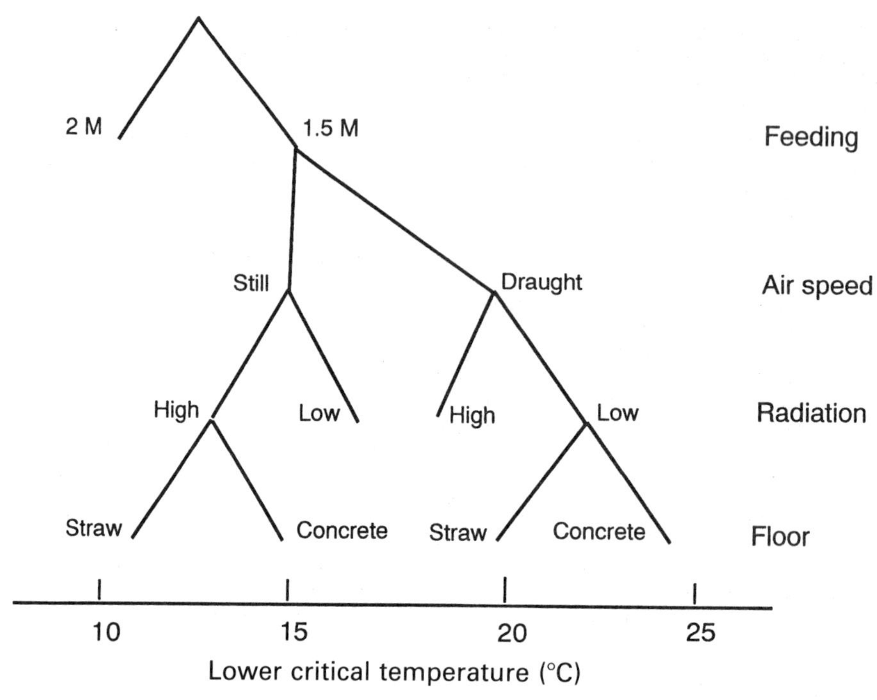

Figure 12.2 A diagrammatic representation of environmental factors influencing the lower critical temperature of the sow (M = maintenance feed intake)

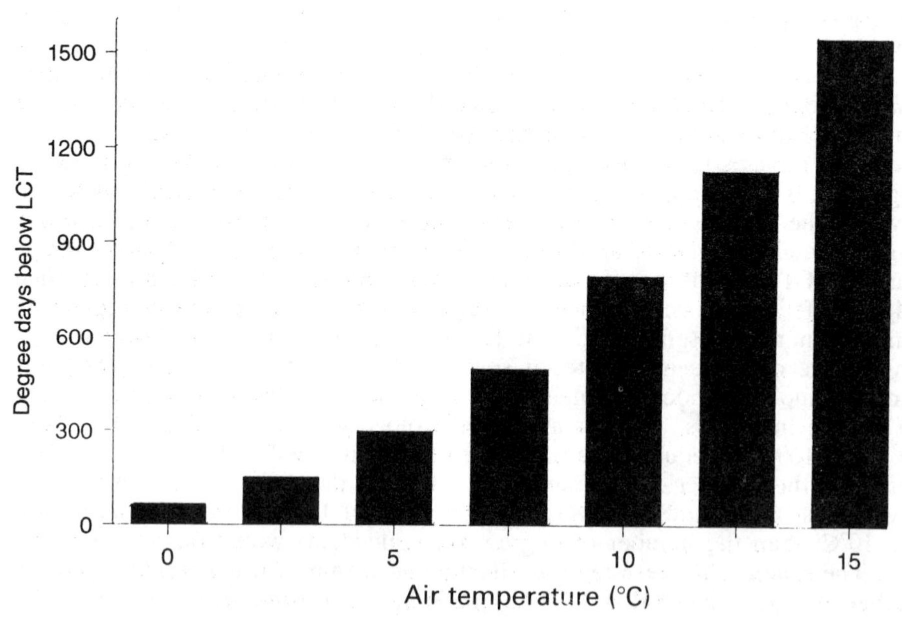

Figure 12.3 Relationship between the air temperature and the number of days below the lower critical temperature of sows (after Smith, 1974)

annual temperature deficit of 5.3°C (1600/300). Since there are only 800 degree days below an LCT of 10°C, then the temperature deficit is only 2.7°C (800/300). Since each 1°C below LCT necessitates an additional energy requirement of 18.8 kJ ME/kg body weight$^{0.75}$/day, the above values represent increased intakes of 100 and 50 kJ ME/kg body weight$^{0.75}$/day, when the LCT of the animals are 15 and 10°C, respectively.

The above values are means throughout the year, whereas most of the days when the environmental temperature is below LCT will occur within the 6-month winter period. Thus, during winter the mean temperature will be 10.7 (1600/150) and 5.3 (800/150) °C below the desired levels of 15 and 10°C, respectively, and the energy requirements correspondingly increased by 200 and 100 kJ ME/kg body weight$^{0.75}$/day. For a 200 kg sow these represent increased intakes of 10.6 and 5.3 MJ ME/day (11.1 and 5.6 MJ DE/day), respectively.

EFFECTS OF EXERCISE

Compared with indoor animals, outdoor sows have considerable opportunity for exercise. This must be considered in the calculation of requirements, although the heat generated during exercise may be considered to meet some of the extra thermoregulatory heat needed to compensate for the cold environment. Petley and Bayley (1988) measured the energy expenditure of pigs running on a treadmill at up to 6 km/hour. The energy expenditure of the exercised pigs was 20% higher than the control animals. This compares with an increased requirement of 16% determined by Jakobsen, Thorbek and Henckel, 1992 (personal communication). When calculated per km, these values represent additional expenditures of 7 kJ/kg body weight/km. Thus, a 200 kg sow walking 1 km/day would dissipate an additional 1.4 MJ heat/day, and assuming a net efficiency of energy utilisation of 0.8, would require an additional 1.7 MJ ME/day (1.8 MJ DE/day).

Table 12.3 CALCULATED ENERGY REQUIREMENTS (MJ ME/DAY) OF OUTDOOR SOWS DURING PREGNANCY

Body weight (kg)	*Net gain (kg)*	*Indoor sows*[a]	*Exercise*[b]	*Requirements for Environment*[c]	*Total*[d]
120	50	28.9	0.9	3.6	33.4
160	40	30.4	1.4	4.3	36.1
200	30	31.7	1.7	4.8	38.2
240	20	32.8	2.1	5.2	40.1
280	10	33.8	2.4	5.5	41.7
320	5	35.9	2.8	5.7	44.4
360	0	37.9	3.2	5.8	46.9

[a] Calculated according to information presented in Figure 12.1 for sows in optimal environment
[b] Based on sow walking 1 km/day
[c] A LCT value of 15°C has been taken at 120 kg body weight, decreasing by 1°C for each 80 kg increase in body weight
[d] Includes wastage

TOTAL ENERGY REQUIREMENTS

On the basis of the above calculations, energy requirements can be determined for the outdoor animal living under a range of environmental conditions (Table 12.3). For animals between 120 and 360 kg body weight, the requirements increase from 33.4 to 46.9 MJ ME/day (35.1 to 49.2 MJ DE/day), respectively, and are some 20% higher than the corresponding animals kept indoors. This is in reasonable agreement with the 13% estimated from feed usage by Meat and Livestock Commission (1992). These values represent the mean throughout the year when the mean environmental temperature of 10°C is approximately 5°C below the animals lower critical temperature. In winter the temperatures are likely to be considerably below this and if a mean temperature of 5°C is taken, then the requirements will be approximately 12% higher, as illustrated in Figure 12.4.

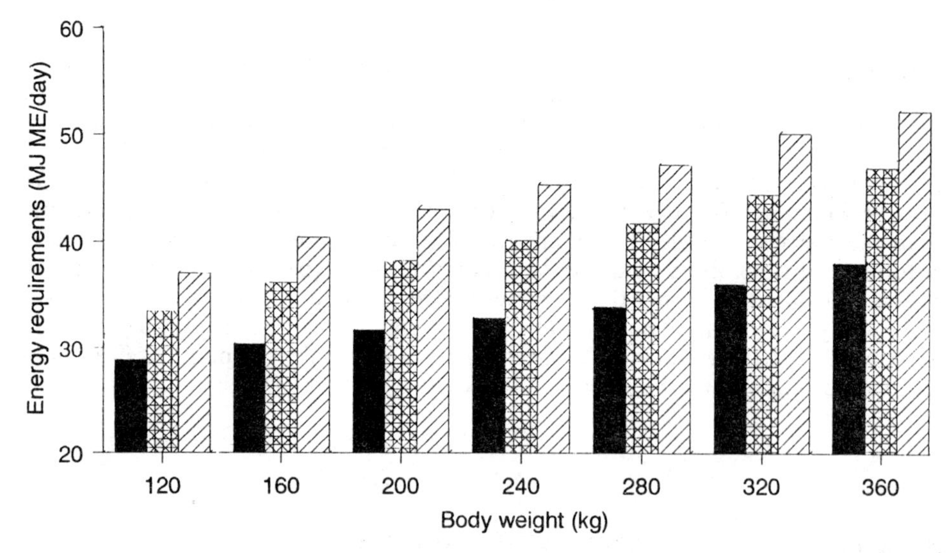

Figure 12.4 Energy requirements of the pregnant sow under different conditions: indoors/summer ■, mean outdoors ▩, outdoors/winter ▨

The above estimates of energy requirements assume that the animal is exposed to the cold conditions throughout the 24-hour period. This may not necessarily be the case, since the animal spends considerable time inside the insulated hut normally provided, where the effective environmental temperature is close to the animal's LCT. On the other hand, when the sow is outside in the open, the temperature may be considerably lower than that estimated. Unfortunately, there is insufficient information to allow for such variables to be considered. Similarly, it has not been possible to consider the effects of rainfall, snow and other environmental circumstances, but the values presented in Table 12.3 and Figure 12.4 represent the best current estimates of energy requirements of sows living out of doors.

The above energy values relate predominantly to sows living outdoors in winter. In summer, the requirements will be similar to those of indoor animals kept within thermal neutrality. These levels of energy should be well tolerated since they are low relative to the voluntary intake of the animal and should not cause problems of heat stress. However, the provision of shade, shelter and wallows will be most useful in alleviating any problems of heat stress during pregnancy.

PROTEIN AND AMINO ACID REQUIREMENTS

Compared with energy, there is less information on the requirements of the pregnant sow for protein and amino acids. Agricultural Research Council (1981) commented on the lack of good information for protein, and especially for amino acid requirements, and little new information has been added since then. Although the factorial approach may be used, major difficulties exist in applying appropriate values to the rate and efficiency of protein deposition in both the conceptus tissue and maternal body.

In this respect, it has been assumed there is little difference in the requirements between indoor and outdoor animals, other than that associated with variations in both body weight and net maternal gain. Using lysine as the reference amino acid in the ideal protein concept (Cole, 1978; Agricultural Research Council, 1981, National Research Council, 1988), the evidence suggests that the requirement for lysine is between 8 and 11 g/day. It is likely to be higher for the young gilt than the mature sow since the rate of maternal gain allowable will be higher. The requirement for lysine should therefore be set at 11 g/day.

Since the requirement for lysine is known, it is possible to determine the requirements for the other essential amino acids (Table 12.4). These are based on Agricultural Research Council (1981), National Research Council (1988) and Agricultural and Food Research Council (1990). There are some small discrepancies between Agricultural Research Council (1981) and National Research Council (1988), which take account of differences in feed ingredients and characteristics.

Table 12.4 ESTIMATES OF THE ESSENTIAL AMINO ACID REQUIREMENTS DURING PREGNANCY AND LACTATION

	Pregnancy	*Lactation*
Lysine (g/day)	11	36–50
Other amino acids (% of lysine):		
Methionine + Cystine	52	53
Threonine	60	65
Tryptophan	17	19
Leucine	100	100
Valine	70	70
Isoleucine	55	55
Phenylalanine + Tyrosine	100	112
Histidine	33	35
Arginine	—	67

Based on Agricultural Research Council (1981) and Agricultural and Food Research Council (1990)

Requirements during lactation

ENERGY

The estimates of the nutritional requirements of the sow during lactation must take account of the varying body weight of the animals (the maintenance component), the yield and composition of milk produced during lactation and must make allowances for any mobilisation of body tissue that occurs. The latter will take place under conditions where the nutrient intake of the animal, even when fed to appetite, does not meet metabolic or nutritional needs.

The factorial procedures to calculate the energy requirements during lactation have been outlined (Mullan, Close and Cole, 1989). These authors suggested that the maintenance energy requirement of the sow during lactation was 471 kJ ME/kg body weight$^{0.75}$/day. Thus, for animals between 160 and 360 kg body weight, the requirement will increase from 21.2 to 38.9 MJ ME/day (22.3 to 40.8 MJ DE/day), respectively. Similarly, they calculated that each 1 kg of piglet gain required some 21.0 MJ ME derived from milk, and since the net efficiency with which dietary energy is utilised by the sow for milk production is 0.72, each 1 kg of piglet gain during lactation requires a maternal intake of 29 MJ ME (30.5 MJ DE). Details of the calculations are presented in Table 12.5 and Figure 12.5 and show the changes in the requirements of the sow during lactation at litter sizes and piglet gains likely to be achieved in practice. These values represent the mean requirements during lactation; they are significantly less in early lactation, and about 20–25% higher in late lactation.

Table 12.5 CALCULATED ENERGY REQUIREMENTS OF OUTDOOR SOWS DURING LACTATION

Body weight (kg)	*Maintenance* (MJ ME/d)	*Milk production*[a] (MJ ME/day)		*Total* (MJ ME/day)	
		10 piglets	*12 piglets*	*10 piglets*	*12 piglets*
160	21.2	58.0	69.6	79.2	90.8
200	25.0	58.0	69.6	83.0	94.6
240	28.7	58.0	69.6	86.7	98.3
280	32.2	58.0	69.6	90.2	101.8
320	35.7	58.0	69.6	93.7	105.3
360	38.9	58.0	69.6	96.9	108.5

a 10 or 12 piglets each gaining 200 g/day during a 28 day lactation period.

During lactation the outdoor sow should consume between 75 MJ ME/day (79 MJ DE/day) for a 160 kg animal suckling 8 piglets to 108 MJ ME/day (113 MJ DE/day) for a 360 kg animal suckling 14 piglets in order to meet her own maintenance needs and those of her suckling piglets. However, in commercial pig production the sow is often unable to consume sufficient feed to meet this need and mobilisation of body reserves occurs. If these losses of body reserves are excessive, and under situations where the sow has limited body stores, milk production is diminished, piglet growth is reduced and reproductive performance is impaired (King and Williams, 1984a, b; Mullan and Williams, 1989).

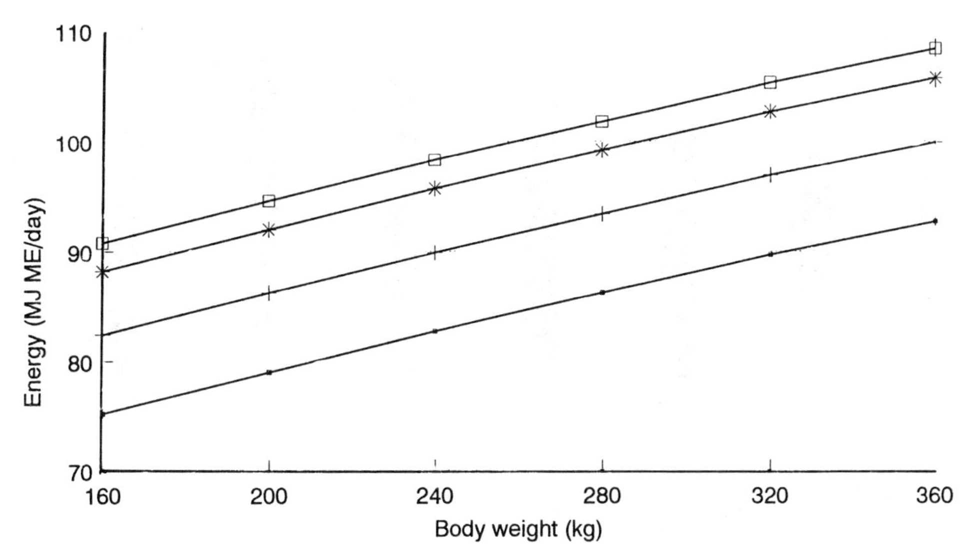

Figure 12.5 Energy requirements of the lactating sow suckling 8 (●), 10 (+), 12 (*) and 14 (□) piglets

EFFECTS OF THE ENVIRONMENT

The environment has a major influence on the voluntary feed intake of the sow (National Research Council, 1987). Under cold conditions (below the lower critical temperature) the animal responds by consuming more feed and thus is capable of meeting her energy requirements. Under hot conditions, however, feed intake is reduced and is often insufficient to meet metabolic demands. Whereas the temperature needs of the piglets may be as high as 30–35°C, the LCT value of the *ad libitum* fed sow may be no higher than 12–15°C.

From a review of the literature, Black, Mullan, Lorschy and Giles (1993), calculated that for each 1°C increase in ambient temperature above 20°C, the daily voluntary energy and feed intake decreased by 2.4 MJ DE and 0.17 kg, respectively. The body weight of the animals was close to 200 kg, and although the relationship between temperature and voluntary feed intake may not be linear, it may be calculated that each 1°C increase in temperature reduces energy intake by 45 kJ DE (43 kJ ME)/kg body weight$^{0.75}$/day. Thus, for sows between 160 and 360 kg body weight, this represents a decrease in energy intake between 1.9 and 3.5 MJ ME/°C (2.0 and 3.7 MJ DE/°C), respectively (Figure 12.6).

From the above information, it may be calculated that between 20 and 30°C, voluntary energy intake will decrease by 19 and 35 MJ ME/day, for sows between 160 and 360 kg body weight, respectively. This assumes that 20°C represents the upper limit of the zone of thermal neutrality of the lactating sow when feed intake will become significantly reduced. However, if it is assumed that only fat tissue is being mobilised, and on the basis that each 1 kg of fat mobilised from body stores supplies 31.8 MJ (39.7 × 0.8) available energy, then over a 28 day lactation, body fat reserves would be reduced by 17 and 31 kg, respectively. Recent experimentation by Schoenherr, Stahly and Cromwell (1989) and Vidal,

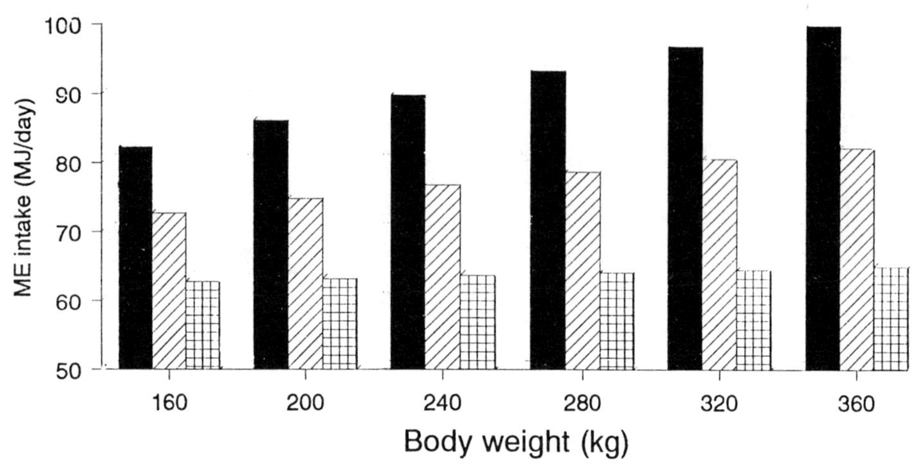

Figure 12.6 The relationship between predicted energy intake and temperature in lactating sows, suckling 10 piglets at ■ 20°C, ▨ 25°C and ▦ 30°C

Edwards, MacPherson, English and Taylor (1991) suggests that each 1 kg loss of body weight during lactation, as a result of an increase in temperature, resulted from a reduction in energy intake of 60 MJ ME. This suggests body weight losses of 9 and 16 kg when applied to the above values and indicates that the loss of body tissue, and especially fat, may be higher than is estimated solely from loss of body weight. This highlights the importance of ensuring adequate body reserves in the outdoor gilt at first mating and during its reproductive life. However, it is now known that both lean and fat reserves may be mobilised during lactation and it is likely that loss of both body substances influences subsequent reproductive performance (Hughes and Pearce, 1989; King, 1987; Mullan, Close and Foxcroft, 1991).

PROTEIN AND AMINO ACID REQUIREMENTS

In determining the animal's requirement for protein, the quality of the protein is of critical importance, and hence its amino acid content. Since lysine is the most limiting amino acid, it is therefore more usual to express requirements in terms of lysine, according to the ideal protein concept (Cole, 1978: Agricultural Research Council, 1981). The requirements of the lactating sow are closely correlated with the quantity of milk produced and upon its amino acid composition. Estimates suggest that the requirement for lysine for milk production may be as high as 90% of the total requirement, with the remainder being associated with the maintenance needs of the animal. This suggests that, similar to energy, the factorial approach may be used to establish requirements.

The procedures developed by Mullan *et al.* (1989) have been used to calculate requirements. These suggest that the nitrogen (N) requirement for maintenance is 0.45 g N/kg body weight$^{0.75}$/day and that each 1 kg of piglet gain requires 57.4 g nitrogen (359 g protein). It has been further assumed that each 1 g of ideal protein is equivalent to 0.07 g lysine and that each 1 g of ideal protein is about 65% of crude protein. The derivation of the requirements is presented in

Table 12.6 CALCULATED LYSINE REQUIREMENTS OF OUTDOOR SOWS DURING LACTATION

Body weight (kg)	*Maintenance* (g/day)	*Milk production*[a] (g/day)		*Total* (g/day)	
		10 piglets	*12 piglets*	*10 piglets*	*12 piglets*
160	5.7	32.6	39.2	38.3	44.9
200	6.8	32.6	39.2	39.4	46.0
240	7.8	32.6	39.2	40.4	47.0
280	8.8	32.6	39.2	41.4	48.0
320	9.7	32.6	39.2	42.3	48.9
360	10.6	32.6	39.2	43.2	49.8

a 10 or 12 piglets each gaining 200 g/day during a 28 day lactation period.

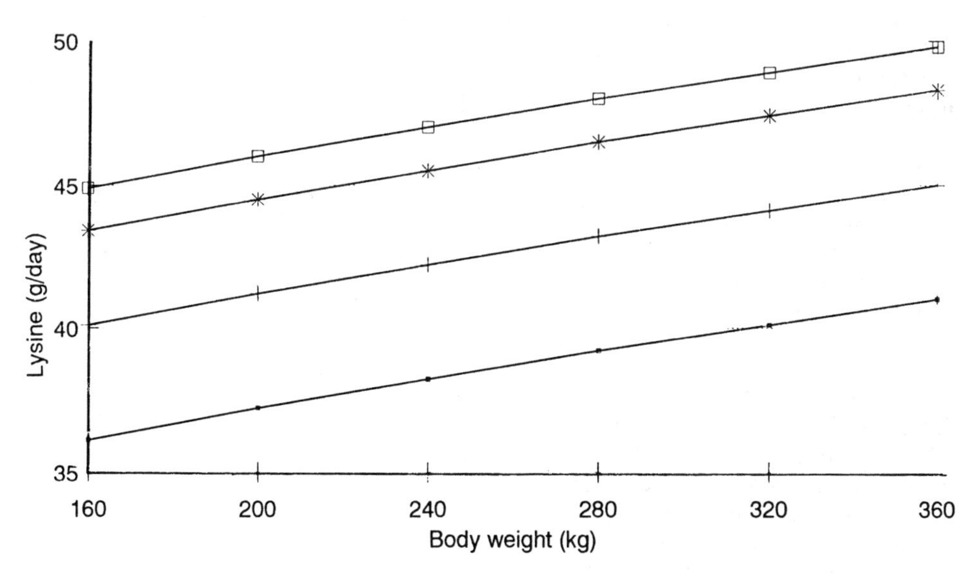

Figure 12.7 Lysine requirements of the lactating sow suckling 8 (●), 10 (+), 12 (*) and 14 (□) piglets

Table 12.6. Figure 12.7 shows that, depending upon the level of milk production, that is litter size and piglet growth, the requirements for lysine range between 36 and 50 g/day for animals between 160 and 360 kg body weight, respectively. When compared with the energy requirements presented in Figure 12.5, these represent lysine : energy values close to 0.5 g/MJ ME.

Knowing the lysine requirements, and using the 'ideal protein' concept, it is possible to determine the requirements of the other essential amino acids, as presented in Table 12.4.

MINERAL AND VITAMIN REQUIREMENTS

Research on the specific requirements of outdoor pigs for minerals and vitamins is lacking. In general, there is little difference in the requirements between indoor and outdoor animals, with some exceptions.

In some ways, outdoor sows have an advantage over indoor animals since they have access to minerals in the soil, as well as vitamins in grass and other vegetable matter. Similarly, they have access to faecal material which can provide minerals and vitamins derived from undigested feed, exogenous sources and as a by-product from bacterial fermentation in the hind gut. Indoor sows may require more vitamin D due to reduced access to sunlight, but outdoor sows may have a higher requirement for zinc and vitamin A in order to overcome sunburn and to facilitate the repair of damaged tissue.

Most outdoor systems of production are situated on free-draining soils, such as chalk, sand and gravel, and this has special significance for meeting the mineral needs of the animal. Outdoor pigs consume significant amounts of soil, and since chalk contains an overabundance of calcium, can lead to an increased requirement of phosphorus if the recommended Ca:P ratio of 1.3:1–1.5:1 is to be maintained. It may also affect the acid/base status of the gut, which influences digestion and nutrient absorption.

An adequate level of vitamin D must also be provided if the absorption and utilisation of both calcium and phosphorus is to be optimised. Similarly, high levels of calcium depress the utilisation of iron and dietary levels greater than 100 mg Fe/kg are recommended if the intake of calcium is greater than the dietary equivalent of 12 g/kg. The utilisation of zinc is also impaired at high levels of calcium intake and dietary zinc levels greater than 100 mg/kg are recommended under such conditions. Habitual ingestion of soil is common and may limit feed intake, with an associated reduction in nutrient intake.

Sows often have to walk on hard, stony ground and this may precipitate lameness and foot lesions. Although Ca and P have been shown to be extremely important in the general development of limb soundness, supplementation of the diet with biotin has been shown to significantly strengthen the hoof (Brooks, Smith and Irwin, 1977; Penny, Cameron, Johnson, Keynon, Smith, Bell, Cole and Taylor, 1980; Webb, Penny and Johnston, 1984). The normal recommendation for biotin is 300 μg/kg, but in an outdoor situation this could be increased to 1 mg/kg in an attempt to reduce the incidence of leg weakness.

During periods of high temperature and in conditions where water intake is limited, feed intake is reduced and this leads to reduced mineral and vitamin intake. Similarly, the physiological adaptations needed to maintain body temperature during wide fluctuations of climatic temperature and in cold, wet and windy conditions, mean that provision should be made to ensure adequate mineral and vitamin supplementation. Because of the difficulty in determining specific requirements of the outdoor sow in the wide range of prevailing environmental circumstances, it is suggested that in certain circumstances the inclusion rate of minerals and vitamins in the diet is greater than for indoor sows, as recommended by Agricultural Research Council (1981) and Agricultural and Food Research Council (1990).

Development of feeding strategies

Although it is possible to determine the nutritional requirements of the outdoor sow during both pregnancy and lactation, appropriate feeding strategies must be developed which meet as exactly as possible the metabolic needs of the animal at all stages of reproduction. Indeed, an important factor is whether the sow will eat the feed provided, or whether there are factors that limit appetite.

VOLUNTARY FEED INTAKE

The establishment of any feeding strategy must ensure that the feed allowances are within the appetite capacity of the animal. The appetite of the sow is influenced by a number of animal, dietary, environmental and husbandry factors and these influence the strategies that are available for the nutrition of the sow. This is especially important during lactation, since the objective is to ensure as high an intake as possible during this period of high metabolic requirements. It is obviously outside the scope of this paper to consider all those factors which influence nutrient intake in the sow, and these have recently been reviewed by Lynch (1989). However, for the outdoor sow, it is important to consider those dietary and environmental factors which can influence appetite.

The diet fed can markedly influence voluntary intake and it is well known that animals eat less of a high energy than a low energy ration (Cole and Chadd, 1989; Lynch, 1989). Zoiopoulos, English and Topps (1982) showed that when diets were diluted with 30% barley straw, daily feed intake was increased by 19%, but DE intake actually decreased by 14%. On the other hand, O'Grady and Lynch (1978) fed diets containing 12.5 or 13.8 MJ DE/kg and showed that sows consumed more of the higher density ration, resulting in a 14% increase in energy intake. The response to low or high energy diets may be temperature dependent, since fat has a low heat increment of feeding and is more suitable for hot environments, whereas fibre has a high heat increment of feeding and part of this heat may be used to compensate for some of the extra thermoregulatory heat demanded in cold conditions (Noblet, Dourmad, Lc Dividich and Dubois, 1989). Similarly, protein also has a high heat increment of feeding and sows consumed less of a high protein diet at a temperature of 28°C than a low protein diet at the colder temperature of 16°C (Lynch, 1989). More recently, Close, Pettigrew, Sharpe, Keal and Harland (1990) demonstrated that higher fibrous diets, containing sugar beet pulp, fed during pregnancy improved the appetite of sows fed a conventional cereal-based ration during lactation. The form of the diet also affects intake, with wet feed or a water/meal mixture encouraging more intake than dry feed (O'Grady and Lynch, 1978). In this respect, the importance of water in influencing voluntary feed intake must be emphasised (Brooks and Carpenter, 1990).

There are important associations both within and between parities which can influence voluntary feed intake. It is well-established that the more feed an animal consumes during pregnancy, the lower its feed intake in lactation (Salmon-Legagneur and Rerat, 1962: Harker and Cole, 1985). It also increases with litter size (Mullan *et al.*, 1989). Similarly, Mahan and Mangan (1975) demonstrated a close relationship between dietary protein content during both pregnancy and lactation; the lower the dietary protein content during both pregnancy and lactation, the lower the feed intake of the sow during lactation. In addition to feed supply, intake is also influenced by body weight, level of performance and fatness of the sow at farrowing (O'Grady, Lynch and Kearney, 1985; Mullan and Williams, 1989). Feeding to enhance the deposition of fat rather than body lean during pregnancy may therefore have an adverse effect upon feed intake in lactation.If it is the level of fatness which is the contributing factor, then the question which arises is whether the modern outdoor sow has the capacity to deposit sufficient fat during pregnancy to adversely affect voluntary feed intake in lactation.

As has previously been indicated, the environment has a major influence on voluntary feed intake. Lynch (1977, 1989) and Stansbury, McGlone and Tribble (1987) demonstrated that each 1°C increase in temperature above 21 and 18°C reduced voluntary feed intake by 0.1 and 0.2 kg/day, respectively, with consequential effects upon the loss of body weight of the sow and the growth of the piglets. Black *et al.* (1993) have suggested that the reduction in milk yield of sows at high temperatures is greater than that solely associated with the reduction in maternal feed intake. Thus, management and feeding strategies must be designed to minimise these effects and to allow sows exposed to hot conditions to maintain milk yield. This involves a reduction in heat production, so that the animals do not become hyperthermic, and can be achieved by dietary manipulation, as previously discussed. On the other hand it is possible to enhance heat dissipation to the environment so that the animals feel cooler, bringing the temperature of exposure within the zone of thermal neutrality. This can be brought about by appropriate husbandry and management procedures, such as the provision of wallows and shades, increasing the rate of air movement around the animal and the use of insulated arcs and cooling surfaces (Culver, Andrews, Conrad and Noffsinger, 1960; Garrett, Bond and Kelly, 1960; Ingram, 1965; McGlone, Stansbury and Tribble, 1988).

NUTRITIONAL STRATEGIES AND DIETARY SPECIFICATIONS

There are numerous raw materials that can be used as ingredients for outdoor sow feeds. However, the secret lies in matching the ingredients to the needs of the animal at each stage of reproduction, especially in relation to the seasons of the year. For example, fibrous feeds have a high heat increment of feeding and this makes them more suitable for feeding during winter than summer. Indeed, it is now known that the mature pig has considerable potential to digest and utilise fibrous materials (Low, 1985; Close, 1993) and there are considerable nutritional, welfare and production advantages associated with the feeding of low density rations to sows (Lee and Close, 1987). These include better behaviour and welfare, increased litter size and weight, improved health, reduced water consumption, increased appetite of the sow and increased milk fat content in the subsequent lactation, together with a decreased lower critical temperature. In addition, bulky crops such as beets, silages and grasses can make a significant contribution to the nutritional needs of the outdoor sow and these have been discussed in detail by Machin (1990). The limitations to their use are: exact information on their nutritive value, matching nutrient intake to requirements and handling and processing the feeds.

Because of the considerable fermentation in the hind gut and the ensuing increase in heat production, the continuation of the feeding of high fibre diets is not recommended during summer. The sow has considerable difficulty in dissipating this additional heat and responds by reducing intake. The objective should therefore be to provide a ration which minimises heat production. The addition of fat or oils, with their low heat increment of feeding, not only increases dietary energy content but also reduces metabolic heat production, improves energy intake and increases sow comfort (Stahly, Cromwell and Aviotti, 1979; Pettigrew, 1981; Coffey, Seerley, Funderburke and McCampbell, 1982; Britt,

Armstrong and Cox, 1988). The protein content of the ration should not be excessive as animals consume less of a high protein diet during summer than winter and the use of synthetic amino acids under such circumstances may be beneficial. Feeding behaviour may also change under hot conditions and in an attempt to reduce the additional heat generated with the digestion and assimilation of nutrients, and activity, animals may eat on a 'little and often' basis rather than a few discrete meals throughout the day. In many instances, most of the feed will be consumed in the cooler periods of the early morning and at night. This suggests that *ad libitum* feeders may be the most appropriate means of supplying feed for the outdoor lactating sow. It is therefore possible to formulate nutritional packages for outdoor sows dependent upon the environmental conditions to which they are exposed.

Animals kept outdoors have the opportunity to graze, but information about the feeding value of grass is difficult to obtain. From the results of Chambers (1987), it can be calculated that sows on grass achieved an extra 4 kg body weight gain during pregnancy, compared with control animals. Sows may also consume substantial quantities of bedding, especially straw, which obviously has some feeding value, but these should be treated as a bonus rather than be considered as part of the feeding strategy.

FEEDING REQUIREMENTS

On the basis of the information provided in this paper, it is suggested that the ration fed to outdoor sows during pregnancy should contain between 12.5 and 13.0 MJ DE/kg. Thus the feed requirements will increase from 2.5–2.7 kg/day to 3.6–3.8 kg/day for sows between 120 and 320 kg body weight, respectively. These represent the mean values throughout the year, but during winter the requirements will be some 10 to 15% higher than the values cited. Although the lysine requirement was calculated to be 11 g/day, it is suggested that the dietary content should be not less than 4.5 to 5.0 g/kg. However, if 'user friendly' raw materials are used, it may be difficult to produce a cost-effective ration with so low a level of lysine.

During lactation, the diet should contain between 13.5 and 14.0 MJ DE/kg; thus the feed requirements, established on the basis of the information presented in Figure 12.5, will increase from 5.4–5.6 kg/day for a 160 kg sow suckling 8 piglets to 7.7–8.0 kg/day for a 360 kg sow suckling 14 piglets. These values represent the mean throughout lactation and will be about 30% higher at the later stages of lactation, when the suckling demand of the piglets is exceptionally high. On the basis of the above intakes, the diets should contain about 6.5 g lysine/kg in order to meet the calculated requirements established in Figure 12.7. However, there are many factors which limit the appetite of the lactating sow in the outdoor situation and it is unlikely that she is able to achieve such feed intakes. If intakes are 20% lower than those predicted, then the dietary lysine requirements should be increased to at least 8 g/kg. In summer, the value may need to be higher than this, up to 10.0 g/kg, if the requirements of the animal is to be met.

FEED PRESENTATION

Outdoor sows have traditionally been fed rolls or biscuits which presents the feed in a large bulky form and has a number of advantages. Sows find them easier to pick up, even on unsuitably wet or muddy conditions. It also helps to minimise wastage and birds find them more difficult to consume. More recently, there has been a move towards presenting feed in roll form since these are free flowing, may be stored in bulk bins and are more convenient to distribute in the field using mechanical distributors such as feed trailers.

Ideally, it would be preferable to feed a gestation ration to pregnant sows on an individual basis and a lactation feed to suckling sows from *ad libitum* hoppers. In practice, it is common for producers to use a single feed for all sows and to provide this once per day. During pregnancy pigs are fed restrictedly in groups, predominantly on the ground. This means that the quantity of feed supplied may not meet the requirements of each individual animal at each specific stage of reproduction.

It is important that the feed for outdoor sows is extremely durable, with good pellet quality and with weather resistant properties. The latter is especially important since the feed may be spread in mud. Certain raw materials contain oil and this aids water-proofing. Similarly, the inclusion of additional fat can help in this process, but if excessive, can reduce both pelletability and palatability. Coating the biscuit or roll with fat is also a possibility, but it sometimes sticks to the surface and this increases the ingestion of soil and this can adversely affect appetite (Poornan, 1990). Certain raw materials absorb water and swell faster than others, and if used incorrectly can cause the biscuit or roll to disintegrate. The choice or raw materials is therefore crucial to the success of feeding the outdoor sow.

Concluding remarks

In the assessment of nutritional requirements and the establishment of practical feeding strategies, it is necessary to consider the environmental and husbandry conditions of the animals living out of doors. Knowledge of these will ensure that the requirements of the animals are met at all stages of reproduction during each season and thus ensure that nutrition does not limit optimum sow productivity. As far as is possible, nutritional management should aim to consider individual animals rather than the whole herd and the essential role of the stockman in the provision of feed to the animals must not be ignored.

References

Agricultural and Food Research Council (1990) Technical Committee on Responses to Nutrients, Report No. 4, Nutrient Requirements of Sows and Boars. *Nutrition Abstracts and Reviews, Series B: Livestock Feeds and Feeding*, **60** 383–406

Agricultural Research Council (1981) *The Nutrient Requirements of Pigs*. Slough, Commonwealth Agricultural Bureaux, 307 pp.

Aherne, F.X. and Kirkwood, R.N. (1985) Nutrition and sow prolificacy. *Journal of Reproduction and Fertility*, Suppl. **33** 169–183

Black, J.L., Mullan, B.P., Lorschy, M.C. and Giles, L.R. (1993) Lactation in the sow during heat stress. 43rd Annual Meeting of EAAP, Madrid. September 1992. *Livestock Production Science* (in press)

Bonnett, F. (1923) The outdoor pig. In *Outdoor Pigs: How to Make Them Pay*, pp. 17–33. The Rolls House Publishing Co. Ltd.

Britt, J.H., Armstrong, J.D. and Cox, N.M. (1988) Metabolic interfaces between nutrition and reproduction in pigs. In *Proceedings of 11th International Congress on Animal Reproduction and Artificial Insemination. University College, Dublin*, **5**, 117–125

Brooks, P.H., Smith, D.A. and Irwin, V.C.R. (1977) Biotin supplementation of diets; the incidence of foot lesions, and the reproductive performance of sows. *Veterinary Record*, **101**, 46–50

Brooks, P.H. and Carpenter, J.L. (1990) The water requirements of growing/finishing pigs — theoretical and practical considerations. In *Recent Advances in Animal Nutrition — 1990*, pp. 115–126. Edited by W. Haresign and D.J.A. Cole. London: Butterworths

Chambers, J. (1987) Feeding outdoor pigs: electronic sow feeders and other methods. *The Pig Veterinary Society Proceedings,* **18**, 62–66

Close, W.H. (1987) The influence of the thermal environment on the productivity of pigs. In *Pig Housing and the Environment. Occasional Publication No. 11 — British Society of Animal Production*, pp. 9–24 Edited by A.T. Smith and T.L.J. Lawrence

Close, W.H. (1989) The influence of the thermal environment on the voluntary feed intake of pigs. In *The Voluntary Food Intake of Pigs. Occasional Publication No. 13 — British Society of Animal Production*, pp. 87–96. Edited by J.M. Forbes, M.A. Varley and T.L.J. Lawrence

Close, W.H. (1993) Fibrous feeds for pigs. In *Animal Production in Developing Countries. Occasional Publication No 16 – British Society of Animal Production*, Edited by M. Gill, E. Owen, G.E. Pollard and T.L.J. Lawrence (in press)

Close, W.H., Pettigrew, J.E., Sharpe, C.E., Keal, H.D. and Harland, J.I. (1990) The metabolic effects of feeding diets containing sugar beet pulp to sows. *Animal Production*, **50**, 559–560

Coffey, M.J., Seerley, R.W., Funderburke, D.W. and McCampbell, H.C. (1982) Effects of heat increment and level of dietary energy and environmental temperature on the performance of growing/finishing swine. *Journal of Animal Science,* **54**, 95–105

Cole, D.J.A. (1978) Amino acid nutrition of the pig. In *Recent Advances in Animal Nutrition — 1978*, pp. 59–72. Edited by W. Haresign and D. Lewis. London: Butterworths

Cole, D.J.A. (1990) Nutritional strategies to optimise reproduction in pigs. *Journal of Reproduction and Fertility, Suppl.* **40**, 67–82

Cole, D.J.A. and Chadd, S.A. (1989) Voluntary food intake of growing pigs. In *The Voluntary Food Intake of Pigs. Occasional Publication No. 13 — British Society of Animal Production*, pp. 61–70. Edited by J.M. Forbes, M.A. Varley and T.L.J. Lawrence

Culver, A.A., Andrews, F.N., Conrad, J.H. and Noffsinger, T.L. (1960) Effectiveness of water sprays and a wallow on the cooling and growth of swine in a normal summer environment. *Journal of Animal Science*, **19**, 421–433

Garrett, W.N., Bond, T. E. and Kelly, C.F. (1960) Environmental comparisons of swine performance as affected by shaded and unshaded wallows. *Journal of Animal Science*, **19**, 921–925

Geuyen, T.P.A., Verhagen, J.M.F. and Verstegen, M.W.A. (1984) Effect of housing and temperature on metabolic rate of pregnant sows. *Animal Production*, **38**, 477–485

Harker, A. and Cole, D.J.A. (1985) The influence of pregnancy feeding on sow and litter performance during the first two parities. *Animal Production*, **40**, 540.

Holmes, C.W. and Close, W.H. (1977) The influence of climatic variables on energy metabolism and associated aspects of productivity in pigs. In *Nutrition and the Climatic Environment*, pp. 51–73. Edited by W. Haresign, H. Swan and D. Lewis. London: Butterworths

Holmes, C.W. and McLean, N.R. (1974) The effect of low ambient temperature on the energy metabolism of sows. *Animal Production*, **19**, 1–12

Holmes, C.W. and McLean, N.R. (1977) The heat production of groups of young pigs exposed to reflective or non-reflective surfaces of walls and ceilings. *Transaction of the American Society of Agricultural Engineers*, **20**, 527–528

Hovell, F.D.de B., Gordon, J.G. and MacPherson, R.M. (1977) Thin sows. 2. Observations on the energy and nitrogen exchange of thin and normal sows in environmental temperatures of 20°C and 5°C. *Journal of Agricultural Science, Cambridge*, **89**, 523–433

Hughes, P.E. and Pearce, G.P. (1989) The endocrine basis of nutrition — reproduction interactions. In *Manipulating Pig Production II. Australian Pig Science Association, Werribee*, pp. 290–295. Edited by J.L. Barnett and D.P. Hennessey

Ingram, D.L. (1965) Evaporative cooling in the pig. *Nature*, **207**, 415–416

Kemp, B., Verstegen, M.W.A., Verhagen, J.M.F. and van der Hel, W. (1987) The effect of environmental temperature and feeding level on energy and protein retention of individual housed pregnant sows. *Animal Production*, **44**, 275–283

King, R.H. (1987) Nutritional anoestrus in growing sows. *Pig News and Information*, **8**, 15–22

King, R.H. and Williams, I.H. (1984a) The effect of nutrition on the reproductive performance of first litter sows. 1. Feeding level during lactation and between weaning and mating. *Animal Production*, **38**, 241–247

King, R.H. and Williams, I.H. (1984b) The effect of nutrition on the reproductive performance of first litter sows. 2. Protein and energy intakes during lactation. *Animal Production*, **38**, 249–256

Lee, P.A. and Close, W.H. (1987) Bulky feeds for pigs: a consideration of some nonnutritional aspects. *Livestock Production Science*, **16**, 395–405

Low, A.G. (1985) Role of dietary fibre in pig diets. In *Recent Advances in Animal Nutrition — 1985*, pp. 87–112. Edited by W. Haresign and D.J.A. Cole, London: Butterworths

Lynch, P.B. (1977) Effect of environmental temperature on lactating sows and their litters. *Irish Journal of Agricultural Research*, **16**, 123–130

Lynch, P.B. (1989) Voluntary food intake of sows and gilts. In *The voluntary food intake of pigs. Occasional Publication No. 13 — British Society of Animal Production*, pp. 71–77. Edited by J.M. Forbes, M.A. Varley and T.L.J. Lawrence

Machin, D.H. (1990) Alternative feeds for outdoor pigs. In *Outdoor Pigs, Principles and Practice*, pp. 103–114. Edited by B.A. Stark, D.H. Machin and J.M. Wilkinson. Marlow: Chalcombe Publications

Mahan, D.C. and Mangan, L.T. (1975) Evaluation of various protein sequences on the nutritional carry-over from gestation to lactation with first-litter sows. *Journal of Nutrition*, **105**, 1291–1298

McGlone, J.J., Stansbury, W.F. and Tribble, L.F. (1988) Management of lactating sows during heat stress. Effects of water drip, snout coolers, floor type and a high energy diet. *Journal of Animal Science*, **66**, 885–891

Meat and Livestock Commission (1992) *Pig Yearbook*. Milton Keynes: Meat and Livestock Commission

Mount, L.E. (1968) *The Climatic Physiology of the Pig*. 271 pp. Edward Arnold, London

Mullan, B.P., Close, W.H. and Cole, D.J.A. (1989) Predicting nutrient responses of the lactating sow. In *Recent Advances in Animal Nutrition — 1989*, pp. 229–243. Edited by W. Haresign and D.J.A. Cole. London: Butterworths

Mullan, B.P. and Williams, I.H. (1989) The effect of body reserves at farrowing on the reproductive performance of first litter sows. *Animal Production*, **48**, 449–457

Mullan, B.P., Close W.H. and Foxcroft, G.R. (1991) Metabolic state of the lactating sow influences plasma LH and FSH before and after weaning. In *Manipulating Pig Production III*. p 31. Edited by E.S. Batterham. Werribee: Australian Pig Science Association

National Research Council (1988) *Nutrient Requirements of Swine*, 93 pp. Washington, DC: National Academy Press

Noblet, J., Dourmad, J.Y., Le Dividich, J. and Dubois, S. (1989) Effect of ambient temperature and addition of straw or alfalfa in the diet on energy metabolism in pregnant sows. *Livestock Production Science*, **21**, 309–324

O'Grady, J.F. and Lynch, P.B. (1978) Voluntary feed intake by sows: influences of system of feeding and nutritional density of the diet. *Irish Journal of Agricultural Research*, **17**, 1–5

O'Grady, J.F., Lynch, P.B. and Kearney, P.A. (1985) Voluntary feed intake by lactating sows. *Livestock Production Science*, **12**, 355–365

Pettigrew, J.E. (1981) Supplemental dietary fat for peripartal sows: a review. *Journal of Animal Science*, **52**, 107–117

Penny, R.H.C., Cameron, R.D.A., Johnson, J., Keynon, P.J., Smith, H.A., Bell, A.W.P., Cole, J.P.L. and Taylor, J. (1981) Influence of biotin supplementation on sow reproductive efficiency. *Veterinary Record*, **109**, 80–81

Petley, M.P. and Bayley, H.S. (1988) Exercise and post-exercise energy expenditure in growing pigs. *Canadian Journal of Physiology and Pharmacology*, **66**, 721–730

Poornan, P. (1990) Formulation, compounding and raw material use in feeds for outdoor pigs. In *Outdoor Pigs, Principles and Practice*, pp. 85–101 Edited by B.A. Stark, D.H. Machin and J.M. Wilkinson. Marlow: Chalcombe Publications

Salmon-Legagneur, E. and Rerat, A. (1962) Nutrition of the sow during pregnancy. In *Nutrition of Pigs and Poultry*, pp. 207–223. Edited by J.T. Morgan and D. Lewis. London: Butterworths

Schoenherr, W.D., Stahly, T.S. and Cromwell, G.L. (1989) The effects of dietary fat or fibre addition on yield and composition of milk from sows in a warm or hot environment. *Journal of Animal Science*, **67**, 482–495

Smith, C.V. (1974) Farm Buildings. In *Heat loss from Animals and Man*, pp. 345–365 Edited by J.L. Monteith and L.E. Mount. London: Butterworths

Stahly, T.S., Cromwell, G.L. and Aviotti, M.P. (1979) The effects of environmental temperature and dietary lysine source and level on the performance and carcass characteristics of growing swine *Journal of Annual Science*, **49**, 1242–1251

Stansbury, W.F., McGlone, J.J. and Tribble, L.F. (1987) Effect of season, floor type, air temperature and snout coolers on sow and litter performance. *Journal of Animal Science*, **65**, 1507–1513

Verhagen, J.M.F., Verstegen, M.W.A., Geuyen, T.P.A. and Kemp, B. (1986) Effect of environmental temperature and feeding level on heat production and lower critical temperature of pregnant sows. *Journal of Animal Physiology and Animal Nutrition*, **55**, 246–256

Verstegen, M.W.A. and van der Hel, W. (1974) The effects of temperature and floor type on metabolic rate and effective critical temperature in groups of growing pigs. *Animal Production*, **18**, 1–11

Verstegen, M.W.A., Verhagen, J.M.F. and den Hartog, L.A. (1987) Energy requirements of pigs during pregnancy: a review. *Livestock Production Science*, **16** 75–89

Vidal, J.M., Edwards, S.A., MacPherson, O., English, P.R. and Taylor, A.G. (1991) Effects of environmental temperature on dietary selection in lactating sows. *Animal Production*, **52**, 597

Webb, N.G., Penny, R.H.C. and Johnston, A.M. (1984) Effect of a dietary supplement of biotin on pig hoof horn strength and hardness. *Veterinary Record*, **114**, 185–189

Zoiopoulos, P.E., English, P.R. and Topps, J.H. (1982) High fibre diets for *ad libitum* feeding of sows during lactation. *Animal Production*, **35**, 25–33

13

RECENT ADVANCES IN PROBIOSIS IN PIGS: OBSERVATIONS ON THE MICROBIOLOGY OF THE PIG GUT

C.S. STEWART[a], K. HILLMAN[b], F. MAXWELL[a], DENISE KELLY[a] and T.P. KING[a]
a *Rowett Research Institute, Bucksburn, Aberdeen, AB2 9SB, UK* and
b *Scottish Agricultural College, King Street, Aberdeen AB9 1UD, UK*

Introduction

The excretion of cellular solutes that occurs in pigs and other animals suffering from diarrhoea is a protective defence mechanism which is elicited in response to the presence of infective agents in the gut. It is triggered by a wide range of pathogenic organisms, and causes immense problems in pig production worldwide. The causative agents of diarrhoea in pre-weaning piglets include enterotoxigenic *Escherichia coli*, clostridia, gastroenteritis virus, rotavirus and *Cryptosporidium* (Jonsson and Conway, 1992). The presence of these organisms in the gut does not in itself guarantee that the disease will develop, and this has led to the view that the disease process can be seen as partly the result of the failure, albeit sometimes highly localized, of a protective effect exerted by the normal (commensal) gut microbial population. For this reason, understanding the microbial ecology of the gut is of prime importance in understanding the disease state and its possible prevention, and this is reflected in numerous reviews (Savage, 1977; Lee, 1985; Tannock, 1992).

Of all pathogens, *E. coli* is by far the most widely studied. Under normal circumstances, peristalsis, flowing ingesta and secreted mucus serve to propel bacteria through the small intestine, thereby curtailing their establishment. Surface fimbriae enable *E. coli* strains to overcome these physiological defences and rapidly proliferate in the small intestine (Holland, 1990). Many of the symptoms of infection result from the production of toxins of various classes (Gyles, 1992). Although the disease state itself is increasingly understood, the way in which *E. coli* and other pathogens out-compete the commensal bacteria that normally inhabit the same or similar niches is still not clear (Freter, 1988).

The origin and development of the scientific probiotic concept, that the presence of certain micro-organisms (especially lactic-acid producing bacteria) or their products may protect against gastrointestinal disease, has been reviewed by Fuller (1992). Some of the most frequently cited potentially beneficial effects of probiotic organisms on the health of the host animal, including effects other than the control of pathogens, are listed in Table 13.1. A schematic representation of the attachment of pathogens and their toxins, and of some interactions between probiotics and pathogens that involve adhesion, competition or inhibition are

Table 13.1 POTENTIALLY BENEFICIAL EFFECTS OF PROBIOTIC ORGANISMS OR THEIR PRODUCTS

Effect	*Reference*
Antagonism towards pathogens	
Bacteriocins	Havenaar *et al.* (1992)
Acids and other compounds	Vandevoorde *et al.* (1992)
Virulent phage	Houghton and Fuller (1980); Smith and Huggins (1983)
Competition with pathogens	
Energy sources	Hentges (1992)
Minerals	Bezkorovainy (1989); Vandevoorde *et al.* (1992)
Adhesion sites	Barrow (1992); Coconnier *et al.* (1992)
Stimulation of enzyme reactions	
Detoxification of xenobiotics	Rowland (1992)
Digestion	Rowland (1992)
Other effects	
Decreased ammonia production	Muting, Eschrich and Mayer (1968)
Decreased production of phenols	Muting *et al.* (1968)
Decreased cytotoxin production	Corthier, Dubos and Raibaud (1985)
Macrophage activation	Tomioka and Saito (1992)
Reduction in serum cholesterol	Fernandes, Chandran and Shahani (1992)
Anti-tumor activity	Adachi (1992)

illustrated diagrammatically in Figure 13.1. Given the changes that occur during the maturation of animals, in the dietary transitions during and after weaning, and the variations in the nature of the pathogenic challenge, it is possible that no one mechanism operates all of the time.

Here, the conditions in the gut as they may affect the species composition and metabolic activity of the gut flora are considered, especially in relation to the availability of oxygen and attachment to the gut wall. The potential for the use of fermenter model systems for screening probiotics and for studies of microbial interactions in the gut is briefly described.

The gut as a bacterial habitat

The composition and metabolic activity of microbial populations reflect the prevailing environmental conditions (Brock, 1966; Alexander; 1971; Latham, 1979). Important factors that are likely to determine the distribution, composition and metabolic activity of the gut flora include the presence of local defence systems (Banks, Board and Sparks, 1986), the presence of gut receptors to which certain bacteria may attach, the nutrient composition and flow of digesta, the presence of endogenous nutrients, the pH and E_h (oxidation/reduction potential on the hydrogen scale) and the availability of molecular oxygen.

DIGESTA FLOW AND BACTERIAL ADHERENCE TO THE GUT

The rate of flow of digesta is greatest in the small intestine, in which digesta particles reside for around 2.5 h (Jonsson and Conway, 1992) and lowest in the caecum and colon (Lee, 1985). As a result, bacterial populations in the small

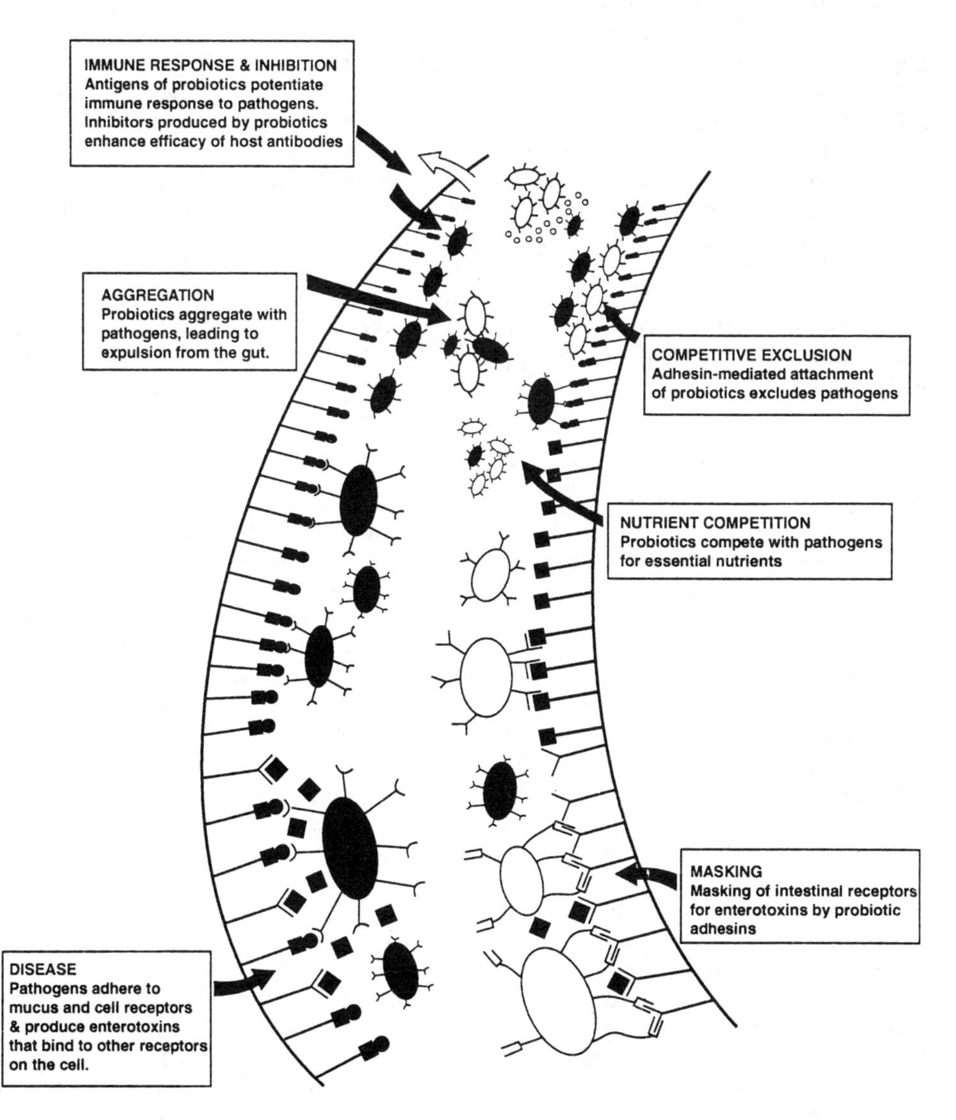

Figure 13.1 Schematic representation of the adhesion of pathogens (solid ovals) and their enterotoxins (◆) to gut receptors and of some mechanisms by which commensal or probiotic bacteria (open ovals) may decrease pathogenicity

intestine ensure their survival by colonizing the intestinal surface, or by associating with the mucin layer covering the wall (Savage, Dubos and Schaedler, 1968; Savage, 1977; Rozee, Cooper, Lam and Costerton, 1982). The invasion of the mucus layer, growth within it, and attachment to epithelial cells protected by mucus are critical steps in pathogenesis by *E. coli* (Smith, 1992).

Some bacteria have adapted their life cycle closely to the turnover of epithelial cells. As these cells migrate from the crypts towards the top of the villi, a process that in mice takes about 2 days (Abrams, Bauer and Sprinz, 1963), some segmented filamentous bacteria detach from the epithelial cells and return to the crypts to colonize new cells (Chase and Erlandson, 1976). The mucin producing goblet cells of the colon of rodents contain spiral organisms attached to the cell surface (Lee, 1985). These bacteria are thought to maintain their populations by being sufficiently motile to counter the velocity of the flow of mucin (Lee, 1985). Spirochaetes have been observed in the intestinal crypts of pigs and many other animals (Harris and Kinyon, 1974; Takeuchi, Jervis, Nakazawa and Robinson, 1974).

ADHESION OF PATHOGENS AND PROBIOTICS

Age-related and diet-induced intestinal glycosylation changes have been observed in the pig which are believed to play an important role in modifying the properties of intestinal receptors for both commensal and pathogenic bacteria (King and Kelly, 1991; Kelly and King, 1991). The nature of these host receptors, their binding affinity and their distribution play a significant role in host specificity and tissue tropism (Wick, Madara, Fields and Normark, 1991; Krogfelt, 1991). Fimbrial and non-fimbrial bacterial adhesins which interact with intestinal surface receptors have been identified in both commensals (e.g. *Lactobacilli*) and pathogens (e.g. *E. coli*).The accumulating evidence suggests that carbohydrate recognition is the principal mechanism of fimbrial-intestinal receptor interaction.

Strains of *E. coli* which cause diarrhoea have been partially classified according to the nature of their fimbrial adhesins. The synthesis of these antigens (e.g. K88, K99, 987P, F41) and the production of enterotoxins are considered essential virulence factors which often enable pathogens to compete successfully with commensals in the intestine. Enterotoxigenic *E. coli* strains bearing the K88 fimbrial antigen are frequently associated with outbreaks of diarrhoea in pigs. Three major antigenic subtypes of K88 fimbrial adhesin have been identified (K88ab, K88ac and K88ad). These antigenic variants have unique adhesion specificities and their binding to intestinal mucosal surfaces (Figure 13.2) provides the basis for a phenotypic classification of pig populations (Bijlsma and Bouw, 1987). The basis of this variation is uncertain; although K88 fimbrial adhesins are known to recognise oligosaccharide sequences, the precise structure of the receptors has still to be ascertained (Conway, Welin and Cohen, 1990; Erickson, Willgohs, McFarland, Benfield and Francis, 1992; Metcalfe, Krogfelt, Krivan, Cohen and Laux, 1991). In recent experiments *in vitro*, suspensions of K88ab, K88ac and K88ad positive *E. coli* strains were applied to resin sections of pig jejunal tissue and attachment of K88 fimbriae to villus surfaces was revealed by immunofluorescence microscopy. Cytochemical examination of the glycoconjugate complexion of the porcine intestine suggested that the K88 fimbriae recognised oligosaccharide precursors of histo-blood group A and O antigens (King, Begbie, Spencer and Kelly, 1992). These observations are in keeping with earlier reports

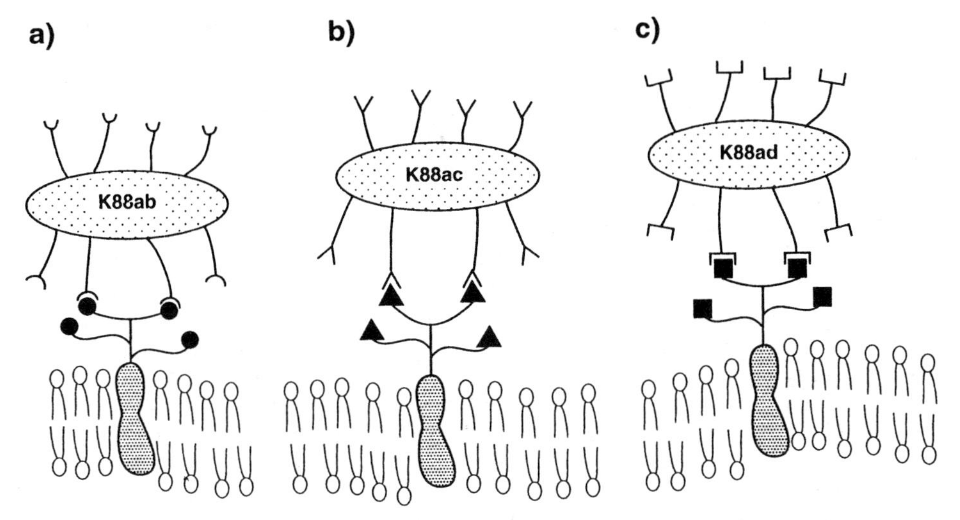

Figure 13.2 Schematic diagram illustrating the hypothetical interactions between different *E. coli* K88 serotypes and enterocyte receptors (solid symbols)

that K88 receptors contain terminal galactosyl (Gibbons, Jones and Sellwood, 1975) or D-galactosamine moieties (Sellwood, 1984; Staley and Wilson, 1983).

Sialic acids present on the terminal position of glycoproteins and gangliosides are attracting increasing interest because of their involvement in various aspects of normal and pathological cellular growth and development. These sugars, which represent a large family of differently N- and O-substituted neuraminic acid derivatives, play a key regulatory role in cellular and molecular recognition. In some instances they act as signals for recognition and in other situations they mask recognition sites on molecules and cell membranes (Schauer, 1991). Serotype K99 *E. coli*, which cause diarrhoea in neonatal but not adult pigs, express fimbrial adhesins which bind to sialylated glycoproteins and glycolipids (Mouricout and Julien, 1987; Lindahl and Carlstedt, 1990). A preferential affinity of K99 fimbriae for neuraminic acid derivatives has been established, N-glycolylneuraminic acid being twofold more potent than N-acetylneuraminic acid (Lindahl, Brossmer and Wadstrom, 1987). Two pig phenotypes have been identified, those expressing high levels of K99 receptors and those exhibiting low levels. Seignole, Mouricout, Duval-Iflah, Quintard and Julien (1991) have determined that piglets susceptible to K99 adhesion express higher levels of sialylated glycolipids than those resistant to K99 attachment. In recent affinity biochemical and cytochemical studies both age-related and individual diversity was observed in the expression of intestinal sialic acid species in sucking and weaned pigs (King, Kelly, Begbie, McFadyen and Slater, unpublished observations). In keeping with the susceptibility to K99 infection, demonstrably higher levels of intestinal membrane sialylation were evident in newborn and sucking pigs than in animals that had been weaned. Other investigations have demonstrated that the piglet intestinal mucosa contains a substantially higher content of acidic glycolipids than that of adult pigs (Teneberg, Willemsen, de Graff and Karlsson, 1990).

Enterotoxigenic *E. coli* strains carrying 987P fimbriae colonize the small intestine and cause diarrhoea in neonatal (> 6 days post partum) pigs but not in older

(> 3 weeks) post-weaned pigs in spite of the fact that intestinal receptors are present in both groups (Dean, 1989). A < 17 kDa 987P-binding component was present in the intestinal mucus of the older animals but not in neonates and Dean (1990) concluded that this receptor in the mucus may prevent *in vivo* adhesion of 987P+ *E. coli* by competing with brush border 987P-receptors.

The chromosomally encoded *E. coli* fimbrial adhesin F41 is often produced by strains which also elaborate K99 adhesins (Morris, Thorns, Wells, Scott and Sojka, 1983; To, 1984). F41 has a high affinity for terminal N-acetylgalactosamine moieties (Lindahl, Brossmer and Wadstrom, 1988). Cytochemical analysis of the membrane and mucin glycoconjugates in piglets reveal that these sugar moieties occur on immature glycoconjugates in newborn animals but are also present on histo-blood group A antigens in some weaned animals (King and Kelly, 1991).

Nagy, Casey, Whipp and Moon (1992) investigated the adhesion of pathogenic enterotoxigenic *E. coli* strains which did not produce K88, K99 F41 or 987P fimbriae. Colonization of these so-called 4P- strains was characterised by fimbriae-mediated adhesion to villi overlying Peyer's patches. Small-intestinal adhesion by these isolates was found to be dependent on receptors that develop progressively with age during the first 3 weeks after birth. The carbohydrate specificity of 4P- fimbriae is unknown.

Salmonella typhimurium causes severe diarrhoeal disease in pigs. Frequent reinoculation of the intestines by the faecal-oral route is considered an important mechanism for the persistent colonization of the porcine intestine although evidence has also been presented that colonization is much enhanced in *Salmonella* strains which possess adhesive fimbriae (Isaacson and Kinsel, 1992). Again, the receptor specificities of the fimbrial adhesins has not been determined.

Lactobacilli are a prevalent group of the commensal flora of the intestine. For almost a century organisms of this genus have been purported to augment the protective barrier of the intestine, however their mode of action remains speculative. Much attention has centred on the ability of these bacteria to compete with pathogens (and perhaps enterotoxins) for binding sites (Figure 13.1). Pedersen and Tannock (1989) demonstrated that adherence of *Lactobacilli* to epithelial surfaces of the digestive tract was a prerequisite for successful colonisation and may also represent a key feature of their probiotic action. In a recent model Coconnier, Klaenhammer, Kerneis, Bernet and Servin (1992) proposed that an extracellular proteinaceous component produced by adhering *Lactobacilli* provides a divalent bridge that links the bacteria to enterocyte surfaces. The extracellular bridging protein interacts with carbohydrate components of the bacterial cell and the intestinal epithelium. The extracellular bridging proteins may also be derived from the host epithelium or introduced as lectin constituents in the diet (Tannock, 1992; Pusztai, Grant, King and Clarke, 1990). Recently, Mukai, Arihara and Itoh (1992) obtained evidence for the existence on the cell surfaces of *Lactobacillus acidophilus* of lectin-like proteinaceous components. Although the precise nature of the lectin receptors remains to be elucidated there is evidence that species specificity of different *Lactobacillus* strains may reflect specific- and age-related variations in the glycoconjugate complexion of intestinal surfaces (Tannock, 1992; Chauviere, Coconnier, Kerneis, Fourniat and Servin, 1992).

Bioreactive dietary constituents can be used to interfere with the attachment of bacteria to intestinal surface either by competitively masking the receptor moieties or through the provision of alternative binding sites. The potential of dietary lectins to advantageously manipulate pathogen mucosal interactions has been fully discussed by Pustzai *et al.* (1990). Oligosaccharides, natural constituents of many foods including milk have been shown to function as bacterial receptor analogues. This probiotic property has been demonstrated to inhibit the attachment of *Streptococcus* and *Haemophilus* to human epithelial cells and diminish K99 colibacillosis in calves (Andersson, Porras and Hanson, 1986; Mouricout, Petit, Carias and Julien, 1990).

NUTRIENT STATUS

For growth, bacteria require energy sources and nutrients, derived either exogenously from the host diet or endogenously from sloughed-off epithelial cells, cell secretions or from the mucus blanket that coats much of the inner surface of the gut. The mucus is colonized by bacteria (Conway *et al.*, 1990), and has been shown to support rapid growth of *E. coli in vitro* (Jonsson and Conway, 1992). Mucin was found to be the major endogenous carbohydrate source excreted from the upper gut when rats were fed fibre-free diets or diets containing gum arabic (Monsma, Vollendorf and Marlett, 1992). The relative contribution of endogenous and dietary nutrients is not understood, although the view that competition for substrates largely determines the composition of the gut population (Freter, Brickner, Botney, Cleven and Aranki, 1983) is widely accepted. The absorption of soluble carbohydrates and other nutrients by the host animal lowers the concentration of these substrates in the liquid phase, so that beyond the small intestine, only the dietary polymers resistant to rapid microbial degradation and endogenous substrates remain to support microbial growth. Guiot (1982) found that cell-free intestinal contents from rats failed to support growth of *E. coli* in slices of non-nutrient agar. It might be argued from such findings that the provision of soluble substrates for bacterial growth is dependent on the degradation of high molecular weight insoluble polymers either of dietary or animal origin. The disappearance of the carbohydrates of swede (*Brassica napus*) anterior to the terminal ileum of adult pigs was studied by Millard and Chesson (1984). They reported that over 98% of the free sugar (glucose and fructose), together with around 20% of the cellulose, and over 40% of the arabinose and galactose units from polysaccharides was lost anterior to the terminal ileum.

The major site for fermentation of the unabsorbed dietary or endogenous polysaccharides is the large intestine. Here, a number of species of *Bacteroides* are thought to play an important role. In humans, *B. thetaiotaomicron*, *B. ovatus* and *B. fragilis* ferment the mucopolysaccharide chondroitin sulphate together with dietary polysaccharides (Salyers, Vercellotti, West and Wilkins, 1977: Salyers, 1984). The presumed role of these and other bacteria may not be the same as their actual role. For example, although many strains of *Bacteroides* were able to degrade polygalacturonic acid (PGA) in culture, direct examination of the PGA lyases present in human faeces showed that they had lower isoelectric points and higher molecular weights than the PGA lyases of predominant colonic *Bacteroides* grown in pure culture (McCarthy and Salyers, 1986). The growth rate of these bacteria in culture and in the gut is likely to differ, however, and some enzymes of *Bacteroides* are subject to growth rate dependent regulation (Strobel and Russell, 1987).

Estimating the proportion of dietary non-starch polysaccharide fermented in the gut has proved difficult. The stoichiometry of the rumen fermentation and that in human faeces appears to be broadly similar (Miller and Wolin, 1979), and it seemed from this that the extent of fermentation could readily be calculated if the amount of methane produced could be determined. However, Zhu, Fowler and Fuller (1993) detected only around 20% of the expected amounts of methane from pigs fed sugar beet pulp. It may be that the actual stoichiometry of the large intestinal fermentation differs markedly from that in the rumen. Prins and Lankhorst (1977) reported that acetate was formed by the reduction of CO_2 in the caecum of rodents. Although acetogenic bacteria have also been isolated from the rumen (Sharak-Genthner, Davis and Bryant, 1981), such bacteria may compete more effectively with methanogens for hydrogen in the hind gut than in the rumen. It is also possible that there are other reduced products of the fermentation (including components of the biomass) yet to be discovered, or that the methane formed may not all escape from the gut. The latter point is discussed below in relation to the presence and role of oxygen.

Sugars that pass undegraded to the colon (Table 13.2) have been used in attempts to influence the composition of the gut bacterial population (reviewed by Rowland, 1992). Several classes of these sugars, including fructo-, transgalactosylated- and soya bean-oligosaccharides are utilized by bifidobacteria, although *Bacteroides*, *Lactobacilli* and coliforms also utilize some of the preparations currently available.

Table 13.2 OLIGOSACCHARIDES USED TO PROMOTE THE ACTION OF PROBIOTICS

Product	*Reference*
Palatinose condensation product	Nakajima, Nishio, Mizutani, Ogasa and Kashimura (1988)
Xylo-oligosaccharides	Suwa, Koga, Fujikawa, Okazaki, Irie and Nakada (1988)
Inulo-oligosaccharides	Hidaka, Eida and Hamaya (1988)
Transgalactosylated oligosaccharides	Rowland (1992)
Fructo-oligosaccharides	Rowland (1992)
Soybean oligosaccharide extracts	Rowland (1992)

pH

The secretion of gastric juice containing HCl, proteases and mucus lowers the pH, especially in the pyloric region near the exit to the small intestine. Reviewing earlier observations, Johnsson and Conway (1992) reported pH values between 3.0 and 4.4 in the stomach of unweaned pigs, and between 2.3 and 4.5 in adult animals. Most studies showed a rise in pH in the small intestine, and in the caecum and colon, pH values were typically between 6.0 and 7.7. Vervaeke, van Nevel, Decuypere and van Assche, (1973) recorded higher pH values in their animals, which were between 3 and 7 weeks of age. The pH in the stomach of these animals was around 5.3, but rose to pH 8.4 in the posterior part of the small intestine. In a study at the Rowett Institute, Hillman and his colleagues (unpublished data) also found alkaline pH values (typically pH 8.5 to 9.2) *in situ* in the caecum and colon. They found that the pH of gut contents fell rapidly on withdrawal from the gut, which could account in part for the lower pH values reported by other investigators.

OXIDATION/REDUCTION POTENTIAL

Perhaps because of the significant impact of rumen microbiology on the development of anaerobic methods and the investigation of gastrointestinal micro-organisms, the gut is often described as being a highly reduced anaerobic habitat (Freter *et al.*, 1983, Robinson, Smolenski, Ogilvie and Peters, 1989; McFarlane, Hay and Gibson, 1989). For the stomach and the small intestine at least, this is not the case. The E_h (oxidation reduction potential on the hydrogen scale) of the gut contents of piglets (25 to 45 days old) was measured *in situ* by Vervaeke *et al.* (1979). As expected, the stomach was found to be oxidized (265 mV) and the small intestine was increasingly reduced from the anterior (150 mV) to posterior (−43 mV) segments. The E_h of the caecum and anterior colon was around −210 mV, rather more oxidized than the commonly cited value for the rumen of around −350 mV (Smith and Hungate, 1958). Posterior to the colon, the E_h rose progressively towards the anus.

DISSOLVED OXYGEN

All bacteria, including those in the gut, can be categorized on the basis of their response to, or utilization of, oxygen. The effects of oxygen result in part from its use by some organisms as a terminal electron acceptor for the oxidation of reduced coenzymes. At high concentrations, however, oxygen is inhibitory to many microorganisms or microbial processes (Morris, 1975; Wimpenny and Samah, 1978; Wilde and Schlegel, 1982). As a result of these properties, the distribution of oxygen is a primary ecological determinant of the nature and distribution of life in the biosphere and all the habitats within it. In the gastrointestinal tract, the strict aerobes such as *Neisseria*, unable to grow in the absence of oxygen, are normally limited to the nasopharynx (Vedros, 1984). Facultative bacteria, such as *E. coli*, possess a functioning tricarboxylic acid cycle, but are also capable of anaerobic growth. Aerotolerant or microaerophilic bacteria lack a functioning tricarboxylic acid cycle, but some species may use oxygen in reactions involving flavoprotein oxidases (Condon, 1987). *Enterococcus faecium* is able to use oxygen to oxidize NADH using NADH oxidase. In the presence of oxygen, these cells produce acetate in place of lactate, and as the production of acetate is linked to the conservation of energy as ATP, the growth yields/mole substrate utilized are increased when oxygen is present (Gottschalk, 1979). Certain oral streptococci including *S. mutans* and *S. sanguis* have been found to consume large amounts of oxygen in reactions involving NADH oxidase and NADH peroxidase (van Beelan, van der Hoeven, de Jong and Hoogendorn, 1986). Similar reactions occur in bifidobacteria (Uesugi and Yamija, 1978). Other reactions found in lactic acid bacteria and involving molecular oxygen include the conversion of pyruvate plus phosphate to acetylphosphate and the conversion of α-glycerophosphate to dihydroxyacetone phosphate (Condon, 1987).

The more frequently isolated strict anaerobes such as *Bacteroides* species are killed by exposure to low concentrations of oxygen. The oxygen sensitivity of aerotolerant and anaerobic bacteria varies between species (Loesche, 1969; Kikuchi and Suzuki, 1986). Following the discovery of superoxide dismutase (SOD), an enzyme which scavenges superoxide free radicals to H_2O_2, the susceptibility of anaerobes to oxygen was thought to be due to the lack of this enzyme (McCord, Keele and Fridovich, 1971; Morris, 1975). However, some anaerobes such as *Selenomonas ruminantium* were later found to possess

SOD (Wimpenny and Samah, 1978). Production of this enzyme was induced at partial pressures of oxygen above 60 mm Hg. Protection against oxygen was also provided by an NADH oxidase in this bacterium. Recently, Adler, Mural and Suttle (1992) reported that the oxygen sensitivity of a mutant of *E. coli* was linked to the production of a metabolic product of the cells that is secreted to the culture medium. The compound is heat labile and of low molecular weight.

The concentration of oxygen in the rumen has been found by membrane-inlet mass spectrometry to be up to around 3 μM (Hillman, Lloyd and Williams, 1985). Even in the most reduced regions of the gut of piglets, Hillman, Whyte and Stewart (1993) found substantially higher oxygen concentrations. Measurements were made immediately after removing piglets from the sow at 3 weeks of age, and after feeding other piglets for 6 days on one of three separate diets (milk replacer, solid weaner or solid grower diets). The dissolved oxygen concentrations measured with the oxygen electrode, which is subject to interference from sulphide, ranged from 108 to 159 μM in the stomach and was not affected by the diet. Piglets receiving the milk replacer diet showed lower concentrations of dissolved oxygen in the ileum (65 μM) than the unweaned piglets, or those receiving the other diets (range 93 to 150 μM). A similar trend was found in the posterior segment of the jejunum. In the caecum, and in 3 sites in the colon, oxygen concentrations ranged from 52 to 104 μM, and were not significantly influenced by the nature of the diet.

Further measurements were made in the duodenum, ileum, caecum and colon of two additional unweaned piglets by membrane-inlet mass spectrometry. This

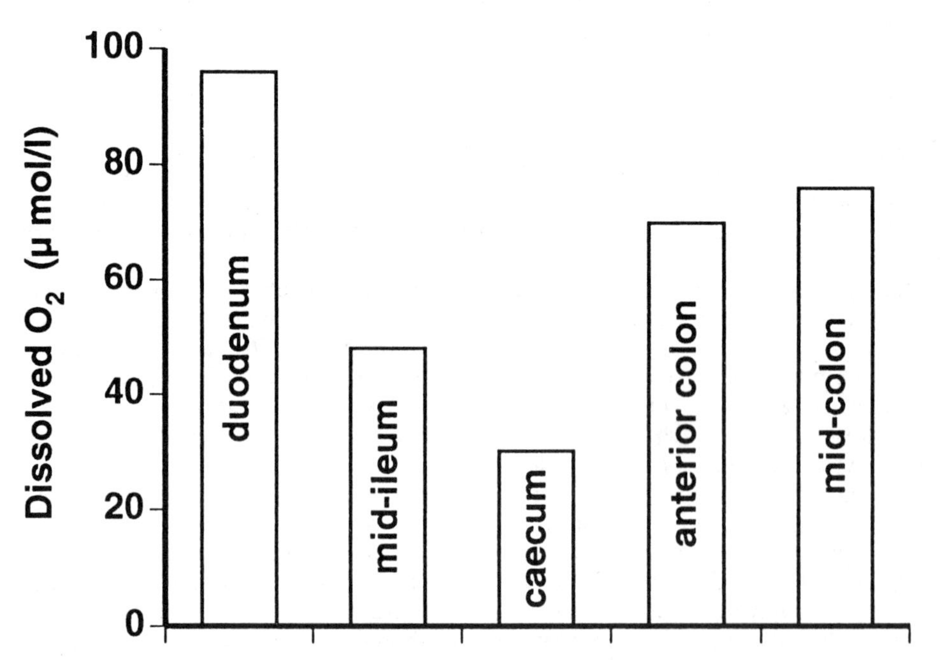

Figure 13.3 Average concentrations of dissolved oxygen in the gut of 2 unweaned piglets determined by mass spectrometry after correction for sulphide

method allows corrections to be made for sulphide, which as a result of ionization presents peaks at m/z (mass charge ratio) 32 in addition to the main peak at m/z 34. The dissolved oxygen concentrations recorded are shown in Figure 13.3. Although these values were lower than those found using the oxygen electrode, the results confirmed that there is a pool of oxygen in the gut of young piglets. The rate of flux of oxygen through this pool is unknown, but as a potential electron acceptor the metabolic effect of oxygen could be considerable. The implications of this finding are discussed below.

Bacterial species from the pig gut

The commensal bacterial flora of the gut acts as a primary defence mechanism, by increasing the colonization resistance of the gut towards pathogens (van der Waaij, Berghuis de Vries and Lekkerkerk, 1971). A list of some of the bacterial species isolated from the gut or faeces of pigs, grouped according to their utilization of or response to oxygen, is presented in Table 13.3. Tannock (1992) provides more detailed citations of the isolation of lactic acid bacteria. Strictly anaerobic *Lactobacilli* have been isolated from the rumen (Sharpe, Latham, Garvie, Zirngibl and Kandler, 1973), and might also be present in pigs. The relative proportions of the bacterial species present varies in different parts of the gut (Savage, 1977). Smith and Jones (1963) found that in healthy pigs, *Lactobacilli* were often predominant in the stomach and small intestine. *Bacteroides* were normally absent from these sections of the gut, although the anaerobe *Veillonella* was present in small numbers. In the large intestine, the numbers of *Lactobacilli* and *Bacteroides* present were broadly comparable. The predominant bacteria in the caeca of normal pigs were found to be *Bacteroides* (*Prevotella*) *ruminicola* (35% of isolates) and *Selenomonas ruminantium* (21% of isolates). Apart from *Lactobacillus acidophilus* (8% of isolates) and *Butyrivibrio* (6% of isolates), none of the other species found contributed more than 3 of the isolates (Robinson, Allison and Bucklin, 1981). In contrast, the same group (Robinson, Whipp, Bucklin and Allinson, 1984) found that the colonic epithelium of normal pigs was colonized mainly by streptococci and *Bacteroides*, together with *Lactobacillus acidophilus*; none of the other species found contributed more than 5% of the isolates. Robinson *et al*., (1984) found that in dysenteric pigs, the predominant bacteria isolated from the colonic epithelium were *Acetivibrio ethanolgignens*, *Selenomonas ruminantium* and *E. coli*.

Not all known species of gut anaerobic bacteria have so far been isolated from pigs. In contrast to the human gut, which contains a wide range of *Bacteroides* species including *B. thetaiotaomicron* and *B. fragilis* (Salyers, 1984), the number of *Bacteroides* species represented in the pig gut is comparatively restricted (Table 13.3: see also Varel, Robinson and Jung, 1987; Hespell and Whitehead, 1990). However, recent advances in understanding the phylogenetic relationships within *Bacteroides* and in *Prevotella* (Mannarelli, Ericsson, Lee and Stack, 1991; Avgustin, Flint and Whitehead, 1992) may lead to significant advances in our knowledge of the distribution of these and related bacteria in the pig gut. Among the aerotolerant organisms, *Enterococcus faecium* and *E. faecalis* are not routinely found in the pig gut according to Tannock (1992), although these species can be isolated from rodents, and both species have been used as probiotics in farm animals (Fuller, 1992). *Bifidobacteria* in general demonstrate a significant degree of host-animal specificity (Scardovi, 1986).

Table 13.3 SOME FACULTATIVE, AEROTOLERANT AND ANAEROBIC BACTERIA FROM THE PIG

Species	*Reference*
Facultative bacteria	
Escherichia coli	Smith and Jones (1963)
Aerotolerant bacteria	
Lactobacillus acidophilus, L. brevis	Tannock, Fuller and Pedersen (1990)
L. crispatus, L. fermentum	"
L. johnsonii, L. agilis, L. amylovorans	Mitsuoka (1992)
L. reuteri	Axelsson and Lindgren (1987)
L. plantarum, L. delbrueckii	Russell (1979)
L. salivarius	Fuller, Barrow and Brooker (1978)
Enterococcus avium	Mundt (1986)
Streptococcus salivarius, S. bovis	Fuller *et al.* (1978)
S. morbillorum	Molitoris, Krichevski, Fagerberg and Quarles (1986)
S. intermedius	Robinson *et al.* (1981)
Bifidobacterium adolescentis	Robinson *et al.* (1984)
B. boum, B. choerinum, B. globosum,	Scardovi (1986)
B suis, B. thermophilum,	"
B. pseudolongum	"
Anaerobes	
Acetivibrio ethanolgignens	Robinson *et al.* (1984)
Bacteroides (Prevotella) ruminicola,	"
B. uniformis, B. furcosus, B capillosus,	"
B. (Mitsuokella) multiacidus	Robinson *et al.* (1981)
B. (Ruminobacter) amylophilus	Robinson *et al.* (1984)
Fusobacterium prausnitzii, F. necrophorum	"
Selenomonas ruminantium,	"
Megasphaera elsdenii,	"
Eubacterium aerofaciens,	"
E. tenue, E. lentum, E. cylindroides	"
Peptostreptococcus anaerobius	"
Clostridium welchii	Smith and Jones (1963)
Butyrivibrio fibrisolvens	Robinson *et al.* (1981)
Methanobrevibacter species	Miller, Wolin and Kusel (1986)

In a recent study of bifidobacteria from pig faeces isolated using the selective medium of Beerens (1990), Maxwell and Stewart (1992) found 19 strains of *B. boum*, 12 strains of *B. thermophilum*, 2 strains of *B. choerinum*, 1 strain of *B. suis* and 1 strain belonging to the *B. pseudolongum/B. globosum* group. Nine strains (6 from Italy and 3 from Scotland) of uncertain taxonomic status were discovered. All nine strains showed the same pattern of fermentation of sugars (Table 13.4), but the Scottish and Italian isolates differed in the electrophoretic mobility of their transaldolases. The Scottish isolates had no detectable 6-phosphogluconate dehydrogenase activity that could be visualized on zymograms. The protein fingerprints of these bacteria on sodium dodecylsulphate polyacrylamide gel electrophoresis were unlike those of the type strains of species previously isolated from pigs, and those of other isolates obtained in this study. These isolates appear to be most similar to species previously isolated from humans (*B. pseudocatenulatum, B. infantis* and *B. breve*). Establishing their taxonomic relatedness awaits the outcome of DNA-DNA hybridization studies.

Table 13.4 DISTINGUISHING FEATURES OF BIFIDOBACTERIA FROM PIG FAECES

	B. boum	B. choerinum	B. thermo-philum	B. suis	B. pseudolongum/ globosum	B. *spp*[a] *S*	*I*
Sugar fermentation							
Sorbitol	–	–	–	–	–	–	–
Arabinose	–	–	–	+	+	–	–
Raffinose	+[b]	+	+	+	+	+	+
Ribose	–	–	–	–	+	+	+
Starch	+	+	+	+	+	+	+
Lactose	+/–	+/–	+/–	+	+	+	
Cellobiose	–	–	+/–	–	–	–	–
Melezitose	+	+/–	+/–	–	–	–	–
Gluconate	–	–	–	–	–	–	–
Enzyme mobility[c]							
Transaldolase	6	3	8	6	2	6	3/4/5
6-Phosphogluconate	ns	ns	ns	5	ns	nd	< 4

Maxwell and Stewart (1992); F. Maxwell (unpublished data)
[a] S = isolated in Scotland: I = isolated in Italy; [b] Except for one raffinose negative strain, otherwise similar to *B. boum*, [c] Determined electrophoretically and classified according to the system of Scardovi, Casalicchio and Vincenzi (1979); ns = not studied; nd = not detected

Although methanogens have been isolated from pig faeces (Table 13.3), the distribution of methanogenesis in the various segments of the pig gut has not been reported. Following an investigation of the bacterial flora of adult pigs fed cereal or vegetable fibre (Chesson, Richardson and Robertson, 1985), the distribution of methanogenic activity in gut samples from the same pigs was studied. For these tests, samples of gut contents were added to an anaerobic nutrient medium. The tubes containing the samples were incubated for up to 14 d at 38°C. The proportion of methane in the headspace gas after incubation is shown in Table 13.5. It was clear that the samples from the large intestine produced most methane, but some methanogenesis was found in samples from all regions of the gastrointestinal tract that were tested (Table 13.5). Although the possibility of the upward flow of digesta during anaesthesis cannot be excluded, the results also suggest that methanogenesis may occur, albeit at low levels, in the small intestine of pigs. Samples from the caecum, colon and rectum of pigs fed bran produced more methane than pigs fed fibre from swede.

In view of the presence of oxygen, it was thought that the pig gut might be a site for the oxidation of various reduced products of anaerobic metabolism such as volatile fatty acids, methane and hydrogen, all of which can support the growth of some aerobic species (Gottschalk, 1989). To test for the presence of methane oxidizers, enrichment cultures were set up using a medium containing minerals, without added carbohydrates or other sources of carbon or energy. When this medium was inoculated with dilutions of gut contents from piglets, then incubated in an atmosphere containing methane and air, five of the bacterial cultures obtained were found to remove methane from the headspace gas (Hillman, Whyte and Stewart, unpublished data). The cultures which grew in the presence of methane were mixed, containing sporing rods and large spores or cyst-like structures.

Table 13.5 METHANE PRODUCTION ON INCUBATION OF INTESTINAL CONTENTS OF ADULT PIGS[a]

	CH_4 in headspace gas following incubation in vitro[b] (%)			
Dietary fibre source	*Swede*		*Bran*	
Animal number	*H5*	*H8*	*H6*	*H7*
Intestinal contents from				
Stomach	3.2	5.8	< 1	< 1
Jejunum	2.1	> 1	< 1	4.3
Anterior ileum	1.6	2.4	0	5.9
Mid-ileum	6.5	2.4	0	3.1
Terminal ileum	6.3	5.9	5.1	4.3
Anterior caecum	4.6	6.9	9.2	6.0
Posterior caecum	4.5	6.0	7.5	12.1
Anterior colon	4.2	5.9	12.1	21.6
Posterior colon	3.8	5.2	17.7	27.1
Rectum	3.8	12.5	19.6	25.7

[a] A.J. Richardson, unpublished data; [b] Total headspace volume approximately 7 ml. Samples were incubated for up to 14 days at 38°C 1.0 ml of gut contents to 10 ml of medium M2 of Hobson (1969) prepared and maintained under anaerobic conditions (Bryant, 1972)

Fermenter simulation of the ileum for studies of bacterial interactions

Although interactions at surfaces are widely believed to be an important feature of the mode of action of probiotics, some lactic acid bacteria demonstrate competitive or antagonistic interactions with *E. coli* and other pathogens when growing in culture broth. A wide range of bacteriocins and bacteriocin-like compounds are produced by lactic acid bacteria, and these compounds together with organic acids and H_2O_2 may inhibit the growth of specific pathogens (reviewed in Vandevoorde, Vande Woestyne, Bruyneel, Christiaens and Verstraate, 1992; Havenaar, Ten Brink and Huis in't Veld, 1992). Some lactic acid bacteria are also thought to compete with *E. coli* for growth substrates (Raibaud, 1992) and essential minerals including iron (Marcelis, den Daas-Slagt and Hoogkamp-Korstanje, 1978; Neilands, 1984; 1992; Griffiths, 1991) and manganese (Bruyneel, Vande Woestyne and Verstraete, 1990). The role of iron in bacterial virulence has received particular attention. In animals and humans, iron is found inside cells in ferritin, haeme or haemosiderin. Extracellular iron is normally complexed to the iron binding glycoproteins lactoferrin or transferrin. Thus the amount of free iron in solution in equilibrium in the presence of iron-binding glycoproteins has been estimated at around 10^{-18} M (Griffiths, 1991). The aquisition of iron by bacteria is not wholly dependent on competition for the uptake of free iron: a number of mechanisms exist for the removal of iron from iron-binding proteins. These include chelators such as the siderophores enterobactin and aerobactin which are present in *E. coli*, and siderophore-independent systems involving membrane receptors specific for transferrin or lactoferrin, present in gonococci, meningococci and *Haemophilus influenzae* (Griffiths, 1991).

Competitive and antagonistic interactions can readily be studied in fermenter model systems and such systems, simulating the conditions in the caecum and colon, have been developed for this purpose (Miller and Wolin, 1981: Veilleux and Rowland, 1981; Freter *et al.*, 1983; Macfarlane *et al.*, 1989). Recently, Hillman and his colleagues (Hillman, Breen, Spencer and Stewart, unpublished results) have constructed a continuous flow fermentation device reproducing the conditions in the pig ileum for studies on the interactions between pathogens and probiotics. It was found that the inclusion of a controlled dissolved oxygen concentration of 50 μM provided a more realistic balance in the gut microflora than that obtained when the system was operated anaerobically. Total counts of anaerobic bacteria (which would include facultative bacteria) obtained using Columbia sheep blood agar, were higher than the numbers of aerobic bacteria using the same medium incubated in air. This suggests that the anaerobes were capable of growth in the presence of 50 μM dissolved oxygen, about 25% of air saturation, presumably as a result of the presence of anaerobic microniches as discussed further below. This simulation proved to be very stable and has been used to demonstrate the resilience of the population to changes in pH and dissolved oxygen concentration. Over the pH range 5.0 to 9.0, the total numbers of bacteria remained stable, although the composition of the population changed. Microbial groups such as the lactic acid bacteria which were expected to disappear at high pH were in fact highly active under all of the conditions tested.

Discussion

The co-existence of aerobic and anaerobic microorganisms in an environment containing oxygen has been seen in many habitats, and has been explained by the existence of anaerobic microniches (Scott, Yarlett, Hillman, Williams, Williams and Lloyd, 1983), or by the ability of some strict anaerobes to withstand exposure to oxygen (Kiener and Leisinger, 1983; Patel, Roth and Agnew, 1984). The presence of methanogenic activity in the small intestine, together with the presence of significant amounts of oxygen, suggests that aerobic and anaerobic bacteria occupy separate niches within this region of the gut. Aerobic/anaerobic interfaces occur in many habitats, but have been most extensively studied in soils and sediments. Here, conditions can change from aerated to anaerobic within several millimetres (Frenzel, Thebrath and Conrad, 1990). The flow of digesta would disrupt any such interface in the gut, but the richly oxygenated blood supply to the gut wall is likely to be a source of oxygen for the gut contents in the vicinity of the wall, whereas the central luminal region of the gut containing digesta particles, could be anaerobic. Stable co-cultures of the aerobic bacterium *Pseudomonas* with the anaerobe *Veillonella alcalescens* were maintained in chemostat culture by Gerritse, Schut and Gottschal (1990) over a range of oxygen supply rates from around 0.4 to 2.4 mol/l/h. Increasing the oxygen supply rate was accompanied by increased growth of *Pseudomonas*, but the oxygen concentration remained below 0.1 μM, and the growth of *Veillonella* was little affected by increasing the oxygen supply. A mixed culture from a fresh water sediment, containing aerobic, fermentative and methanogenic bacteria, continued to produce methane in the presence of oxygen providing the oxygen supply rate was increased slowly, allowing the population to adapt. Low rates of oxygen supply actually increased methane production, presumably as a consequence of increased supply of (unnamed) growth factors from the aerobic population to the methanogens. In mixed culture therefore, some aerobes and methanogens can co-exist without the provision of obvious microniches. However, there are as yet no known reports of the survival in co-culture of methanogenic and methane-oxidizing bacteria in stirred culture: it may be that such co-existence requires the presence of microniches such as food particles. If the food particles are the niche for the fermentative bacteria and methanogens, then the distribution of these bacteria on and within these particles is likely to determine whether or not hydrogen diffuses from these populations, or is largely consumed by methanogens and acetogens within the anaerobic population (Schink, 1992). A scheme, illustrating diagrammatically how fermentative bacteria, acetogens methanogens and methane oxidizing bacteria might be located in aerobic and anaerobic microniches in the gut, is illustrated in Figure 13.4.

The apparent stability of the bacterial population in the fermenter simulating conditions in the ileum is probably attributable to the presence within the system of microniches providing habitats for the different bacterial groups. The nutrient medium used contained particles of insoluble polysaccharides (Macfarlane *et al.*, 1989). The presence of particles would allow the formation of localized micro-environments. Thus, the inner part of these particles might possess a lower pH and dissolved oxygen content than the bulk of the medium, allowing the growth of anaerobic bacteria in an apparently aerated system. The presence of oxygen would favour the growth of facultative and aerobic micro-organisms within the liquid and at the surface of the particles. Within the particles, a range of microbial habitats from microaerophilic to anaerobic are likely to exist and as a result, the fermenter provides niches for a physiologically diverse range of bacteria. However, it is clear that there are limits to our ability to reproduce conditions in the gut. Thus, no satisfactory way of providing surfaces with specific receptors similar to those possessed by epithelial cell has yet been devised. In addition, since the bacterial competition for iron involves the removal of iron from host glycoproteins, modelling this process would require the addition of such iron chelators to the system.

The putative methane oxidizing bacteria from the pig gut have not yet been identified. Methane is oxidized by *Methylococcus* and *Methylomonas* species (Whittenbury and Krieg, 1984). These bacteria possess complex internal membrane structures which have not been observed in the cultures isolated here and it seems unlikely that the present isolates will prove to belong to the known species of methanotrophs. The use of enrichment processes often results in the isolation of stable consortia, and it can be difficult to assign functions to the components of such mixtures. For example, Barker (1940) isolated a culture (*Methanobacillus omelianski*) capable of producing acetate and methane from ethanol and bicarbonate. Over a quarter of a century later, this culture was shown to be a syntrophic association of two species of bacteria, an S organism that oxidizes ethanol to hydrogen and acetate, and a methanogen that uses this hydrogen to reduce carbon dioxide to methane (Bryant, Wolin, Wolin and Wolfe, 1967). Interestingly, growth of the S organism is very poor in pure culture because the accumulation of hydrogen inhibits the oxidative reaction that provides energy for its growth.

Apart from the possible oxidation of reduced fermentation products, the presence of oxygen in the piglet gut raises other possibilities. It may be that some carbohydrates of dietary or endogenous origin are oxidized to carbon dioxide and water by bacteria growing aerobically. Although it has been shown (Zhu *et al.*, 1993) that the recovery of hydrogen from the fermentation of fibre in the pig gut is low, the fate of the 'missing' hydrogen might be clarified if a carbon balance were available for microbial activity in the pig gut. Such calculations might be possible

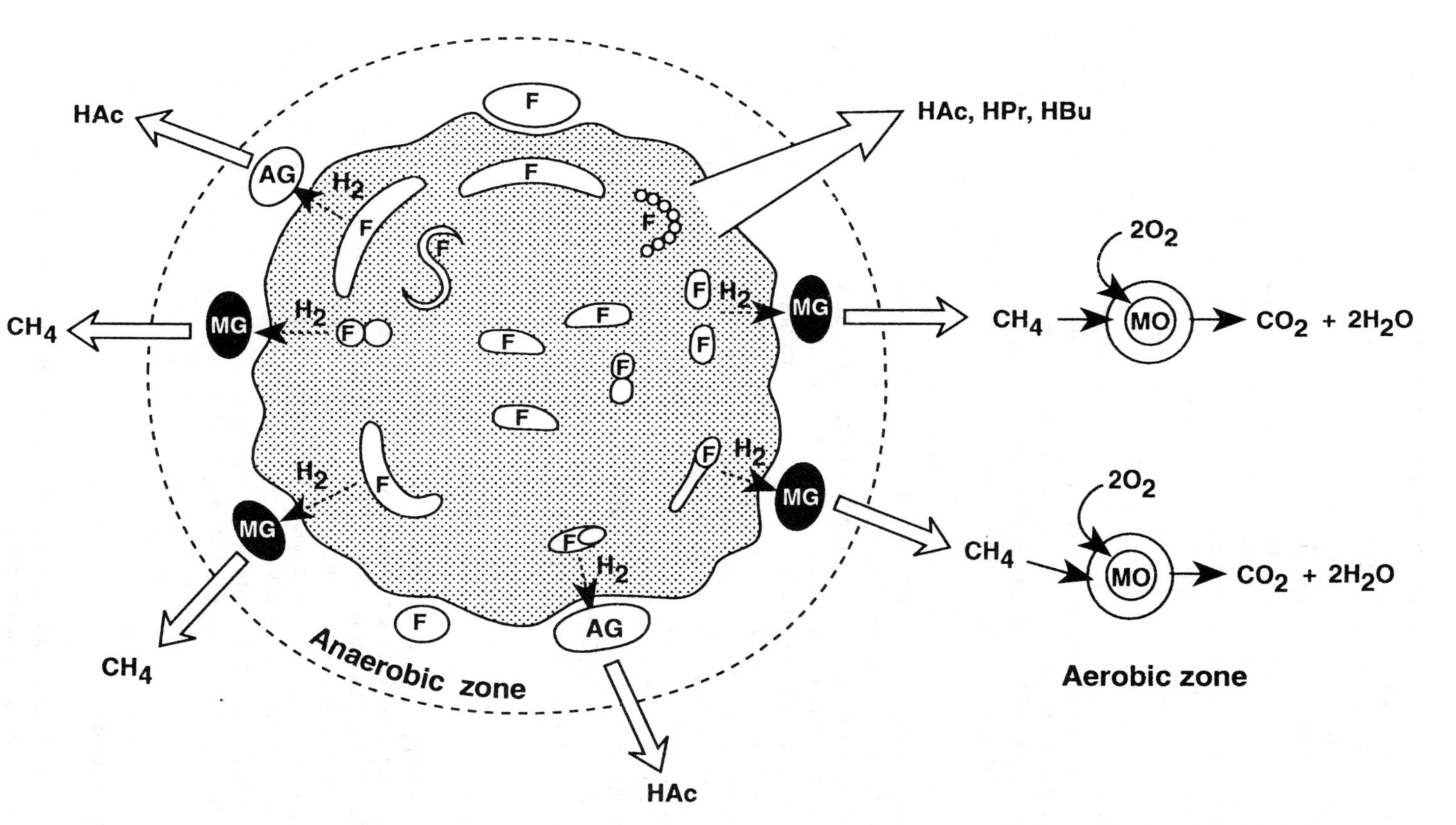

Figure 13.4 Schematic diagram illustrating possible interactions between anaerobic and aerobic bacteria in the pig gut. Feed particles (shaded area) are colonized by fermentative bacteria (open symbols labelled F) which produce acetate (HAc) propionate (HPr) and butyrate (HBu), CO_2 and H_2. Interspecies hydrogen transfer occurs to methane (CH_4) producers (MG) and also to acetogenic bacteria (AG). It is hypothesized that some of this methane may diffuse to oxygen-containing niches where it is oxidized by methane oxidizing bacteria (MO)

if realistic simulations of events in the gut can be reproduced in the fermenter systems referred to here.

Although some bacteria such as *Paracoccus denitrificans* are capable of dissimilatory reduction of nitrate to nitrogen gas under anaerobic conditions (Kocur, 1984), this reaction does not appear to occur in the rumen, the gastro-intestinal habitat which has been most widely investigated. This is thought to be due to the presence of excess carbon in the form of fermentable substrates, and a comparative shortage of suitable electron acceptors in this ecosystem (Tiedje, Sexstone, Myrold and Robinson, 1982). Nitrate is largely converted to ammonia in the rumen, and this reaction can compete with methanogenesis for reducing equivalents (Allison and Macfarlane, 1988). It has previously been assumed that the conditions in the gut of pigs and other monogastric animals would be broadly similar to those in the rumen. However, the availability of oxygen might allow ammonia oxidizing bacteria to survive in the pig gut, particularly in those regions in which the high pH would favour the formation of ammonia from ammonium ions. There exists a wide range of bacterial reactions involving the use of molecular oxygen (Gottschalk, 1979) and there is a clear need for a reassessment of bacterial metabolism in the gut in view of the findings reported here.

Interference with bacterial attachment to surface oligosaccharide receptors has been a key focus for the development of probiotic strategies to prevent enteric disease. Competitive exclusion of enteropathogens from the mucosal surface through the application of innocuous strains of bacteria has proven difficult largely because of the inappropriate match of probiotics and pathogens with regard to their receptor specificities and also the problems of introducing new strains of bacteria into the complex microcosm of the intestine. Research on the molecular structure of receptors for pathogens such as *E. coli* should circumvent these problems and lead to new strategies involving the application of chemical probiotics in the form of dietary and synthetic oligosaccharides and bacterial and dietary lectins. The temporal and spatial targeting of these largely natural food additives will be facilitated by further study of age-related and diet-induced intestinal glycosylation changes in pigs.

Conclusions

The gut is a complex habitat in which micro-organisms varying in their utilization of and sensitivity to oxygen coexist, presumably as a consequence of the presence of specialized microhabitats. The use of live probiotics involves deliberate modification of the ecology of the gut, and this is not a simple task. However, the prospects for the use of bacteriocins, oligosaccharides, lectins or other agents to prevent colonization by pathogens is considerable. The factors produced by *Lactobacilli* that prevent binding of *E. coli* to epithelial cells (Coconnier *et al.*, 1992; Blomberg, Henriksson and Conway, 1993) are of particular interest in this regard.

Acknowledgements

We thank A.J. Richardson for permission to include his data on methane production by samples from the pig gut.

References

Abrams, G.D., Bauer, H. and Sprinz, H. (1963) *Laboratory Investigation*, **12**, 355–364

Adachi, S. (1992) In *The Lactic Acid Bacteria in Health and Disease, Volume 1*, pp. 233–262. Edited by B.J.B. Wood. London: Elsevier Applied Science

Adler, H., Mural, R. and Suttle, B. (1992) *Journal of Bacteriology*, **174**, 2072–2077

Alexander, M. (1971) *Microbial Ecology*. New York: J. Wiley and Sons

Allison, C. and Macfarlane, G.T. (1988) *Journal of General Microbiology*, **134**, 1397–1405

Andersson, B., Porras, O. and Hanson, L.A. (1986) *The Journal of Infectious Diseases*, **153**, 232–237

Avgustin, G., Flint, H.J. and Whitehead, T.R. (1992) *FEMS Microbiology Letters*, (in press)

Axelsson, L. and Lindgren, S. (1987) *Journal of Applied Bacteriology*, **62**, 433–440

Banks, J.G., Board, R.G. and Sparks, N.H.C. (1986) *Biotechnology and Applied Biochemistry*, **8**, 103–147

Barker, H.A. (1940) *Antonie van Leeuwenhoek*, **6**, 201–220

Barrow, P.W. (1992) In *Probiotics*, pp. 225–259. Edited by R. Fuller. London: Chapman and Hall Beerens, H. (1990) *Letters in Applied Microbiology*, **11**, 155–157

Bezkorovainy, A. (1989) In *Biochemistry and Physiology of* Bifidobacteria, pp. 147–176. Edited by A. Bezkorovainy and R.M. Catchpole. Boca Raton: CRC Press

Bijlsma, I.G.W. and Bouw, J. (1987) *Veterinary Research Communications*, **11**, 509–518

Blomberg, L., Henriksson, A. and Conway, P.L. (1993) *Applied and Environmental Microbiology*, **59**, 34–39

Brock, T.D. (1966) *Principles of Microbial Ecology*. New Jersey: Prentice Hall

Bruyneel, B., Vande Woestyne, M. and Verstraete, W. (1990) *Antonie van Leeuwenhoek*, **57**, 119–122

Bryant, M.P. (1972) *American Journal of Clinical Nutrition*, **25**, 1324–1328

Bryant, M.P., Wolin, E.A., Wolin, M.J. and Wolfe, R.S. (1967) *Archive fur Mikrobiologie*, **59**, 20–31

Chauviere, G., Coconnier, M.H., Kerneis, S., Fourniat J. and Servin, A.L. (1992) *Journal of General Microbiology*, **138**, 1–7

Chase, D.G. and Erlandson, S.L. (1976) *Journal of Bacteriology*, **127**, 572–583

Chesson, A., Richardson, A.J. and Robertson, J.A. (1985) *Beretning fra Statens Husdyrbrugsforsog*, **580**, 272–275

Coconnier, M.H., Klaenhammer, T.R., Kerneis, S., Bernet, M.F. and Servin, A.L. (1992) *Applied and Environmental Microbiology*, **58**, 2034–2039

Condon, S. (1987) *FEMS Microbiology Reviews*, **46**, 269–280

Conway, P.L., Welin, A. and Cohen, P.S. (1990) *Infection and Immunity*, **58**, 3178–3182

Corthier, G., Dubos, F. and Raibaud, P. (1985) *Applied and Environmental Microbiology*, **49**, 250–252

Dean, E.A. (1989) *Microecology and Therapy*, **19**, 101–104

Dean, E.A. (1990) *Infection and Immunity*, **58**, 4030–4035

Erickson, A.K., Willgohs, J.A, McFarland, S.Y., Benfield, D.A. and Francis, D.H. (1992) *Infection and Immunity*, **60**, 983–988

Fernandes, C.F., Chandan, R.C. and Shahani, K.M. (1992) In *The Lactic Acid Bacteria in Health and Disease, Volume 1*, pp. 297–342. Edited by B.J.B. Wood. London: Elsevier Applied Science
Frenzel, P., Thebrath, B. and Conrad, R. (1990) *FEMS Microbiology Ecology*, 73, 149–158
Freter, R. (1988) In *Virulence Mechanisms of Bacterial Pathogens*, pp. 45–60. Edited by J.A. Roth. Washington: American Society for Microbiology
Freter, R., Brickner, H., Botney, M., Cleven, D. and Aranki, A. (1983) *Infection and Immunity*, **39**, 676–685
Fuller, R. (1992) In *Probiotics*, pp. 1–8. Edited by R. Fuller. London: Chapman and Hall
Fuller, R., Barrow, P.A. and Brooker, B.E. (1978) *Applied and Environmental Microbiology*, **35**, 582–591
Gerritse, J., Schut, F. and Gottschal, J.C. (1990) *FEMS Microbiology Letters*, **66**, 87–94
Gibbons, R.A., Jones, G.W. and Sellwood, R. (1975) *Journal of General Microbiology*, **86**, 228–240
Gottschalk, G. (1979) *Bacterial Metabolism*. New York: Springer-Verlag
Griffiths, E. (1991) *Biology of Metals*, **4**, 7–13
Guiot, H.F.L. (1982) *Infection and Immunity*, **38**, 887–892
Gyles, C.L. (1992) *Canadian Journal of Microbiology*, **38**, 734–746
Harris, D.L. and Kinyon, J.M. (1974) *American Journal of Clinical Nutrition*, **27**, 1297–1304
Havenaar, R., Ten Brink, B. and Huis in't Veld, H.J. (1992) In *The Lactic Acid Bacteria in Health and Disease, Volume 1*, pp. 151–170. Edited by B.J.B. Wood. London: Elsevier Applied Sciences
Hentges, D.J. (1992) In *Probiotics*, pp. 87–110. Edited by R. Fuller. London, Chapman and Hall
Hespell, R.B. and Whitehead, T.R. (1990) *Journal of Dairy Science*, **93**, 3013–3022
Hidaka, H., Eida, T. and Hamaya, T. (1986) *European Patent Application 85109590.1*, publication number 0 171 026
Hillman, K., Lloyd, D. and Williams, A.G. (1985) *Current Microbiology*, **12**, 335–312
Hillman, K., Whyte, A. and Stewart. C.S. (1993) *Letters in Applied Microbiology*, in press
Hobson, P.N. (1969) In *Methods in Microbiology, Volume 3B*, pp. 133–149. Edited by J.R. Norris and D.W. Ribbons. London; Academic Press
Holland, R.E. (1990) *Clinical Microbiology Reviews*, **3**, 345–375
Houghton, S.B. and Fuller, R. (1980) *Applied and Environmental Microbiology*, **39**, 1054–1058
Isaacson, R.E. and Kinsel, M. (1992) *Infection and Immunity*, **60**, 3193–3200
Jonsson, E. and Conway, P. (1992) In *Probiotics*, pp. 259–316. Edited by R. Fuller. London: Chapman and Hall
Kelly, D. and King, T.P. (1991) *Histochemical Journal*, **23**, 55–60
Kiener, A. and Leisinger, T. (1983) *Systematic and Applied Microbiology*, **4**, 305–312
Kikuchi, H.E. and Suzuki, T. (1986) *Applied and Environmental Microbiology*, **52**, 971–973
King, T.P. and Kelly, D. (1991) *Histochemical Journal*, **23**, 43–54

King, T.P., Begbie, R., Spencer, R. and Kelly, D. (1992) In *Proceedings of the Nutrition Society*, (in press)
Kocur, M. (1984) In *Bergey's Manual of Systematic Bacteriology, Volume 1*, pp. 399–402. Edited by N.R. Krieg. Baltimore: Williams and Wilkins
Krogfelt, K.A. (1991) *Reviews of Infectious Diseases*, **13**, 721–735
Latham, M.J. (1979) In *Microbial Ecology A Conceptual Approach*, pp. 115–137. Edited by J.M. Lynch and N.J. Poole. Oxford: Blackwell Scientific Publications
Lee, A. (1985) *Advances in Microbial Ecology*, **8**, 115–162
Lindahl, M., Brossmer, R. and Wadstrom, T. (1987), *Glycoconjugate Journal*, **4**, 51–58
Lindahl, M., Brossmer, R. and Wadstrom, T. (1988) In *The Molecular Immunology of Complex Carbohydrates*, pp. 123–152. Edited by A.M. Wu and L.G. Adams. New York, London: Plenum Press
Lindahl, M. and Carlstedt, I. (1990) *Journal of General Microbiology*, **136**, 1609–1614
Loesche, W.J. (1969) *Applied Microbiology*, **18**, 723–727
Mannarelli, B.M., Ericsson, L.D., Lee, D. and Stack, R.J. (1991) *Journal of Bacteriology*, **173**, 2975–2980
Marcelis, J.H., den Daas-Slagt, H.H. and Hoogkamp-Korstanje, J.A.A. (978) *Antonie van Leeuwenhoek*, **44**, 257–267
Maxwell, F. and Stewart, C.S. (1992) In *Les Bacteries Lactiques*, pp. 142. Caen: Adria Normandie
McCarthy, R.E. and Salyers, A.A. (1986) *Applied and Environmental Microbiology*, **52**, 9–16
McCord, J.M., Keele, B.B. and Fridovich, I. (1971) In *Proceedings of the National Academy of Sciences of the United States of America*, **68**, 1024
McFarlane, G.T., Hay, S. and Gibson, G.R. (1989) *Journal of Applied Bacteriology*, **66**, 407–417
Metcalfe, J.W., Krogfelt, K.A., Krivan, H.C., Cohen, P.S. and Laux, D.C. (1991) *Infection and Immunity*, **59**, 91–96
Millard, P. and Chesson, A. (1984) *British Journal of Nutrition*, **52**, 583–594
Miller, T.L. and Wolin, M.J. (1979) *American Journal of Clinical Nutrition*, **32**, 164–172
Miller, T.L. and Wolin, M.J. (1981) *Applied and Environmental Microbiology*, **42**, 400–407
Miller, T.L., Wolin, M.J. and Kusel, E.A. (1986) *Systematic and Applied Microbiology*, **8**, 234–238
Mitsuoka, T. (1992) In *The Lactic Acid Bacteria in Health and Disease*, pp. 69–114. Edited by B.J.B. Wood. London: Elsevier Applied Sciences
Molitoris, E., Krichevski, M.I., Fagerberg, D.J. and Quarles, C.L. (1986) *Journal of Applied Bacteriology*, **60**, 111–120
Monsma, D.J., Vollendorf, N.W. and Marlett, J.A. (1992) *Applied and Environmental Microbiology*, **58**, 3330–3336
Morris, J., Thorns, C., Wells, G., Scott, A. and Sojka, W. (1983) *Journal of General Microbiology*, **129**, 2753–2759
Morris, J.G. (1975) *Advances in Microbial Physiology*, **12**, 169–246
Mouricout, M.A. and Julien, R.A. (1987) *Infection and Immunity*, **55**, 1216–1223
Mouricout, M.A., Petit, J.M. Carias, J. R. and Julien, R. (1990) *Infection and Immunity*, **58**, 98–106

Mukai, T., Arihara, K. and Itoh, H. (1992) *FEMS Microbiology Letters*, **98**, 71–74
Mundt, J.O. (1986) In *Bergey's Manual of Systematic Bacteriology*, pp. 1063–1065. Edited by P.H.A. Sneath. Baltimore: Williams and Wilkins
Muting, D., Eschrich, W., and Mayer, J.B. (1968) *American Journal of Proctology*, **19**, 336
Nagy, B., Casey, T.A., Whipp, S.C. and Moon, H.W. (1992) *Infection and Immunity*, **60**, 1285–1294
Nakajima, Y., Nishio, K., Mizutani, T., Ogasa, K. and Kashimura, J. (1989) *UK Patent Application 8813095*, GB 2 206 582 A
Neilands, J.B. (1984) *Microbiological Sciences*, **1**, 9–14
Neilands, J.B (1992) *Canadian Journal of Microbiology*, **38**, 728–733
Patel, G.B., Roth, L.A. and Agnew, B.J. (1984) *Canadian Journal of Microbiology*, **30**, 228–235
Pedersen, K. and Tannock, G.W. (1989) *Applied and Environmental Microbiology*, **55**, 279–283
Prins, R.A. and Lankhorst, A. (1977) *FEMS Microbiology Letters*, **1**, 252–258
Pusztai, A., Grant, G., King, T.P., and Clarke, E.M.W. (1990) In *Recent Advances in Animal Nutrition — 1990*, pp. 47–60. Edited by W. Haresign and D.J.A. Cole. London: Butterworths
Raibaud, P. (1992) In *Probiotics*, pp. 9–28. Edited by R. Fuller. London: Chapman and Hall
Robinson, I.M., Allison, M.J. and Bucklin, J.A. (1981) *Applied and Environmental Microbiology*, **41**, 950–955
Robinson, I.M., Whipp, S.C., Bucklin, J.A. and Allison, M.J. (1984) *Applied and Environmental Microbiology*, **48**, 964–969
Robinson, J.A., Smolenski, W.J., Ogilvie, M.L. and Peters, J.P. (1989) *Applied and Environmental Microbiology*, **55**, 2460–2467
Rowland, I.R. (1992) In *Probiotics*, pp. 29–54. Edited by R. Fuller. London: Chapman and Hall
Rozee, K.R., Cooper, D., Lam, K. and Costerton, J.W. (1982) *Applied and Environmental Microbiology*, **43**, 1451–1463
Russell, E.G. (1979) *Applied and Environmental Microbiology*, **37**, 187–193
Salyers, A.A. (1984) *Annual Review of Microbiology*, **38**, 293–313
Salyers, A.A., Vercellotti, J.R., West, S.E.H. and Wilkins, T.D. (1977) *Applied and Environmental Microbiology*, **33**, 319–322
Savage, D.C. (1977) *Annual Review of Microbiology*, **31**, 107–133
Savage, D.C., Dubos, R. and Schaedler, R.W. (1968) *Journal of Experimental Medicine*, **127**, 67–76
Scardovi, V. (1986) In *Bergey's Manual of Systematic Bacteriology, Volume 2*, pp. 1418–1424. Edited by P.H.A. Sneath. Baltimore: Williams and Wilkins
Scardovi, V., Casalicchio, F. and Vincenzi, N. (1979) *International Journal of Systematic Bacteriology*, **29**, 312–327
Schauer, R. (1991) *Glycobiology*, **1**, 449–452
Schink, B. (1992) In *The Prokaryotes*, pp. 276–299. Edited by A. Balows *et al.* Berlin: Springer Verlag
Scott, R.I., Yarlett, N., Hillman, K., Williams, T.N., Williams, A.G. and Lloyd, D. (1983) *Journal of Applied Bacteriology*, **55**, 143–149
Seignole, D., Mouricout, M. Duval-Iflah, Y., Quintard, B. and Julien, R. (1991) *Journal of General Microbiology*, **137**, 1591–1601

Sellwood, R. (1984) In *Attachment of Organisms to The Gut Mucosa, Volume 2*, pp. 167–175. Edited by E.C. Boedecker. Boca, Raton, Fla.: CRC Press
Sharak-Genthner, B.R., Davis, C.L. and Bryant, M.P. (1981) *Applied and Environmental Microbiology*, **42**, 12–19
Sharpe, M.E., Latham, M.J., Garvie, E.I., Zirngibl, J. and Kandler, O. (1973) *Journal of General Microbiology*, **77**, 37–49
Smith, H.W. (1992). *Canadian Journal of Microbiology*, **38**, 747–752
Smith, H.W. and Jones, J.E.T. (1963) *Journal of Pathology and Bacteriology*, **86**, 387–412
Smith, H.W. and Huggins, M.B. (1983) *Journal of General Microbiology*, **129**, 2659–2675
Smith, P.H. and Hungate, R.E. (1958) *Journal of Bacteriology*, **75**, 713–718
Staley, T.E. and Wilson, I.B. (1983) *Molecular and Cellular Biochemistry*, **52**, 177–189
Strobel, H.J. and Russell, J.B. (1987) *Applied and Environmental Microbiology*, **53**, 2505–2510
Suwa, Y., Koga, K., Fujikawa, S., Okazaki, M., Irie, T., Nakada, T. (1988) *European Patent Application 87115999.2*, publication number 0 265 970 A2
Takeuchi, A., Jervis, H.R., Nakazawa, H. and Robinson, D.M. (1974) *American Journal of Clinical Nutrition*, **27**, 1287–1296
Tannock, G.W. (1992) In *The Lactic Acid Bacteria in Health and Disease, Volume 1*, pp. 21–48. Edited by B.J.B. Wood. London: Elsevier Applied Science
Tannock, G.W., Fuller, R. and Pedersen, K. (1990) *Applied and Environmental Microbiology*, **56**, 1310–1316
Teneberg, S., Willemsen, P., de Graff, F.K. and Karlsson, K.A. (1990) *FEBS Letters*, **263**, 10–14
Tiedje, J.M., Sexstone, A.J., Myrold, D.D. and Robinson, J.A. (1982) *Antonie van Leeuwenhoek*, **48**, 569–583
To, S. (1984) *Infection and Immunity*, **43**, 549–554
Tomioka, H. and Saito, H. (1992) In *The Lactic Acid Bacteria in Health and Disease, Volume 1*, pp. 263–296. Edited by B.J.B. Wood. London: Elsevier Applied Science
Uesugi, I. and Yamija, M. (1978) *Zeitschrift für Allgemeine Microbiologie,* **18,** 593–601.
Van Beelan, P., van der Hoeven, J.S., De Jong, M.H. and Hoogendorn, H. (1986) *FEMS Microbiology Ecology*, **38**, 25–30
Van der Waaij, D., Berghuis de Vries, J.M. and Lekkerkerk, J.E.C. (1971) *Journal of Hygiene (Cambridge)*, **69**, 405–441
Vandevoorde, L., Vande Woestyne, M., Bruyneel, B., Christiaens, H. and Verstraate, W. (1992) In *The Lactic Acid Bacteria in Health and Disease, Volume 1*, pp. 447–476. Edited by B.J.B. Wood. London: Chapman and Hall
Varel, V.H., Robinson, I.M. and Jung, H.J. (1987) *Applied and Environmental Microbiology*, **53**, 22–26
Vedros, N.A. (1984) *Neisseria*. In *Bergey's Manual of Systematic Bacteriology, Volume 1*, pp. 290–296. Edited by J.G. Holt and N.R. Kreig. Baltimore: Williams and Wilkins
Veilleux, B.G. and Rowland, I.R. (1981) *Journal of General Microbiology*, **123**, 103–115
Vervaeke, I.J., van Nevel, C.J., Decuypere, J.A. and van Assche, P.F. (1973) *Journal of Applied Bacteriology*, **36**, 397–405

Whittenbury, R. and Kreig, N.R. (1984) In *Bergey's Manual of Systematic Bacteriology, Vol 1*, pp. 256–261. Edited by N.R. Krieh and J.G. Holt. Baltimore: Williams and Wilkins

Wick, M.J., Madara, J.L., Fields, B.N. and Normark, S.J. (1991) *Cell*, **67**, 651–659

Wilde, E. and Schlegel, H.G. (1982) *Antonie van Leeuwenhoek*, **48**, 131–143

Wimpenny, J.W.T. and Samah, O.A. (1978) *Journal of General Microbiology*, **108**, 329–332

Zhu, J.Q., Fowler, V.R. and Fuller, M.F. (1993) *British Journal of Nutrition*, in press

14

NEW APPROACHES WITH PIG WEANER DIETS

G.G. PARTRIDGE
BP Nutrition (UK) Ltd, Wincham, Northwich, Cheshire, CW9 6DF, UK and
B.P. GILL *The Scottish Agricultural College, 581 King Street, Aberdeen, AB9 1UD, UK*

Introduction

Most weaning of pigs in the UK currently takes place between 3 and 5 weeks of age, with the majority of producers opting for 'early' weaning (19–25 days, Table 14.1). Weaning early to increase sow productivity has to be justified on the basis that any increase in sow output (i.e. pigs/sow/year) outweighs the cost of replacing what the sow is able to deliver naturally in the care of her piglets.

Table 14.1 TRENDS IN WEANING AGE 1983–1991

Age at Weaning (days)	*1983*	*1985*	*1987*	*1989*	*1991*
< 19	8	5	3	2	< 1
19–25	54	62	66	59	59
26–32	19	22	22	30	33
33–39	15	9	8	7	6
> 39	4	2	1	2	1

Courtesy of Meat and Livestock Commission (1992)

The emphasis on early weaning systems in recent years in the UK has favoured the development of a market place for relatively high cost, specialised weaner diets which, almost invariably, rely on a significant use of milk products and highly processed raw materials (e.g. denatured skimmed milk powder; whey powder and derivatives; specially processed fish meals; isolated vegetable proteins; cooked cereals etc.). Such diets are, of necessity, a compromise between what is nutritionally desirable for the young pig and what is economically justifiable to the pig farmer. Nutrition is not routinely regarded as a primary welfare issue but in the case of the young pig separated from the sow at weaning it assumes major importance.

The primary objective at weaning is to ensure that the transition from sow's milk, the ideal piglet food (Table 14.2), to post-weaning diet is as smooth as possible without compromising growth or predisposing the young animal to disease. A

Table 14.2 COMPOSITION OF SOW'S MILK AT AROUND 21 DAYS *POST-PARTUM*

	Composition
Dry matter (g/kg)	193.9
Principle constituents (g/kg DM)	
Crude protein	289.9
Fat	393.0
Lactose	271.6
Ash	45.6
Digestible energy (MJ/kg DM)	26.0
Amino acid content (g/kg DM)	
Lysine	22.0
Methionine + cystine	9.5
Threonine	12.0
Tryptophan	3.8
Fatty acid composition (%)	
Myristic C14:0	4.2
Palmitic C16:0	35.6
Linoleic C18:2	9.4
Minerals (g/kg DM)	
Calcium	10.9
Phosphorous	8.1
Sodium	2.5
Potassium	4.2
Vitamins	
Vitamin A (IU/kg DM)	5000.0
Vitamin E (mg/kg DM)	2.8
Vitamin C (mg/kg DM)	440.0
Riboflavin (mg/kg DM)	11.0

Note: This is not an exhaustive list. Composition can vary according to genotype, nutrition, health and environment. Compiled from various sources including: Braude, Coates, Henry, Kon, Rowland, Thompson and Walker (1947), Elsley (1970), Elliott, van der Noot, Gilbreath and Fisher (1971), Fahmy (1972) and Agricultural Research Council (1981)

number of nutritional strategies can be used in pursuing this objective and the major ones will be examined in this review.

The weaning process — gut function and health

In an ideal world of pig production, weaning would be a gradual process rather than an abrupt event in the young animal's life. Unfortunately the constraints imposed by many management systems (e.g. financial, nutritional and environmental) can result in poor health and performance in the immediate post-weaning period. Various explanations and hypotheses have been forwarded to account for the variable growth check and loss of health experienced by many piglets at weaning and some have been summarised in Table 14.3.

It is beyond the scope of this chapter to examine all of these aspects but those which are of current interest and are directly relevant to the health and nutritional management of the newly weaned piglet are briefly discussed.

Table 14.3 PRINCIPAL FACTORS ASSUMED TO BE ASSOCIATED WITH THE GROWTH CHECK EXPERIENCED BY YOUNG, NEWLY WEANED PIGS, GIVEN AN ADEQUATE THERMAL ENVIRONMENT AND LOW DISEASE CHALLENGE

1. Insufficient digestive enzymes for new food substrates
2. Reduced absorptive capacity due to changes in villus architecture
3. Poorly developed gastric secretions
4. Removal of beneficial factors present in sows' milk (e.g. natural antibacterials, immunoglobulins)
5. Inadequate feed intake
6. Form of post-weaning diet offered, dry meal/pellet cf. liquid (sows' milk)
7. Environmental changes, 'stress'

MATERNAL FACTORS

Sow's milk contains important protective maternal antibodies which are lost to the piglet at weaning. The major immunoglobulin component of sow's milk is IgA (Husband and Bennell, 1980) and between 3 and 4 weeks of age the sucking piglet would be receiving about 1.6 g IgA/day (Svendsen and Brown, 1973). Unlike colostral antibodies, IgA and other milk immunoglobulins are not absorbed from the gut and thus form a continuous defence along the lumen against infectious organisms to which the sow has developed resistance (Svendsen and Larsen, 1977; Kidder 1982). This protection is essential while the piglet's own production of IgA remains undeveloped, that is from birth to about 6 weeks of age (Svendsen and Brown, 1973).

The literature therefore suggests that piglets weaned between 3 and 4 weeks of age lose the valuable protection of maternal antibodies at a time when their own immature immune system cannot resist a significant disease challenge. Furthermore susceptibility to disease may be increased since the immune status of the piglet is compromised by an immunosuppressive reaction to weaning (Coenen and Kruse, 1986).

CHANGES IN DIGESTIVE ENZYMES AND GUT MORPHOLOGY

Several studies have shown a consistent depression in levels of pancreatic and intestinal enzymes in the newly-weaned piglet (e.g. Kidder and Manners, 1980; Shields, Ekstrom and Mahan, 1980; Hampson and Kidder, 1986; Lindemann, Cornelius, El Kandelgy, Moser and Pettigrew, 1986).

Some observations, however, would have been negatively influenced by poor or variable feed intake in the pre-weaning and immediate post-weaning period. The technique of gavage feeding (gastric intubation) has allowed the effects of continuous supplies of dietary substrate on enzyme induction to be more critically examined (e.g. Kelly, Smyth and McCracken, 1990; 1991a; b). Such studies indicate that high levels of creep feed intake can induce marked increases in specific and total enzyme activities at weaning (Kelly *et al.*, 1990). In the post- weaning period high feed intakes (via gastric intubation) also stimulated gut development and enzyme levels in the small intestine. Indeed it appears numerically that the small intestine of the newly weaned animal still has spare capacity in terms of enzyme complement, to deal with cereal-based weaner diets (Kelly *et al.*, 1991b).

Creep feed intakes in animals weaned at 3 to 4 weeks of age are notoriously variable both within and between litters (Table 14.4). Factors such as weaning age;

Table 14.4 MEASUREMENTS OF CREEP FEED INTAKE WHEN WEANING TOOK PLACE AT 3 OR 4 WEEKS OF AGE (VARIOUS STUDIES)

Period that creep feed was offered	*Average creep feed intake/pig before weaning (g) or range where individual intakes were measured*	*Type of diet*	*Source*
7–21	93	Highly digestible skimmed milk/ cooked cereals	Partridge (1988b, unpublished observations)
10–21	71	'Complex'; including milk powders/cooked starches	Okai, Aherne and Hardin (1976)
	23	'Semi-complex'	
	13	'Simple'; wheat/soya	
10–28	13–194	Maize/soya/whey	Barnett *et al.* (1989)
7–28	600	Highly digestible skimmed milk/cooked cereals	English, Robb and Dias (1980)
21–28	450	Maize/soya/whey	Appleby, Pajor and Fraser (1992)
10–28	13–1911	Maize/soya/whey	Pajor, Fraser and Kramer (1991)
14–28	150	'Complex', skimmed milk powder, 25% soya	Aherne, Danielsen and Nielsen (1982)

sow milk output (i.e. drive of the litter to eat); palatability and digestibility of the creep feed offered; wet or dry feeding; form of feed (i.e. meal or pellets); access to the creep (e.g. floor fed or in hoppers) are all likely to influence the quantities consumed and thereby the degree of induction of digestive enzymes.

One consistent effect on digestive physiology in the immediate post-weaning period is the abrupt change in villus morphology, from long slender villi characteristic of the suckling animal to leaf-shaped villi with varying degrees of atrophy (Miller, James, Smith and Bourne, 1986; Cera, Mahan, Cross, Reinhart and Whitmoyer, 1988). Villus height is often halved and crypt depth doubled in the newly weaned pig, a situation which results in immature enterocytes reaching the villus tip where they show a reduced capacity to express enzyme activity (Miller *et al.*, 1986). Continuous supply of dietary substrate via gavage feeding can reduce, but does not completely prevent, these changes in villus height and crypt depth occurring (Kelly *et al.*, 1991a).

In practical pig nutrition it is frequently advised that lightweight pigs at weaning should ideally be fed a more digestible ration than their 'normal' litter mates. This rationale is supported by studies looking at digestive enzyme activities where chronological age was a poor indicator of physiological maturity. Heavier pigs at weaning tended to have a more developed gastrointestinal tract than their smaller litter mates (de Passillé, Pelletier, Menard and Morisset, 1989).

DIETARY ANTIGENS

It has been proposed that many of the morphological changes that take place in the young pigs' gut in the post-weaning period are a function of a transient

Table 14.5 EFFECT OF DIFFERENT SOYA PRODUCTS ON IMMUNE RESPONSE, ABSORPTION (XYLOSE) AND POST-WEANING GROWTH RATE IN YOUNG PIGS[a] (WEANING AGE 21 DAYS)

	Milk protein	*Soyabean meal*	*Soya protein concentrate I*	*Soya protein concentrate II*	*Extruded soya protein concentrate*
Residual antigens in products					
Glycinin	—	2.4	0.9	0.5	0.8
β-*Conglycinin*	—	3.6	1.5	1.2	1.3
Villus height (μm)	364	234	309	280	319
Crypt depth (μm)	198	222	214	190	196
Skinfold thickness (mm)	0.82	3.33	2.65	2.59	2.50
Xylose absorption (mg/100 ml)	0.82	0.42	0.61	0.78	0.67
Coliforms (% total bacteria)	2	37	24	23	4
Postweaning growth rate (14 d, g/d)	326	182	208	211	227

[a] Pigs were orally 'sensitised' to these products before weaning. After Li *et al.* (1991b)

hypersensitivity to antigenic components present in the diet (e.g. Miller, Newby, Stokes and Bourne 1984). Soya antigens have particularly been implicated. The theory proposed that immunopathological damage to the intestine would vary in severity between abruptly weaned, 'sensitised' (low creep feed intakes) and immunologically 'tolerant' animals (high or 'adequate' levels of creep feed intake). The last group would be least likely to incur immunologically mediated gut damage.

A number of recent studies have examined this hypothesis. Kelly *et al.* (1990) and Hampson *et al.* (1988) both failed to reproduce the hypersensitivity response when precisely controlling 'sensitising' and tolerising' levels of creep food intake by gavage feeding. Similarly Sarmiento, Runnels and Moon (1990) working with a K88 (positive) *Escherichia coli* challenge model failed to induce morphological changes in the small intestine which were characteristic of an allergic response. Heppell, Sissons and Banks (1989) comparing the immune response stimulated by soya protein in both calves and pigs concluded that the animals' differed considerably in their ability to develop oral tolerance. Calves were particularly prone to developing adverse immune responses to dietary antigens, i.e. they failed to develop oral tolerance. In contrast piglets quickly became hyporesponsive to injections of soya protein.

In contrast Li, Nelssen, Reddy, Blecha, Klemm and Goodband (1991a) 'sensitised' young pigs prior to weaning with dried skim milk, or a variety of soya products, and found a negative linear correlation between skin-fold thickness (cutaneous hypersensitivity response measured six days post weaning) and average daily gain in the post-weaning period (Table 14.5). More highly processed soya products (e.g. soya protein concentrate) elicited smaller changes in skinfold thickness and showed less detrimental effects on villus morphology than standard soya bean meal. Milk protein based diets (soya free) supported the highest weight gains post-weaning and maintained a superior gut structure when compared with all the soya products (Table 14.5). The authors proposed that enzyme-linked immunosorbent assay (ELISA) could be used as a routine screening process to determine levels of potentially antigenic protein in soya bean products prior to their inclusion in young pig diets.

Although the antigenic response to feed proteins (e.g. soya) in the young pig remains equivocal it is apparent that feed processing has a key role to play in young animal response and this facet is examined in more detail later in this review.

Nutritional approaches for enhancing health and welfare

Nutritionists have a considerable responsibility for safeguarding the welfare of the newly weaned piglet by constructing diets which will not stress the physiology of the immature gut and therefore compromise health. In practice, options for producing the 'ideal' baby pig diet are constrained by cost and the nutritionist must explore alternative strategies that will maintain piglets in satisfactory health. This can be a major task since the nutritional defences must be sufficiently sound to replace the protective maternal factors lost at weaning.

Sometimes good nutritional practice in itself is inadequate when the piglet is exposed to undue levels of stress and disease because of poor housing, hygiene and husbandry standards. The feed manufacturer then has to include antibiotics at levels which have been specifically prescribed for a disease situation or rely on routine additions of antibacterial feed additives at the growth promotion level.

Although antibiotics remain the most reliable option for controlling diarrhoea and associated loss in growth following weaning, increasing consumer resistance to the use of these products has encouraged alternative approaches for enhancing health. The efficacy of alternative approaches must be completely evaluated before the complete withdrawal of antibiotics from feeds is considered. For example, in Sweden the incidence of post-weaning diarrhoea increased after 1986 when the general use of antibiotics in feed was banned (Holmgren, Franklin, Wallgren, Bergstrom, Martinsson and Rabbe, 1990).

DIET ACIDIFICATION

The acidification of diets and drinking water for weaned piglets has received much interest over several years. Current research effort has focused on the efficacy of adding organic acids to diets with a view to enhancing health and performance following weaning. The literature has been recently summarised in a review paper by Easter (1988) who concluded that the weight of evidence is marginally in favour of adding organic acids.

The rationale behind diet acidification seems very sound. That is the maintenance of an acidic environment in the stomach by correcting any increase in pH resulting from a reduction in gastric acid output at weaning and from the acid binding effects of feed ingredients. Acidification would thus stimulate proteolytic activity, assisting the digestion of non-milk proteins and restricting the growth of undesirable acid intolerant bacteria such as pathogenic coliforms. The nutritionist may aim to influence pH beyond the stomach with the hope of protecting the entire tract from harmful bacteria.

Overall, the response in growth and performance to acid supplemented diets is very variable and on average is not of the magnitude expected from what appears to be good nutritional strategy. Furthermore, any beneficial outcome in gut health may be due to more subtle effects than the usual explanation of a lowered gut pH.

There are few studies which have attempted to correlate changes in the digestive tract in response to diet acidification. Recently, Risley, Kornegay, Lindemann, Wood and Eigel (1992), were unable to detect any changes in pH or in the population of lactobacilli and *E. coli* along the entire digestive tract in weaned piglets offered either 15 g/kg fumaric or 15 g/kg citric acid (Table 14.6). On the other hand Burnell, Cromwell and Stahly (1988) decreased digesta pH in the stomach and small and large intestine of weaned piglets fed diets containing 10 g/kg citric acid. The organic acid supplement included the probiotic organisms *Lactobacillus casei* and *Streptococcus faecium* thus confounding evaluation of the independent effects of either.

In the artificially reared piglet, White, Wenham, Sharman, Jones, Rattray and McDonald (1969) were able to reduce gastric pH by 1 unit using a milk-based liquid diet reduced to a pH of 4.8 with lactic acid. A corresponding reduction in haemolytic coliform count and duration of scour was observed. Similar effects have been reported by Thomlinson and Lawrence (1981) in which 1% additions of lactic acid to the drinking water of weaned piglets reduced gastric pH, delayed the multiplication of enterotoxigenic *E. coli* and lowered mortality rate. Earlier studies with lactic acid treated drinking water have also shown favourable responses in

Table 14.6 EFFECT OF ORGANIC ACIDS[a] ON pH, *LACTOBACILLUS* AND *ESCHERICHIA COLI* COUNTS[b] OF THE INTESTINAL CONTENTS OF WEANED PIGLETS

	Stomach	*Jejunum*	*Colon*	*Lower colon*
pH				
Control	4.07	6.76	6.36	7.06
Fumaric acid	3.87	6.42	6.16	6.89
Citric acid	3.82	6.69	6.19	6.93
Lactobacillus				
Control	7.5	7.4	8.9	8.9
Fumaric acid	7.7	7.6	9.0	9.0
Citric acid	8.0	7.9	9.2	9.2
Escherichia coli				
Control	5.4	6.0	7.0	7.3
Fumaric acid	6.0	6.4	7.0	7.4
Citric acid	6.0	5.7	6.8	7.2

[a] Organic acids used at the rate of 15 g/kg diet. Note diets also contained 5.8 to 6.0 g/kg limestone; [b] Microbial counts expressed in $\log_{10}$ of the cultural count/g of wet content. Courtesy of Risely *et al.* (1992)

terms of coliform numbers and subsequent health and performance (Kershaw, Luscombe and Cole, 1966; Cole, Beal and Luscombe, 1968).

The use of water as a vehicle for delivering organic acids, in particular lactic acid, appears to have given more consistent advantages than diet acidification. Unfortunately the two methods of delivery have not been evaluated within a single study. Because the volume of water intake by weaned piglets is about 2.5 fold higher than feed intake and the ratio of water to feed intake is about 10% higher in the first than in subsequent weeks after weaning (Gill, 1989), water acidification provides a greater scope for delivering a higher and a more regular dose of acid to the piglet. Responses to diet supplementation between different studies may be confounded by differences in diet formulation, in particular the inclusion of limestone which has a very high acid neutralising capacity.

It is not entirely clear how organic acids enhance gut health. A number of explanations have been put forward including those concerning gut pH and its effects on reducing coliform counts. Lactic acid, however, may have a specific bacteriostatic effect against *E. coli* which is independent of any change in pH. For example, in laboratory cultures, the lactate anion has been shown to suppress the growth of *Staphylococcus aureus* in the presence of buffers which maintained a neutral pH (Seaman and Woodbine, 1977).

Additionally, lactic acid may create a gut milieu favourable for the growth of lactobacillus. These beneficial micro-organisms are considered to restrict the colonisation of the gut by pathogenic *E. coli*, which is the basis of the probiotic concept. A symbiotic relationship between lactobacillus and the piglet already exits before weaning. In the act of suckling, the piglet delivers a regular supply of lactose to lactobacillus. This is readily fermented to lactic acid, which may be sufficient to compensate for the deficiency in hydrochloric acid production by immature gastric secretory cells.

Lastly, organic and inorganic acids may help to preserve freshness of feed and partly cover for any deficiencies in the hygiene of the feeding environment.

FERMENTABLE SUBSTRATES

The fermentation of lactose to lactic acid in the digestive tract of the sucking piglet has been widely reported and its importance to milk digestion and health have been briefly discussed. The application of this concept to enhance health after weaning are of current interest. As well as lactose, the benefits of including feed ingredients rich in fermentable non-starch polysaccharides are receiving increasing attention.

Lactose

The mode of action by which lactose exerts a beneficial effect on piglet gut health has been briefly considered in this paper. In the immediate pre-weaning period, the intake of lactose at three weeks will average 40 g/day. Although much of this will be readily hydrolysed by mucosal lactase in the small intestine, a significant amount will be utilised by beneficial bacteria along the entire length of the digestive tract. Lactose differs from other sugars in that it is utilised by a far smaller range of bacteria. In particular it is the most appropriate nutrient for lactobacilli (Kidder, 1982). The nutritional significance of lactose to lactobacilli was demonstrated in a study by Pollmann, Danielson and Peo (1980). It was found that early weaned piglets inoculated with a preparation containing viable *Lactobacillus acidophilus* had higher growth rates when their diets were supplemented with lactose (Table 14.7).

Table 14.7 THE EFFECT OF DIETARY LACTOSE SUPPLEMENTATION ON THE GROWTH (g/day) OF WEANED[a] PIGLETS INOCULATED WITH *L. ACIDOPHILUS*

Lactose (g/kg)	*Inoculum* Water	L. acidophilus[b]
0	195	176
100	166	225

[a] Average initial weight 6.0 kg. Duration of study was 28 days; [b] *L. acidophilus* culture (2 × 10^{11} organisms/ml); 10 ml given daily by stomach tube for 2 weeks after weaning.
Courtesy of Pollmann *et al.* (1980)

Following weaning, the intake of lactose from a starter diet will depend on the level of inclusion of various milk products but in general will fall between 20 and 60% of pre-weaning intake. However, any apparent deficiency in lactose intake is compensated by the reduction in intestinal lactase activity around 3 to 4 weeks of age (Manners and Stevens, 1972). Thereafter lactase activity cannot be restimulated to any significant extent in the pig by the addition of lactose to the diet (Ekstrom, Benevenga and Grummer, 1975; Table 14.8). Consequently a greater proportion of dietary lactose will be available for bacterial fermentation, suggesting that even moderate supplementation of weaner diets with lactose will be of some benefit.

In future, synthetic sugars which are exclusively utilised by naturally occurring

Table 14.8 LACTASE ACTIVITY OF MUCOSAL HOMOGENATES FROM THE SMALL INTESTINE OF PIGS FED VARIOUS LEVELS OF DRIED WHEY

	Dried whey[a] (g/kg)				
	0	*100*	*250*	*400*	*s.e.d*
Duodenal lactase[b]	0.16	0.22	0.18	0.18	0.02
Jejunal lactase	2.06	2.44	2.00	2.24	0.24
Ileal lactase	0.18	0.20	0.18	0.18	0.02

[a] Whey containing 470 g/kg lactose; [b] μmol of galactose released/mg protein/hour. Courtesy of Ekstrom *et al.* (1975)

beneficial microbes or by genetically engineered micro-organisms may offer more effective methods of manipulating gut microbial patterns and health.

Non-starch polysaccharides

As with lactose, the inclusion of ingredients rich in readily fermentable non-starch polysaccharides (NSP) in weaner diets may enhance gut health by maintaining or stimulating a suitable microbial population.

Current research has focused on dried sugar beet pulp (SBP) since it is a good source of soluble NSP containing a high proportion of readily fermentable pectic substances. In a recent study, Low, Carruthers and Longland (1990) showed that weaned piglets fed diets supplemented with SBP were able to maintain a satisfactory level of gut health without recourse to antibiotic treatment. Other less familiar products which are also high in fermentable fibre, such as dried apple pomace, have given similar results when fed in association with high-fibre and low-protein diets (Bertschinger, Jucker, Pfirter and Pohlenz, 1983).

It is surprising the extent to which weaned piglets are able to utilise dietary fermentable fibre. For example, Edwards, Taylor and Harland (1991) found no reductions in growth and performance in weaned piglets fed diets containing up to 150 g/kg SBP. This would suggest that the microbial population in the gut of weaned piglets is quite well established and can be easily encouraged by the addition to diets of fermentable NSP. Low *et al.* (1990) found that by 32 days of age the apparent digestibility of SBP in weaned piglets approached 85%.

The role of NSPs and their mode of action in the gut require further investigation. Research in this field of study is likely to benefit alternative drug-free methods of preventing or even treating post-weaning diarrhoea.

LIQUID FEEDING

It is generally accepted that the intake, growth, feed conversion and health of weaned piglets on liquid feeds is very much better than those fed the equivalent nutrients in a dry form. This is based on several comparative studies on dry and liquid feeding conducted with artificially reared and very early weaned piglets (Braude and Newport, 1977; Lecce, Armstrong, Crawford and Ducharme, 1979; English, Anderson, Davidson and Dias, 1981; Taverner, Reale and Campbell, 1987).

The reasons for improved health and performance of piglets on liquid feeds are not clearly understood. An immediate advantage for liquid fed piglets is that they do not have to learn separate patterns of feeding and drinking behaviour, their needs for both water and nutrients are met at a single point of delivery. Consequently dehydration from an inadequate water intake and from increased fluid and mineral losses in scouring piglets may be prevented by increased water and nutrient intake with liquid feeding.

Liquid feeding may also enhance gut health and function by providing appropriate conditions for enzyme activity, digestion, nutrient absorption and microbial growth. Deprez, Deroose, van den Hende, Muylle and Oyaert (1987) showed that the villus height in the small intestine of liquid fed piglets remained unchanged after weaning, but in dry fed piglets it was significantly reduced (Table 14.9).

Table 14.9 MEAN VILLUS HEIGHT (μm) IN THE SMALL INTESTINE OF WEANED PIGLETS ON DRY AND LIQUID FEEDS

		Days after weaning				
		0	*4*	*6*	*8*	*11*
Distal jejunum	Liquid	329	364	312	388	375
	Dry	329	283	269	220	254
Ileum	Liquid	297	324	267	368	375
	Dry	297	305	211	238	243

Piglets were weaned at 5 weeks of age. Courtesy of Deprez *et al.* (1987)

Despite the considered benefits, liquid feeding is not widely used for weaned piglets. This is due to practical difficulties concerning feed hygiene and increased labour requirements for equipment cleaning and refilling. Such problems are currently being addressed by Partridge, Fisher, Gregory and Prior (1992), in their development of an automated wet feeding system for weaned piglets. Preliminary results showed significant improvements in intake and growth compared with a conventional dry feeding regime (Table 14.10).

Table 14.10 INTAKES AND GROWTH RATES OF WEANED PIGLETS ON A CONVENTIONAL DRY OR AN AUTOMATED WET FEEDING SYSTEM

	Dry hopper	*Wet Autofeeder*	*s.e.d.*	*P*
		Intake (g/day)		
Week 1	149	176	14.0	<0.1
Week 2	327	357	23.6	NS
Week 3	453	518	36.3	<0.1
Overall	310	351	16.8	<0.05
		Growth rate (g/day)		
Week 1	133	147	15.5	NS
Week 2	330	355	14.4	<0.05
Week 3	380	434	17.2	<0.001
Overall	281	312	12.1	<0.01

Piglets weaned around 23 days of age. Courtesy of Partridge *et al* (1992)

Meeting the energy requirements of the newly weaned piglet

Currently, the average age at weaning in the UK falls between 19 and 25 days (Meat and Livestock Commission, 1992). At this stage of lactation, the nursing sow is able to support a piglet live-weight growth rate of about 300 g/day on a milk yield (Elsley, 1970) providing about 4.0 MJ DE/piglet/day. Energy intake may be supplemented by an extra 1.0 to 1.5 MJ DE/day if piglets are allowed access to a highly digestible milk-based creep feed containing 19.0 MJ DE/kg (English, Robb and Dias, 1981). The piglet uses about 2 MJ DE for body fat deposition (50 g), about 1.5 MJ DE for protein deposition (40 g) and the remainder for maintenance (calculated from data presented by Wood and Groves (1965) and Le Dividich, Vermorel, Noblet, Bouvier and Aumaitre (1980)). It is not surprising therefore that about 60% of the gross energy in sow milk is in the form of fat. The sow has evolved a very effective strategy for delivering milk fat in a highly emulsified state with an apparent digestibility in the piglet approaching 100%. The fat is uniformly suspended throughout the milk in the form of globules, but is prevented from interfering with other constituents (e.g. lactose) by being encapsulated in a lipo-protein membrane. Fat globules in sow milk are much finer than those found in cow's milk, presenting a very large surface area for digestion (Fahmy, 1972).

FAT SUPPLEMENTATION

Fat is used in starter diets to enhance energy intake following weaning, when stress depresses appetite and consequently growth rate. Irrespective of dietary nutrient density, voluntary intake in the week after weaning will normally average 200 g/day. Thus for the piglet to sustain an intake of 5.5 MJ DE/day the diet would need to provide 27.5 MJ DE/kg. In commercial practice, starter diets will be constrained to a target between 16 and 18 MJ DE/kg, depending on cost and manufacturing technology, which means that energy intake will be about 40% below the requirement to support a growth rate of 300 g/day.

The amount of fat required to raise the energy to this level will depend on the DE value of other ingredients used in the diet, but in general total fat content of cereal based diets will range between 100 and 200 g/kg. This compares with a value for the fat content in the dry matter of sow milk of about 400 g/kg. Fat may be added directly to the diet in the form of extracted vegetable and/or animal fats or it may feature as a major constituent of feed ingredients as in processed oilseeds.

UTILISATION OF FAT

Although growth and performance averaged over several weeks after weaning can be improved on high fat diets, a number of studies suggest that the response to fat supplementation during the first week can be very small or even negative (Lawrence and Maxwell, 1983; Howard, Forsyth and Cline, 1990; Mahan, 1991).

Utilisation of fat may be limited by the reduction in pancreatic lipase activity found immediately after weaning. The concentration of lipase in the pancreas and digesta have been reported to fall to about 30% and 60% of preweaning levels respectively (Scherer, Hays, Cromwell and Kratzer, 1973; Lindemann *et al.*, 1986; Cera, Mahan and Reinhart, 1990; e.g. Table 14.11). It may take several days for pancreatic concentrations to recover to a level found in the sucking piglet at

Table 14.11 EFFECT OF AGE AND WEANING ON PANCREATIC LIPASE ACTIVITY

Age (weeks)	*Body weight* (kg)	*Pancreas weight* (g)	*Lipase activity*[a] /g pancreas	Total
Preweaning				
Birth	1.29	1.06	965	934
1	1.97	2.74	1235	2965
2	3.28	3.70	2938	12142
3	4.79	5.54	2906	15421
4	6.56	7.36	6268	48756
Postweaning				
5	6.50	10.47	1778	15711
6	7.99	12.28	1180	15163

[a] μmoles of fatty acid produced/min. Courtesy of Lindemann *et al.* (1986)

Table 14.12 EFFECT OF DIETARY FAT SUPPLEMENTATION ON LIPASE ACTIVITY IN WEANED PIGLETS

Day after weaning[a]	*Control*	*High fat diet*[b]
	Lipase units[c]/g pancreas	
7	193	196
14	200	224
21	328	265
28	318	358
	Total lipase units	
7	1440	1445
14	2575	2509
21	7542	5577
28	8720	11685

[a] Pigs weaned at 21 days of age; [b] Supplemented with 60 g/kg maize oil; [c] In Tietz units (one unit equivalent to 0.05 N NaOH). Courtesy of Cera *et al.* (1990)

weaning. Furthermore the rate of recovery does not appear to respond to the addition of fat to starter diets (Cera *et al.*, 1990; Table 14.12). Consequently adding fat in an attempt to induce or stimulate lipase activity may be of limited value. This was further demonstrated in the study by Howard *et al.* (1990) in which the subsequent and overall growth and performance of weaned piglets did not respond to a starter diet supplemented with and without soya bean oil (Table 14.13).

Even though the reduction in pancreatic lipase activity is proportional to the reduction in fat intake at weaning, the surface area of dietary fat which is available for lipase attack is considerably smaller than that for milk fat. Consequently the digestibility of dietary fat is low in the week after weaning (Cera, Mahan and Reinhart, 1988) but improves with time in direct correlation with the recovery in lipase output (Scherer *et al.*, 1973).

Another reason why the addition of crude fat may not benefit growth in the early stages after weaning is the possible negative interaction of supplementary

Table 14.13 EFFECT OF AN ADAPTATION PERIOD WITH SOYA BEAN OIL SUPPLEMENTATION ON FEED INTAKE, GROWTH AND PERFORMANCE OF WEANED PIGLETS

Soya bean oil (g/kg)	*Control* 0	*Adaptation* 60	*No adaptation*[a] 60
Feed intake (kg/day)			
Week 1	0.33	0.30	0.29
2	0.63	0.60	0.63
3	0.95	0.96	0.84
4	1.18	1.14	1.10
Growth rate (kg/day)			
Week 1	0.19	0.18	0.20
2	0.42	0.41	0.42
3	0.53	0.58	0.55
4	0.65	0.67	0.68
Feed conversion			
Week 1	2.01	2.28	1.65
2	1.54	1.52	1.55
3	1.80	1.72	1.59
4	1.82	1.77	1.70

[a] Soyabean oil not added to diets in the first 2 weeks after weaning. Piglets were weaned between 4 and 5 weeks of age. Courtesy of Howard *et al.* (1990)

fat with the microbial utilisation of fermentable constituents such as NSPs and lactose. There is much information on dietary fat and microbial fermentation in the ruminant but this subject has received little attention in the pig. In the context of this paper, it is proposed that the addition of extracted fats, as oils or liquids, may limit the availability of fermentable substrates to beneficial microbes, at a time when the recovery of a suitable gut microbial population is critical to the maintenance of health following weaning. This effect may be exacerbated in specialised diets based on milk replacers and other refined ingredients lacking appropriate fermentable matter.

Since dietary fat utilisation in relation to fat quality and fatty acid composition has been comprehensively reviewed elsewhere (Stahly, 1984), this subject will be considered only briefly in this chapter. Earlier studies in this field were conducted with artificially reared piglets which proved a very sensitive model for evaluating different fat sources (Alexander, 1969; Braude and Newport, 1973; Braude, Keal and Newport, 1976). More recent work with neonatal and early weaned piglets has produced similar results where utilisation was enhanced in sources containing a higher proportion of short-chain saturated and long-chain unsaturated fatty acids than in sources rich in long-chain saturated fatty acids (Lawrence and Maxwell, 1983; Cera *et al.*, 1988; Cera, Mahan and Reinhart, 1989). In general, coconut oil, butter fat and lard have been well utilised, soya and corn oil have given satisfactory results and diets containing tallow have given the poorest performance.

It is considered that the digestibility of dietary fat is directly related to the ease with which it can form micelles with bile salts. Short-chain fatty acids have a high affinity for micelle formation (Stahly, 1984) but can also be rapidly absorbed directly into the blood stream following hydrolysis from the triglyceride (McDonald, Edwards and Greenhalgh, 1981). Long-chain unsaturated fatty acids are also readily able to form micelles and their presence can assist the emulsification of fat sources high in long-chain saturated fatty acids. For this reason, mixtures of vegetable oils and animal fats tend to give better results than additions of animal fats only (Stahly, 1984; Cera *et al.*, 1989; Li, Thaler, Nelssen, Harmon, Allee and Weeden, 1990). This must be balanced against any increased health risk associated with the oxidation of unsaturated fats in starter diets. Nevertheless, as the digestive system of the weaned piglet adapts to non-milk fats, differences in the utilisation of fats according to source tend to diminish over time (Cera *et al.*, 1988; 1989; Table 14.14).

Table 14.14 COEFFICIENT OF APPARENT DIGESTIBILITY OF DIFFERENT FATS IN THE WEANED PIGLET

Week after weaning	*Maize oil*	*Lard*	*Tallow*
1	0.790	0.681	0.648
2	0.805	0.718	0.724
3	0.888	0.836	0.818
4	0.888	0.849	0.825

Pigs weaned at 21 days of age. Courtesy of Cera *et al.* (1988)

OILSEEDS

Oilseeds are a valuable source of fat for young pigs. Unless adequately processed, the undesirable effect of antinutritional and toxic factors in the seed will be of greater significance than any contribution to the energy value of the diet.

Some seeds, though potentially cost-effective sources of fat energy, may not be rendered safe by normal processing methods for use in weaner diets. Rapeseed falls into this category of ingredients, though rich in oil (400 g/kg seed) it contains goitrogenic glucosinolates which cannot be easily neutralised by either heat or chemical processing (Fenwick, Spinks, Wilkinson, Heaney and Legoy, 1986). In a recent study with weaned piglets, Gill and Taylor (1989) demonstrated the extent to which the inclusion of milled whole rapeseed, though from a variety selected for low-glucosinolate content (17 μmol/g seed), is limited by these toxic factors. Diets supplemented with either 200 g/kg or 400 g/kg of milled rapeseed depressed intake and growth in weaned piglets by a factor of 0.70 and 0.55 respectively, relative to those fed comparable diets containing extruded full-fat soya beans.

The sensitivity of the newly weaned piglet to rapeseed toxins can be further demonstrated by choice-feeding conditions allowing nutritional freedom from undesirable factors. For example, McIntosh and Aherne (1982) found that selection of a diet without rapeseed meal offered with another containing only 50 g/kg rapeseed meal represented 70% of total voluntary intake. This study would suggest that, given a choice, piglets are willing to tolerate no more than 15 g/kg rapeseed meal in their diet or an equivalent of 30 g/kg of whole rapeseed. However tolerance to rapeseed may vary depending on glucosinolate concentration.

Opportunities for using rapeseed as a high-energy feed ingredient for weaned piglets are improving with continued reductions in the content of glucosinolates in newer varieties. A recent evaluation in weaned piglets assigned an apparent DE value of 33.5 MJ/kg to the oil in milled low-glucosinolate rapeseed, suggesting an apparent oil digestibility of about 0.85 (Aumaitre, Bourdon, Peiniau and Bengala Freire, 1989).

Finally, full-fat oilseed meals may offer advantages over extracted fats. Processing the seed (e.g. milling or extrusion) effectively homogenises the fat whilst retaining it within a vegetable matrix and thus away from other constituents of the diet.

The importance of feed processing to young pig performance

THE EFFECTS OF MILLING ON GRIST SIZE

Relatively few studies have critically examined the effects of raw material particle size on digestibility and growth performance in the young pig. In those trials where it has been studied, the emphasis has been on the cereal fraction of the diet and the general consensus is that fine grinding is desirable, both in terms of feed intake and feed efficiency. Wu and Fuller (1974) using raw maize-based diets, in a meal form, found that feed intake and feed efficiency significantly improved when the maize was hammer milled through screens with decreasing aperture sizes of 9.5, 4.0 and 1.6 mm respectively. Similarly Healy, Hancock, Bramel-Cox, Behnke and Kennedy (1992) observed with raw maize or sorghum-based diets, incorporated into pellets, that decreasing average particle size from 900 μm through to 300 μm resulted in significant linear improvements in average daily gain and feed conversion efficiency. These effects were most apparent in the first two weeks after weaning (weaning weight 5.3 kg) and started to disappear when the data were considered over a 5 week period. This tends to support the view of Lawrence (1979) who found that grinding of cereals at or below 5 mm screen sizes had little effect on growth and digestive utilisation in growing and finishing pigs.

When feeding meal diets to newly-weaned pigs, however, the possible disadvantages of very fine ground cereals (particularly wheat) should be considered, i.e. its tendency to become ‘sticky and pasty’ in the animals’ mouth (Lawrence, 1979). Partridge (1988a, unpublished observations) working with wheat-based meal diets for 3 week weaned pigs found reduced feed intake (13%) in a group offered ‘simulated wheat’ (i.e. gluten plus pre-gelatinised wheat starch). This effect was no longer apparent when the same formulations were incorporated into a pellet.

At a production level the economic feasibility of very fine grinding would also have to be carefully considered. For example, in the studies of Healy *et al.* (1992) grinding corn at 900 μm resulted in a modest throughput of 1.76 t/h but at mean particles sizes of 300 μm this throughput was reduced by 63%.

It seems likely that any beneficial effects of fine grinding of cereals for the young pig are a consequence of the increase in surface area for enzyme action, gut fermentation and other digestive processes. Mercier and Guilbot (1974) working with maize found that grinding through 2 mm rather than 7.5 mm screens only had slight influences on the structure of the starch granule and did not significantly affect its susceptibility to amylases *in vitro*. Additional hydrothermic treatments appear to be required to bring about starch gelatinisation and improved *in vitro* digestibility.

PELLETING

In the pelleting process dietary raw materials are routinely preconditioned using steam and then subjected to pressure as they are forced through the die of the cubing machine, giving a frictional heating effect. These processes, if strictly controlled, are usually beneficial to the production of diets for young pigs (Mercier and Guilbot 1974; Lawrence 1979; Hardy 1992). In commercial production the types of raw materials typically used in pig starter rations (e.g. milk products, cooked cereals, oils, sugars) are usually pressed at relatively low temperatures (60–70°C) to avoid the risk of caramelization and excessive pellet hardness, both of which can reduce food intake in the young animal. This will limit the extent of extra 'cooking' that can potentially take place in the process (cf. section 'Extrusion and expansion'). Small pellet sizes (2.5 mm) appear to be particularly beneficial to feed intake in the newly-weaned animal, giving improved performance over meal and crumbs (Partridge, 1989).

Short time, high temperature (STHT) processing: nutritional implications to the young pig

EXTRUSION

For over 50 years extrusion has been used as a method for processing human food (Ferket 1991). During the last 10–15 years, however, significant advances have been made in the development and application of extrusion technology, particularly in the pet food and aquaculture sector. Consequently it is now becoming economically feasible to consider its use for manufacturing feeds for certain classes of farm livestock. Complete diets for young pigs undoubtedly fall into this category but at the present time they are relatively rare in the market place. In contrast, individual raw materials processed by extrusion have been used extensively in diets for young pigs for many years (e.g. extruded cereals, soya beans).

Extruders are categorised as single-screw or twin screw types using a 'wet' or 'dry' extrusion process (moisture content of input material being 45–50%, or less than 25% respectively). Wet extrusion requires more sophisticated equipment and therefore is more expensive. Similarly single screw extruders are of simpler design, less versatile and adaptable for specialised application, but cost less than twin-screws (Hancock, 1992).

In use extruders generally create friction, pressure and attrition to cause distinct changes in the physical and chemical features of the foodstuff passing through them. The temperatures generated in an extruder barrel generally reach 135–60°C at pressures of 15–40 atmospheres in as little as 30 s. As the extruded material exits the barrel, through a fixed die which shapes the product, the sudden drop in pressure results in violent expansion as steam escapes.

EXPANSION

In 1989 the expander (or STHT conditioner) was introduced to the feed industry. In essence the expander is a form of single-screw extruder usually set ahead of a conventional pellet mill (Pipa and Frank, 1989; Daw, 1991). Partially conditioned meals enter the expander and the passage of meals through the machine, driven by the screw, is regularly resisted by a series of hardened 'interrupter' bolts inserted through the side of the extruder barrel. This resistance creates extreme 'shear' on the meal and the friction developed combined with extra injected steam, raises the temperature. At the outlet end of the machine, in contrast to the extruder, a cone-shaped resistance plate is mounted to obstruct the flow of the material thereby creating extra pressure and friction (temperature range 110–140°C typically, 30 bar pressure). At discharge through the resistance cone the sudden release of this high pressure results in the flash-off of moisture as steam and 'expansion' of the product occurs. Product consistency at this stage resembles a stiff, dry porridge and a cutter/breaker is usually employed to ensure uniform flow to the pellet mill.

NUTRITIONAL IMPLICATIONS OF EXTRUSION AND EXPANSION OF DIETS AND RAW MATERIALS FOR THE YOUNG PIG

Both processes, depending on process conditions and raw material usage, offer a number of potential advantages for the production of complete diets for young pigs. Equally there are certain limitations or disadvantages which need to be addressed. Both are summarised in Table 14.15.

Table 14.15 POTENTIAL ADVANTAGES AND DISADVANTAGES OF EXTRUSION AND EXPANSION PROCESSING OF YOUNG PIG DIETS

Factor	*Advantages*	*Disadvantages*
Protein	Denaturation, reduction of anti-nutritional factors.	Possible destruction of amino acids; Maillard reaction, complexing of reducing sugars and free amino groups.
Starch	Gelatinisation; de-activation of naturally occurring α-amylase inhibitors in raw cereals.	Formation of amylose – fatty acid complexes.
Fats and oils	Release of encapsulated oil from raw materials, inactivation of lipolytic enzymes.	Oxidation of lipids and flavour components.
'Fibre'	Increases the proportion of soluble non-starch polysaccharides at the expense of insolubles.	Not applicable.
Vitamins	Not applicable.	Potential loss of vitamin activity, especially: A, K, C, B1, folic.
Minerals	Not applicable.	Possible phytates complexing with Zn, Mg. Destruction of natural phytase.
Feed hygiene	Reducing levels of bacteria, moulds, fungi, aflatoxins.	Not applicable.
Feed additives	Extruded feeds can provide an absorbent kernel for coatings e.g. oils, flavours, heat labile vitamins, enzymes, probiotics, syrups etc.	Loss of some feed additives if processed through the extruder e.g. some enzymes, some antibiotics.
Palatability	Improved 'mouth feel', texture.	Development of undesirable flavours, reaction products.

There are relatively few reports in the literature comparing the performance of pigs on pelleted diets with complete extruded (or expanded) diets. Several studies have concentrated on the effects of extrusion on the cereal component in the diet but many often fail to characterise both in vitro digestibility and/or the raw material processing conditions used, making interpretation and comparison of animal response difficult. In some studies the inclusion of sub-therapeutic antibacterials also may mask potential responses associated with STHT processing.

EFFECTS ON CARBOHYDRATES

Starch

Approximately 60–70% of the energy ingested by the weaned piglet is derived from carbohydrates, so processing of this portion of the diet is of major significance to young animal performance.

Starch, the predominant carbohydrate in the diet is derived mainly from the cereal component and is ordinarily a granular mass comprising glucose units, amylose and amylopectin molecules. The extrusion process causes expansion and gelatinisation of the starch granules with an opening up of these complex molecular chains. Given optimum process conditions extrusion has the potential to gelatinise high levels of starch in most feed raw materials (85–100%, Björck and Asp, 1982; Asp and Björck, 1989).

Although complete starch gelatinisation at atmospheric pressure requires at least 30–40% moisture many authors have shown that extrusion cooking produces a virtually complete starch gelatinisation at low moisture content when the temperature exceeds 110–135°C (Asp and Björck, 1989). The extruded starch has increased solubility and decreased viscosity, favouring availability to amylase and rapid absorbtion of resultant glucose in vivo.

The implicit assumption is usually made that gelatinisation of dietary starches in diets for young pigs is a desirable feature in view of their immature digestive physiology and enzyme complement at weaning (Kidder and Manners, 1978). Reports from the scientific literature on this aspect however, tend to be equivocal (e.g. Aumaitre and Dumond, 1975; Aumaitre, 1976; Dammers, 1981; van der Poel, den Hartog, van den Abeele, Boer and van Zuilichem, 1989; Vestergaard, Danielsen, Jackobsen and Rasmussen, 1990; van der Poel, den Hartog, Stiphout, Bremmers and Huisman, 1990; Sauer, Mosenthin and Pierce, 1990; Hancock, 1992). Factors such as different weaning ages and weights of the pigs used, disease challenge and different processing conditions (both in terms of raw material grinding and subsequent heat treatment), digestibility measurements (e.g. usually whole tract rather than ileal) and influence of other dietary raw materials all interact and confound the establishment of a consensus view. In commercial practice diet formulation for the young pig from birth to, say, 10 kg almost invariably entails a degree of dietary 'insurance' to cope with a range of different management systems and environmental and disease 'challenges' which will be experienced by the newly-weaned animal. Correctly heat processed cereals are always likely to form an integral part of this nutritional and management package. The boundary line over this weight range, however, where gelatinisation is desirable and economically justified is never likely to be clearly established in view of the tremendous variability in enzyme development of the young pig at this age (Kidders and Manners, 1978; de Passillé *et al.*, 1989).

Table 14.16 LACTIC ACID PRODUCTION IN THE STOMACH OF PIGLETS OFFERED A SIMPLE DIET (BASED ON WHEAT/BARLEY/WHEAT GERM) BEFORE (C) OR AFTER EXPANSION (E). VALUES ARE IN mmol/100 g DIGESTA DRY MATTER

	1 hour after feeding		*2 hours after feeding*	
	C[a]	*E*[b]	*C*[a]	*E*[b]
Lactic acid	2.07[x]	5.27[y]	1.59[x]	8.22[z]

[x,y,z,] Values with different superscripts show significant differences ($p<0.05$).
[a] Extent of starch hydrolysation in diet =15% (amyloglucosidase test);
[b] Extent of starch hydrolysation in diet = 47% (amyloglucosidase test).
After Peisker (1992b)

Some recent studies (Peisker, 1992b) looking at the use of expanded diets for newly-weaned pigs have emphasised another potential benefit of hydrolysing the starch component of the diet by cooking, aside from its increased access to the animals' digestive enzymes. Measurements of lactic acid both *in vitro* and *in vivo* have shown higher production rates in the stomach when using expanded feed (Table 14.16). In the weaner pig lactic acid is an important metabolic product produced by the lactobacteriacea in the stomach and intestines which acts as a barrier against pathogenic bacteria. In parallel growth trials the incidence of diarrhoea was substantially reduced in the animals offered the expanded feed and growth rates were enhanced.

One potentially detrimental effect of extrusion cooking of complete diets is the formation of amylose – lipid complexes. These complexes are fairly resistant to α amylase *in vitro*, however studies *in vivo* in the rat indicate that they are completely digested and absorbed, albeit at a slightly lower rate than free amylose (Björck and Asp, 1982). The formation of these complexes underlies the necessity for using oil (method B) analysis to accurately reflect the oil content of extruded diets (Asp and Björck, 1989).

Effects on dietary fibre and non-starch polysaccharides

Few studies have been done on the effects of high attrition extrusion/expansion on naturally occurring fibre in animal feeds. Fadel, Newman, Newman and Graham (1988), working with barley based diets, found that extrusion caused a shift in insoluble to soluble NSP and insoluble β-glucans to soluble β-glucans. In adult pigs this had the effect of significantly improving ileal digestibility of starch (by 13%), dry matter (by 7%) and energy (by 7%). None of these effects were apparent when measured in the faeces, illustrating the importance of this methodology. The shift of insoluble to soluble NSP after extrusion cooking resulted in increased digestion of soluble NSP in the ileum (19% more) and increased fermentation of insoluble NSP in the lower tract (13% more). Asp and Björck (1989) summarising a number of studies on extrusion cooking of various raw materials found that solubilisation of some fibre components was a consistent feature. 'Resistant starch' (strongly retrograded amylose, Berry, 1986) does not appear to be formed in extrusion cooking.

Although total NSP levels in young pig diets will usually be relatively low it is clear that the extrusion/expansion cooking process is likely to have favourable effects on the young animals' ability to deal with these complex carbohydrates.

Effects on low molecular weight carbohydrates

Loss of low molecular weight carbohydrates has been reported on several occasions after extrusion cooking (Asp and Björck, 1989). This may be due to extraction difficulties, or possibly due to hydrolysis and subsequent Maillard reactions. These effects may be particularly pertinent in young pig feeds where sugar additions are often used. Although sucrose itself is a non-reducing sugar, reactions in the extrusion process may lead to its conversion to reducing monosaccharides, capable of participating in Maillard reactions.

EFFECTS ON FATS AND OILS

Another major interest in extrusion cooking of complete diets for young pigs is in its potential to improve availability and utilisation of in-seed fats by rupturing cell walls and releasing encapsulated oil (e.g. soya beans, rapeseed, sunflower seeds). Adams and Jensen (1985), using weaner pigs, showed that extrusion cooking improved the digestibility of in-seed fats in soya beans by 23% when compared with standard roasted soya beans (86% cf. 63% respectively). Levels of digestibility of in-seed fats in extruded products under ideal process conditions should therefore approach those of extracted seed plus additional free oil (85–95%, as measured by Adams and Jensen (1985) in piglets around 6 kg bodyweight). Aumaitre *et al.* (1989) working with full fat rapeseed in newly weaned piglets showed much smaller increases in fat digestibility after extrusion (+ 3%, 86% cf. 83%), when compared with fine ground double zero rapeseed.

Overall it would appear that extrusion/expansion offers nutritional advantages in terms of fat digestibility to the newly weaned animal but actual process conditions, as ever, are major determinants of animal performance.

EFFECTS ON PROTEIN

Of all the nutrient classes, extrusion processing of protein (predominantly soyabean) dominates the scientific literature. Hancock (1992) reviewed a number of studies comparing extruded soyabeans with roasted soyabeans plus equivalent free oil, across a wide range of pig weights and ages. He concluded that, overall, extruded soyabeans were at least equal to, and up to 5% better than soyabean meal plus added fat. The advantage of the extruded whole bean product presumably relates to the good accessability to the intracellular fat (as described above) plus an important effect on denaturation of the protein, increasing its susceptibility to enzymatic breakdown. Extrusion processing causes not only denaturation but radical disruption of the protein configuration by breaking bonds, shearing protein molecules and actually stimulating formation of new bonds and configurations. Thus, the protein molecules not only become more susceptible to proteases but any biological activity that the proteins might have would be destroyed. The latter, of course, is beneficial in terms of inactivating trypsin inhibitors, lectins and other antinutrients associated with the soyabean seed. The major storage proteins in soyabean, glycinin and β–conglycinin, have also been shown to have antigenic properties in calves (Smith and Sissons, 1975; Kilshaw and Sissons, 1979) and piglets (Sissons, Churchman, Heppell, Hardy, Banks and Miller, 1988). Li, Nelsson, Reddy, Blecha, Klemm, Giesting, Hancock, Allee and Goodband (1991b) and Li *et al.* (1991a) found that wet extrusion processing of soya protein concentrate (produced from ethanol extracted soya flakes) gave additional advantages in terms of growth performance in the immediate post-weaning period, despite eliciting similar immunological responses in the animals to 'standard' soya protein concentrate. They attributed this to improved nitrogen digestibility imparted by the extrusion process. Similar benefits of extrusion were reported by Friesen, Nelssen, Behnke and Goodband (1992). These authors found improved pig performance when the soya bean meal in the diets was 'moist' extruded (rather than 'dry') and both treatments were far superior to 'standard' soya bean meal.

Studies by Prince, Miller, Bailey, Telemo, Patel and Bourne (1988) have illustrated how variable routine soya processing can be in terms of reducing certain anti-nutritional factors e.g. lectins. Samples of 'processed' soya products ranged in lectin content from 2 μg/g (for 'Soycomil' milk replacer) to 1468 μg/g (for one particular extruded full-fat soya sample). Although the physiological significance of soya lectin to the young pig has yet to be determined it is clear that processing conditions are going to have a major bearing on resultant animal performance.

Certain process conditions in extrusion cooking (high temperatures in association with low water content) are, however, known to be disadvantageous by favouring the Maillard reaction. This reaction between reducing sugars and free amino groups in protein leads both to a decrease in protein digestibility and in amino acid availability. Increasing moisture content in the extrusion process significantly improves this situation (Björck and Asp, 1982). Lysine is the most reactive protein bound amino acid due to its free ϵ-amino group, but arginine, tryptophan, cysteine and histidine may also be affected (Hurrell and Carpenter, 1977). Expansion, probably because of its relatively high moisture content and generally lower process temperatures, seems to have negligible effects on amino acid availability (Peisker, 1992a). New, improved methods of monitoring overcooked or undercooked soya products in vitro are also likely to have an important influence on assessing soyabean quality for young pigs (Table 14.17).

Table 14.17 EFFECT OF VARIOUS AUTOCLAVING TIMES (I.E. OVER- PROCESSING) OF HULLED, SOLVENT EXTRACTED SOYABEAN MEAL ON YOUNG PIG PERFORMANCE (7.5 kg START WEIGHT, 13 DAY TRIAL)

Autoclaving time (min)	*Urease index* (units of pH increase)	*Protein solubility*[a] (%)	*Weight gain* (kg)	*FCR*
0	0.17	89	4.87	1.47
10	0.02	71	4.54	1.52
20	0.00	66	3.83	1.54
40	0.00	56	3.75	1.67

[a] Protein solubility in potassium hydroxide (Araba and Dale 1990). Courtesy of Parsons *et al.* (1991)

EFFECTS ON HEAT LABILE NUTRIENTS AND ADDITIVES

The process temperatures and friction generated during extrusion/expansion are likely to have a negative influence on certain vitamins, feed enzymes, antibacterials and probiotics which will feature to varying degrees in diets for young pigs. Coelho (1991) has recently reviewed some aspects of vitamin stability and its relationship to extrusion temperature and extruder barrel retention time. Similarly Asp and Björck (1989) summarised a range of studies done on vitamin stability in extruded human foodstuffs. Vitamins A, C, B_1, and K and folic acid are particularly sensitive to high temperature feed processing and subsequent storage. Process conditions, obviously, are the major determinants of the extent of this loss (Table 14.18) and account for differences in apparent stability between different studies.

Table 14.18 AVERAGE INDUSTRY VITAMIN STABILITY VALUES DURING EXTRUSION AND SUBSEQUENT STORAGE IN COMPLETE FEEDS

Vitamin	*Vitamin retention* (%)			*Loss/month during storage in complete feeds* (%)
	Extrusion temp/barrel retention time (°C/min)			
	110/3 *116/2* *121/1* *127/0.5*	*138/3* *143/2* *149/1* *154/0.5*	*165/3* *171/2* *177/1* *182/0.5*	
A 650 (beadlet)	93	86	71	9.5
D_3 325 (beadlet)	95	91	83	7.5
E acetate	98	95	93	2.0
MSBC[a]	70	45	20	17.0
Thiamine HCl	91	80	55	11.0
Riboflavin	93	87	75	3.0
Pyridoxine	94	90	80	4.0
B_{12}	98	95	92	1.4
Capantothenate	95	91	82	2.4
Folic	94	83	71	5.0
Biotin	94	83	70	4.4
Niacin	93	83	71	4.6
Ascorbic acid	65	42	15	30.0
Choline	99	97	94	1.0

[a] MSBC = menadione sodium bisulphite complex (vitamin K). After Coelho (1991)

Most in-feed enzymes, targeted primarily against dietary non-starch polysaccharides, will be vulnerable to high attrition processing. Spraying on of these enzymes in an oil base is already a well established technology and would be pertinent to post-extrusion / expansion treatment of diets for young pigs. A similar approach would be required for some antibacterials (e.g. penicillin products) and probiotics which are notoriously heat sensitive.

Possible future trends in pig weaner diets

New approaches in diet construction for the young pig over the next few years are likely to be influenced primarily by trends and advances made in the following areas;

1. Weaning age and weight, against a background of continual genetic selection for improved litter size and liveability.
2. Possible legislative recommendations on weaning age on welfare grounds.
3. The availability and price of milk products and their highly processed alternatives.
4. Complete diet processing using specialized equipment to upgrade conventional, relatively low cost, raw materials.
5. Husbandry methods employed e.g. where applicable, the use of wet or liquid feed systems for supplementary feeding on the sow, and post-weaning.
6. Legislation in relation to the widespread use of antibiotics and other medicinal products in diets for young pigs.
7. Alternative nutritional strategies to enhance gut health e.g. feed sterilization, biotechnological methods including the use of in-feed enzymes, probiosis (both chemical and using suitable in-feed micro-organisms), in feed use of beneficial sow milk factors, e.g. lactoferrin and surface-acting antibodies.

References

Adams, K.L. and Jensen, A.H. (1985) *Animal Feed Science and Technology*, **12**, 267–274

Agricultural Research Council (1981) *The Nutrient Requirements of Pigs*. Slough: Commonwealth Agricultural Bureaux

Aherne, F.X., Danielsen, V. and Nielsen, H.E. (1982) *Acta Agriculturae Scandinavica*, **32**, 155–160

Alexander, V.A.W. (1969) *Studies on the Nutrition of the Neonatal Pig*. PhD Thesis. University of Edinburgh

Appleby, M.C., Pajor, E.A. and Fraser, D. (1992) *Animal Production*, **55**, 147–152

Asp, N-G. and Björck, I. (1989) In *Extrusion Cooking*, pp. 399–434. Edited by C. Mercier, P. Linko and J.M. Harper, St Paul, Minnesota: American Association of Cereal Chemists

Aumaitre, A. (1976) *Annales de Zootechnie*, **25**, 41–51

Aumaitre, A. and Dumond, R. (1975) *Journees de Recherche Porcine en France* 151–160

Aumaitre, A., Bourdon, D., Peiniau, J. and Bengala Freire, J. (1989) *Animal Feed Science and Technology*, **24**, 275–287

Barnett, K.L., Kornegay, E.T., Risley, C.R., Lindemann, M.D. and Schurig, G.G. (1989) *Journal of Animal Science*, **67**, 2698–2708

Berry, C.S. (1986) *Journal of Cereal Science*, **4**, 301–314

Bertschinger, H.U., Jucker, H., Pfirter, H.P. and Pohlenz, J. (1983) *Annales de Recherche Veterinarie*, **14**, 469-472

Björck, I. and Asp, N.G. (1982) In *Proceedings of an Extrusion Cooking Symposium, 7th World Cereal and Bread Congress, Prague. June/July*, 181–208

Braude, R., Coates, M.E., Henry, K.M., Kon, S.K., Rowland, S.J., Thompson, S.Y. and Walker, D.M. (1947) *British Journal of Nutrition*, **1**, 64–77
Braude, R. and Newport, M.J. (1973) *British Journal of Nutrition*, **29**, 447–455
Braude, R., Keal, H.D. and Newport, M.J. (1976) *British Journal of Nutrition*, **35**, 253–258
Braude, R. and Newport, M.J. (1977) *Animal Production*, **24**, 271–274
Burnell, T.W., Cromwell, G.L. and Stahly, T.S. (1988) *Journal of Animal Science*, **66**, 1100–1108
Cera, K.R., Mahan, D.C., Cross, R.F., Reinhart, G.A. and Whitmoyer, R.E. (1988) *Journal of Animal Science*, **66**, 574–584
Cera, K.R., Mahan, D.C. and Reinhart, G.A. (1988) *Journal of Animal Science*, **66**, 1430–1437
Cera, K.R., Mahan, D.C. and Reinhart, G.A. (1989) *Journal of Animal Science*, **67**, 2040–2047
Cera, K.R., Mahan, D.C. and Reinhart, G.A. (1990) *Journal of Animal Science*, **68**, 384–391
Coelho, M.B. (1991) *Feed International*, **December**,39-45
Coenen, G.J. and Kruse, P.E. (1986) In *Proceedings of the 9th International Pig Veterinary Congress, Barcelona*, 306
Cole, D.J.A., Beal, R.M. and Luscombe, J.R. (1968) *The Veterinary Record*, **83**, 459–464
Dale, N. and Araba, M. (1991) *Zootechnica*, **March**, 43–49
Dammers, J. (1981) *Kraftfutter*, **64**, 308–313
Daw, D. (1991) *Feed Compounder*, **November**, 42–43
De Passillé, A.M.B., Pelletier, G., Menard, J. and Morisset, J. (1989) *Journal of Animal Science*, **67**, 2921–2929
Deprez, P., Deroose, P., van den Hende, C., Muylle, E. and Oyaert, W. (1987) *Journal of Veterinary Medicine*, **34**, 254–259
Easter, R.A. (1988) In *Recent Advances in Animal Nutrition – 1988*, pp. 61–71. Edited by W. Haresign and D.J.A. Cole. London: Butterworths
Edwards, S.A., Taylor, A.G. and Harland, J.I. (1991) *British Society of Animal Production, Winter Meeting, Scarborough 1991*. Paper No. 131
Ekstrom, K.E., Benevenga, N.J. and Grummer, R.H. (1975) *Journal of Nutrition*, **105**, 851–860
Elliot, R.F., van der Noot, G.W., Gilbreath, R.L. and Fisher, H. (1971) *Journal of Animal Science*, **32**, 1128–1137
Elsley, F.W.H. (1970) In *Lactation*, pp. 393–411. Edited by I.R. Falconer. London: Butterworths
English, P.R., Anderson, P.M., Davidson, F.M. and Dias, M.F.M. (1981) *Research and Development Note*. The Scottish Agricultural Colleges, Aberdeen. No. 1
English, P.R., Robb, C.M. and Dias, M.F.M. (1980) *Animal Production*, **30**, 496
English, P.R., Robb, C.M. and Dias, M.F.M. (1981) *Research and Development Note*. The Scottish Agricultural Colleges, Aberdeen. No. 2
Fadel, J.G., Newman C.W., Newman, K.W and Graham, H. (1988) *Canadian Journal of Animal Science*, **68**, 891–897
Fahmy, M.H. (1972) *Canadian Journal of Animal Science*, **52**, 621–627
Fenwick, G.R., Spinks, E.A., Wilkinson, A.P., Heaney, R.K. and Legoy, M.A. (1986) *Journal of the Science of Food and Agriculture*, **37**, 735–741

Ferket P.R. (1991) *Feedstuffs*, **63 (9)**, 19–21
Friesen K.G., Nelssen J.L., Behnke K.C. and Goodband R.D. (1992) *Feed International*, **September**, 50-55
Gill, B.P. (1989) *Water Use by Pigs Managed Under Various Conditions of Housing, Feeding and Nutrition*. PhD Thesis. Plymouth Polytechnic
Gill, B.P. and Taylor, A.G. (1989) *Animal Production*, **49**, 317–321
Hampson, D.J.and Kidder, D.E. (1986) *Research in Veterinary Science*, **40**, 24–31
Hampson, D.J., Fu, Z.F. and Smith, W.C. (1988) *Research in Veterinary Science*, **44**, 309–314
Hancock, J.D. (1992) In *Proceedings of the Distillers Feed Conference*, 33–49
Hardy, B. (1992) In *Neonatal Survival and Growth*, pp. 99–107. Edited by M.A. Varley, P.E.V. Williams and T.L.J. Lawrence, B.S.A.P. Occasional Publications No.15
Healy B.J., Hancock J.D, Bramel-Cox P.J., Behnke K.C. and Kennedy G.A. (1992) *Journal of Animal Science*, **70** (1), p 59
Heppell, L.M.J., Sissons, J.W. and Banks, S.M. (1989) *Research in Veterinary Science*, **47**, 257–262
Holmgren, N., Franklin, A., Wallgren, P., Bergstrom, G., Martinsson, K. and Rabbe, J. (1990) *Svensk Veterinartidning*, **42**, 407–413
Howard, K.A., Forsyth, D.M. and Cline, T.R. (1990) *Journal of Animal Science*, **68**, 678–683
Hurrell, R.F. and Carpenter, K.J. (1977) In *Physical, Chemical and Biological Changes in Food Caused by Thermal Processing*, pp. 168–184. Edited by T. Høyem and O. Kvåle, London: Applied Science Publishers Ltd
Husband, A.J. and Bennell, M.A. (1980) *Pig News and Information*, **1**, 211–213
Kelly, D., Smyth, J.A. and McCracken, K.J. (1990) *Research in Veterinary Science*, **48**, 350–356
Kelly, D., Smyth, J.A. and McCracken, K.J. (1991a) *British Journal of Nutrition*, **65**, 169–180
Kelly, D., Smyth, J.A. and McCracken, K.J. (1991b) *British Journal of Nutrition*, **65**, 181–188
Kershaw, G.F., Luscombe, J.R. and Cole, D.J.A. (1966) *The Veterinary Record*, **79**, 296
Kidder, D.E. (1982) *Pig News and Information*, **3**, 25–28
Kidder, D.E. and Manners, M.J. (1978) *Digestion in the Pig*. Bristol: Scientechnica
Kidder, D.E. and Manners, M.J. (1980) *British Journal of Nutrition*, **43**, 141–153
Kilshaw, P.J. and Sissons, J.W. (1975) *Research in Veterinary Science*, **27**, 361–365
Lawrence, N.J. and Maxwell, C.V. (1983) *Journal of Animal Science*, **57**, 936–942
Lawrence, T.L.J. (1979) In *Recent Advances in Animal Nutrition – 1978*, pp. 83-98. Edited by W. Haresign and D. Lewis. London: Butterworths
Le Dividich, J., Vermorel, M., Noblet, J., Bouvier, J.C. and Aumaitre, A. (1980) *British Journal of Nutrition*, **44**, 313–323
Lecce, J.G., Armstrong, W.D., Crawford, P.C. and Ducharme, G.A. (1979) *Journal of Animal Science*, **48**, 1007–1014
Li, D.F., Thaler, R.C., Nelssen, J.L., Harmon, D .L., Allee, G.E. and Weeden, T.L. (1990) *Journal of Animal Science*, **68**, 3694–3704
Li, D.F., Nelssen, J.L., Reddy, P.G., Blecha, F., Klemm, R.D., Giesting, D.W., Hancock, J.D., Allee, G.L. and Goodband, R.D. (1991a) *Journal of Animal Science*, **69**, 3299- 3307

Li, D.F., Nelssen, J.L., Reddy, P.G., Blecha, F., Klemm, R. and Goodband, R.D. (1991b) *Journal of Animal Science*, **69**, 4062–4069

Lindemann, M.D., Cornelius, S.G., El Kandelgy, S.M., Moser, R.L. and Pettigrew, J.E. (1986) *Journal of Animal Science*, **62**, 1298–1307

Low, A.G., Carruthers, J.C. and Longland, A.C. (1990) *British Society of Animal Production, Winter Meeting, Scarborough 1990*. Paper No. 149

Mahan, D.C. (1991) *Journal of Animal Science*, **69**, 1397–1402

Manners, M.J. and Stevens, J.A. (1972) *British Journal of Nutrition*, **28**, 113–127

McDonald, P., Edwards, R.A. and Greenhalgh, J.F.D. (1981) *Animal Nutrition. 3rd Edition*. Harlow, England: Longman

McIntosh, M.K. and Aherne, F.X. (1982) *The 61st Annual Feeder's Day Report. Agriculture and Forestry Bulletin. Special Issue*. Department of Animal Science, University of Alberta. 76–77

Meat and Livestock Commission (1992) *Pig Yearbook*. Milton Keynes: Meat and Livestock Commission

Mercier, C. and Guilbot, A. (1974) *Annales de Zootechnie*, **23**, 241–251

Miller, B.G., Newby, T.J., Stokes, C.R. and Bourne, F.J. (1984) *Research in Veterinary Science*, **36**, 187–193

Miller, B.G., James, P.S., Smith, M.W. and Bourne, F.J. (1986) *Journal of Agricultural Science*, **107**, 579–589

Okai, D.B., Aherne, F.X. and Hardin, R.T. (1976) *Canadian Journal of Animal Science*, **56**, 573–586

Pajor, E.A., Fraser, D. and Kramer, D.L. (1991) *Applied Animal Behaviour Science*, **32**, 139–155

Parsons, C.M., Hashimoto, K., Wedekind, K.J. and Baker, D.H. (1991) *Journal of Animal Science*, **69**, 2918–2924

Partridge, I.G. (1989) In *Manipulating pig production II*, pp. 160–169. Edited by J.L. Barnett and D.P. Hennessy. Victoria, Australia: Australasian Pig Science Association

Partridge, G.G. (1988a) Unpublished observations

Partridge, G.G. (1988b) Unpublished observations

Partridge, G.G., Fisher, J., Gregory, H. and Prior, S.G. (1992) *British Society of Animal Production, Winter Meeting, 1992*. Paper No. 136

Peisker, M. (1992a) *Feed International*, **February**, 16–34

Peisker, M. (1992b) *International Milling Flour and Feed*, **September**, 44–47

Pipa, F. and Frank, G. (1989) *Advances in Feed Technology*, **2**, 22–30

Pollmann, D.S., Danielson, D.M. and Peo, E.R. (1980) *Journal of Animal Science*, **51**, 638–644

Prince, R.I., Miller, B.G., Bailey, M., Telemo, E., Patel, D and Bourne, F.J. (1988) *Animal Production*, **46**, 509

Risley, C.R., Kornegay, E.T., Lindemann, M.D., Wood, C.M. and Eigel, W.N. (1992) *Journal of Animal Science*, **70**, 196–206

Sarmiento, J.I., Runnels, P.L. and Moon, H.W. (1990) *American Journal of Veterinary Research*, **51**, 1180–1183

Sauer, W.C., Mosenthin, R. and Pierce, A.B. (1990) *Animal Feed Science and Technology*, **31**, 269–275

Scherer, C.W., Hays, V.W., Cromwell, G.L. and Kratzer, D.D. (1973) *Journal of Animal Science*, **37**, abstract 241, 290

Seaman, A. and Woodbine, M. (1977) In *Antibiotics and Antibiosis in Agriculture*, pp. 139–156. Edited by M. Woodbine. London: Butterworths

Shields, R.G, Ekstrom, K.E. and Mahan, D.C. (1980) *Journal of Animal Science*, **50**, 257–265

Sissons, J.W., Churchman, D, Heppell, L.M.J., Hardy, B., Banks, S.M. and Miller, B.G. (1988) In *Proceedings of a Workshop on Antinutritional Factors*, 73–76

Smith, R.H. and Sissons, J.W. (1975) *British Journal of Nutrition*, **33**, 329–349

Stahly, T.S. (1984) In *Fats in Animal Nutrition*, pp. 313–331. Edited by J. Wiseman. London: Butterworths

Svendsen, J. and Brown, P. (1973) *Research in Veterinary Science*, **15**, 65–69

Svendsen, J. and Larsen, J.L. (1977) *Nordisk Veterinaer Medicin*, **29**, 533–538

Taverner, M.R., Reale, T.A. and Campbell, R.G. (1987) In *Recent Advances in Animal Nutrition in Australia*, pp. 338–346. Edited by D.J. Farrell

Thomlinson, J.R. and Lawrence, T.L.J. (1981) *The Veterinary Record*, **109**, 120–122

van der Poel, A.F.B., den Hartog, L.A., van den Abeele, Th., Boer, H. and van Zuilichem, D.J. (1989) *Animal Feed Science and Technology*, **26**, 29–43

van der Poel, A.F.B., den Hartog, L.A., van Stiphout, W.A.A., Bremmers, R. and Huisman, J. (1990) *Animal Feed Science and Technology*, **29**, 309–320

Vestergaard, E.M., Danielsen, V, Jacobsen, E.E. and Rasmussen, V (1990) *Beretning fra Statens Husdyrbrugsforsog* **No. 674**, 71 pp

White, F., Wenham, G., Sharman, G.A.M., Jones, A.S., Rattray, E.A.S. and McDonald, I. (1969) *British Journal of Nutrition*, **23**, 847–858

Wood, A.J. and Groves, T.D.D. (1965) *Canadian Journal of Animal Science*, **45**, 8–13

Wu, J.F. and Fuller, M.F. (1974) *Animal Production*, **18**, 317–320

LIST OF PARTICIPANTS

The twenty-seventh Feed Manufacturers Conference was organized by the following committee:

Miss P.J. Brooking (W.J. Oldacre Ltd)
Mr J. Fordyce (West Midlands Farmers)
Dr K. Jacques (Alltech Inc)
Dr S. Marsden (Dalgety Agriculture Ltd)
Mr D.R. McLean (W.L. Duffield & Sons Ltd)
Mr R.T. Pass (United Distillers)
Mr F.G. Perry (BP Nutrition (UK) Ltd)
Mr J.R. Pickford
Dr A. Reeve (ICI Nutrition)
Mr M.H. Stranks
Dr B. Vernon (Pauls Agriculture)
Dr D.R. Williams (BOCM Silcock Ltd)
Dr K.N. Boorman
Prof P.J. Buttery
Dr D.J.A. Cole (Chairman)
Dr P.C. Garnsworthy (Secretary)
Dr W. Haresign
Prof G.E. Lamming
Dr J. Wiseman
} University of Nottingham

The conference was held at the University of Nottingham School of Agriculture, Sutton Bonington, 6th-8th January 1993 and the committee would like to thank all the authors for their valuable contributions. The following persons registered for the meeting:

Adams, Dr C.A.	Kemin Europa NV, Industriezone Wolfstee, 2200 Herentals, Belgium
Alderman, Mr G.	University of Reading, Department of Agriculture, Earley Gate, Reading RG6 2AT
Allder, Mr M.	Eurotec Nutrition Ltd, Elendale House, St Martins Lane, Witcham, Ely, Cambridge CB6 2LB
Allen, Dr J.D.	Frank Wright Ltd, Blenheim House, Blenheim Road, Ashbourne, Derbyshire DE6 1HA
Anderson, Mr K.R.	Scientific Products for Agriculture Ltd, Avenue 3, Station Lane, Wirney, Oxon OX8 6BB
Angold, Mr M.	Roche Products Ltd, Heanor Gate, Heanor, Derbys
Antoniella, Dr M.	Niccolai S.p.a., V.le Rimembranza, 14, 53011 Castellina in Chianti, Italy
Appleby, Mr G.	Lilly Industries Ltd, Dextra Court, Chapel Hill, Basingstoke, Hants
Ashington, Mr B.	Cubitt & Walker Limited, Ebridge Mills, North Walsham, Norfolk NR28 9NH
Atherton, Dr D.	Thomson & Joseph Ltd, 119 Plumstead Road, Norwich, NR1 4YT

Auran, Dr T.	Telleskjopet Forutvikling, Boks 3771, Granaslia, 7002 Trondheim, Norway
Banton, Mr C.L.	Dairyfood (UK) Ltd, Englesea House, Barthomley Road, Crewe, Cheshire CW1 1UF
Barrie, Mr M.J.	Elanco Animal Health, Dextra Court, Chapel Hill, Basingstoke
Bates, Mrs A	Vitrition Ltd, Ryhall Road, Stamford, Lincs
Beardsworth, Dr P.	Colborn Dawes Nutrition, Heanorgate, Heanor, Derbys
Beaumont, Mr D.	BP Nutrition (UK) Ltd, Wincham, Northwich, Cheshire CW9 6DF
Bedford, Dr M.	Finnfeeds International Ltd, Ailesbury Court, High Street, Marlborough SN8 1AA
Beer, Mr J.H.	W & J Pye Ltd, Fleet Square. Lancaster LA1 1HA
Beever, Dr D.E.	Reading University, Department of Agriculture, Earley Gate, Reading
Bell, Dr B.	Cranswick plc, The Airfield, Hutton Cranswick, Nr Driffield
Belyavin, Dr C.	2 Pinewoods, Church Aston, Newport, Shrops TF10 9LN
Bercovici, Dr. D.	Eurolysine, 16 rue Ballu, 75009 Paris, France
Berry, Mr M.H.	Berry Feed Ingredients Ltd, Chelmer Mills, New Street, Chelmsford, Essex CM1 1PN
Best, Mr P.	Feed International, 18 Chapel Street, Petersfield, Hampshire GU32 3DZ
Black, Dr H.J.	Colborn Dawes Nutrition, Stockmans Way, Belfast BT9 7JX
Blake, Dr J.S.	Highfield, Little London, Andover, Hants SP11 6JE
Bliss, Mrs H.	Carrs Agriculture Ltd, Old Croft, Stanwix, Carlisle CA3 9BA
Bole, Mr J.	David Patton Ltd, Milltown Mills, Monaghan, Rep. of Ireland
Boorman, Dr K.N.	University of Nottingham, School of Agriculture, Sutton Bonington, Loughborough, Leics LE12 5RD
Booth, Miss A.	Yorkshire Country Feeds Ltd, Darlington Road, Northallerton, N Yorkshire
Bourne, Mr S.	Alltech UK, 16/17 Abenbury Way, Wrexham Industrial Estate, Wrexham, Clwyd
Boyd, Dr P.A.	Cranswick plc, The Airfield, Hutton Cranswick, Nr Driffield
Brenninkmeijer, Dr C.	Postbus 1, 5830 MA Boxmeer, The Netherlands
Brooking, Ms P.J.	W J Oldacre Ltd, Cleeve Hall, Bishops Cleeve, Cheltenham, Glos GL52 4RP
Brophy, Mr A.	Alltech Ireland, Unit 28, Cookstown Ind. Estate, Tallaght, Dublin 24
Brosnan, Mr J.P.	Volac Ltd, Orwell, Royston, Herts SG8 5QX
Brown, Mr G.J.P.	Colborn Dawes Nutrition, Heanorgate, Heanor, Derbys
Brown, Mr M.	Britphos Ltd, Rawdon House, Yeadon, Leeds
Bruce, Dr D.W.	Devenish Feed Supplements Ltd, 96 Duncrue Street, Belfast BT3 9AR
Burt, Dr A.W.A.	Burt Research Ltd, 23 Stow Road, Kimbolton, Huntingdon, Cambs PE18 0HU
Bush, Mr M.	FF, Room 531, Ergon House, 17 Smith Square, London SW1P 3JR

Buttery, Prof P.J.	University of Nottingham, School of Agriculture, Sutton Bonington, Loughborough, Leics LE12 5RD
Buysing Damste, Ir B.	Trouw International BV, Research & Development, PO Box 50, 3880 AB Putten, Holland
Byron, Mr J.R.	British Denkavit, Patrich House, West Quay Road, Poole, Dorset BH15 1JF
Campani, Dr I.	F.LLi Martini & C. S.p.a, 9614 Emilia Street, Budrio di Longiano (FO), Italy
Chandler, Mr N.J.	85 Meols Drive, West Kirby, Merseyside
Charles, Dr D.	ADAS, Government Buildings, Chalfont Drive, Nottingham HG8 3SN
Charlton, Mr P.	Alltech UK, Units 16/17, Abenbury Way, Wrexham Industrial Estate, Wrexham, Clwyd
Charteris-Hough, Mr J.E.	Trident Feeds, British Sugar plc, Dundle Road, Peterborough
Clark, Mr P.	ADAS, Staplake Mount, Starcross, Devon
Clarke, Mr A.N.	Four-F Nutrition, Darlington Road, Northallerton, N Yorks
Clay, Mr J.	Alltech UK, Units 16/17 Abenbury Way, Wrexham Industrial Estate, Wrexham, Clwyd
Close, Dr W.H.	Close Consultancy, 129 Barkham Road, Wokingham, Berks RG11 2RS
Cole, Dr D.J.A.	University of Nottingham, School of Agriculture, Sutton Bonington, Loughborough, Leics LE12 5RD
Cole, Mr M.A.	BOCM Pauls Ltd, 47 Key Street, Ipswich, Suffolk IP4 1BX
Colenso, Mr J.	BP Nutrition (UK) Ltd, Wincham, Northwich, Cheshire CW9 6DF
Cooke, Dr B.	Dalgety Agriculture Ltd, 180 Aztec West, Almondsbury, Bristol BS12 4TH
Cooper, Mr A.	University of Plymouth, Seale-Hayne Department of Agriculture, Newton Abbott, Devon TQ12 6NQ
Cooper, Miss S.E.	Butterworth-Heinemann, Linacre House, Jordan Hill, Oxford OX2 8DP
Cottrill, Dr B.R.	ADAS Starcross, Staplake Mount, Starcross, Devon EX6 8PE
Cowan, Dr D.	Novo Nordisk UK, 4 St Georges Yard, Castle Street, Farnham, Surrey GU9 7LW
Cox, Mr N.	S C Associates Ltd, The Limes, Sowerby Road, Sowerby, Thirsk, N Yorks YO7 1HX
Creasey, Mrs A.	BASF plc, Earl Road, Cheadle Hulme, Cheshire SK8 6QG
Darashah, Mr P.	Dairygold Co-op Soc Ltd, Ballymakeepa, County Cork
Darmody, Mr W.	Avonmore Foods plc, Patrick Street, Kilkenny, Ireland
Davies, Mrs S.	Sun Valley Feed Mill, Tram Inn, Allensmore, Hereford, HR2 9AW
Dawson, Mr W.	Britphos Ltd, Rawdon House, Yeadon, Leeds
De Bruyne, Ir K.	EMC Belgium, Square de Meeus 1, B-1040 Brussels
De Man, Dr T.J.	Kerkstraat 40, 3741 AK BAARN, The Netherlands
De Visser, Dr H.	IVVO-DLO, PO Box 160, 8200 AD Lelystad, The Netherlands
Dickins, Mr A.	Unitrition International, Olympia Mills, Barlby Road, Selby, N Yorks YO8 7AF

Dixon, Mr D.H.	Brown & Gillmer Ltd, Box 3154, The Lodge, Florence House, 199 Strand Road, Merrion, Dublin 4
Edwards, Miss S.	BP Nutrition (UK) Ltd, Wincham, Northwich, Cheshire CW9 6DF
Ewing, Dr W.	Cargill plc, Knowle Hill Park, Fairmile Lane, Cobham, Surrey
Farley, Mr R.L.	BP Nutrition (UK) Ltd, Wincham, Northwich, Cheshire CW9 6DF
Filmer, Mr D.	David Filmer Ltd, Wascelyn, Brent Knoll, Somerset TA9 4DT
Fisher, Dr C.	Leyden Old House, Kirknewton, Midlothian EH27 8DQ
Fletcher, Mr C.J.	Aynsome Laboratories, Kentsford Road, Grange-over-Sands, Cumbria LA11 7BA
Forbes, Mr J.K.H.	Ross Breeders Ltd, Newbridge, Midlothian EH28 8SZ
Fordyce, Mr J.	West Midland Farmers, Bradford Road, Milksham, Wilts SN12 8LQ
Foster, Mr D.	Feedpharm (Pty) Ltd, 13 Sergeant Street, Somerset West, 7130, South Africa
Frank, Mr T.L.	MAFF, Room 529, Ergon House, 17 Smith Square, London SW1P 3JR
Fraser, Mrs S.P.	International Milling, 171 High Street, Rickmansworth, Herts
Freeman, Dr C.P.	Unilever Research, Colworth House, Sharnbrook, Bedford
Fullarton, Mr P.J.	Forum Chemicals Ltd, Forum House, Brighton Road, Redhill, Surrey
Gaisford, Mr M.	Farmers Weekly, Quadrant House, Sutton, Surrey
Garnsworthy, Dr P.C.	University of Nottingham, School of Agriculture, Sutton Bonington, Loughborough, Leics LE12 5RD
Geary, Mr B.	Hoechst Animal Health, Walton Manor, Walton, Milton Keynes MK7 7AJ
Geddes, Mr N.	Nutec Ltd, Eastern Avenue, Lichfield, Staffs
Geerse, Ir C.	Gist-Brocades Agro Business, Group BV, Postbus 1820, 2280 DV Rijswijk, The Netherlands
Gibson, Mr J.E.	Parnutt Foods Ltd, Hadley Road, Woodbridge Industrial Estate, Sleaford, Lincs. NG34 7EG
Gilbert, Mr R.	World Feed Extrusion Comminique, Asbury Publications Ltd, Stoke Road, Bishops Cleeve, Gloucs GL52 4RW
Gill, Dr P.	Scottish Agricultural College, 581 King Street, Aberdeen AB9 1UD, Scotland
Gillespie, Miss F.	United Molasses, Stretton House, Derby Road, Stretton, Burton-on-Trent, Staffs
Givens, Dr D.I.	ADAS Drayton Research Centre, Feed Evaluation Unit, Alcester Road, Stratford on Avon
Goefsen, Dr T.	Stormollen, 5270 Vaksdal, Norway
Goff, Mr S.	Business Management Systems Ld, Sproughton House, Sproughton, Ipswich IP8 3AW
Gooderham, Mr B.J.	W & J Pye Ltd, Fleet Square, Lancaster LA1 1HA
Gould, Mrs M.	Volac Ltd, Orwell, Royston, Herts SG8 5QX
Grace, Mr J.	Elanco Animal Health, Dextra Court, Chapel Hill, Basingstoke

Gray, Mr B.	Orm House, 2 Hookstone Park, Harrogate, Yorks HG2 8QT
Green, Dr S.	Rhone Poulenc Chemicals, Oak House, Reeds Crescent, Watford, Herts WD1 1QH
Griffiths, Mr W.D.E.	Midland Shires Farmers Ltd, Defford Mill, Earls Croome, Worcester
Guenther, Dr C.	BASF AG, Marketing Vitamins, 6700 Ludwigshafen, Germany
Haggar, Mr C.W.	1 Park Crescent, Addingham, Ilkley, W Yorks
Hall, Mr G.R.	Kemin UK Ltd, Becor House, Green Lane, Lincoln LN6 9DL
Hanly, Mr B.J.	Quest International, 1833 57th Street, Sarasota, Florida 34240, USA
Hannagan, Mr M.J.	West Coates, 11 Durbin Park Road, Clevedon, Avon BS21 7EU
Hardy, Dr B.	Dalgety Agriculture Ltd, 180 Aztec West, Almondsbury, Bristol BS12 4TH
Haresign, Dr W.	University of Nottingham, School of Agriculture, Sutton Bonington, Loughborough, Leics LE12 5RD
Harker, Dr A.J.	Finnfeeds International Ltd, Market House, Ailesbury Court, High Street, Marlborough SN8 1AA
Harland, Dr J.	British Sugar plc, PO Box 11, Oundle Road, Peterborough PE2 9QU
Harrison, Mrs J.	Sciantec Analytical Services, Main Site, Dalton, Thirsk, North Yorkshire YO7 3JA
Harrison, Mr M.	Farmlab Ltd, Whetstone Magna, Lutterworth Road, Leicester
Harvey, Mr T.	Bio Agri Mix (Western) Ltd, 217 Douglas Woopds Place SE, Calgary, Alberta, Canada T2Z 1K6
Haythornthwaite, Mr A.	NU Wave Health Products Ltd, 45 Church Road, Warton, Preston PR4 1BD
Hazzledine, Mr M.J.	Dalgety Agriculture Ltd, 180 Aztec West, Almondsbury, Bristol
Hegeman, Mr F.	Borculo Whey Products Ltd, Bryman Four Estate, River Lane, Saltney, Chester CH4 8RQ
Hepworth, Mr N.	Daylay Foods Ltd, Thornton Hill, Strathore Road, Thornton, Kirkaldy, Fife
Higginbotham, Dr J.D.	United Moasses, Stretton House, Derby Road, Stretton, Burton-on-Trent, Staffs DE13 0DW
Hill, Mr B.E.	Midland Shires Farmers Ltd, County Mills, Worcester WR1 3NU
Hillmann, Mr M.	Farmlab Ltd, Whetstone Magna, Lutterworth Road, Leicester
Hitchens, Mr C.	Favor Parker Ltd, The Hall, Stoke Ferry, Kings Lynn, Norfolk
Hogg, Mr A.A.	Elanco Animal Health, Dextra Court, Chapel Hill, Basingstoke
Holbrooke, Mr G.	2 Lodge Park, Whittlebury, Towcester, Northants NN12 8XG
Holmes, Mr G.R.	S E Johnson Ltd, Old Road Mills, Darley Dale, Derbys DE4 2ES
Hopkins, Mr D.P.	ICI Nutrition, Alexander House, Crown Gate, Runcorn WA7 2UP

Huggett, Miss C.D.	University of Nottingham, School of Agriculture, Sutton Bonington, Loughborough, Leics lE12 5RD
Hvelplund, Dr T.	National Institute of Animal Science, Research Centre Foulun, 8830 Tjele, Denmark
Ilola, Mrs M.	Lannen Tehtaat OY, 27820 Iso-Vimma, Finland
Inborr, Mr J.	Finnfeeds International Ltd, Market House, Ailesbury Court, High Street, Marlborough, Wiltshire SN8 1AA
Ince, Mr R.	Business Management Systems Ld, Sproughton House, Sproughton, Ipswich IP8 3AW
Jacklin, Mr D.	ADAS Newcastle, Kenton Bar, Newcastle-upon-Tyne, NE1 2YA
Jackson, Mr J.C.	Nutec Ltd, Eastern Avenue, Lichfield, Staffs
Jacques, Dr K.	Alltech Inc, Biotechnology Centre, 3031 Catnip Hill Pike, Nicholasville, Ky 40356, USA
Jamison, Dr W.	Oregon State University, Dept. Political Science, Social Science Hall 307, Corvallis, OR 97331-6206
Janes, Mr R.	Criddle Billington Feeds, Warrington Road, Glazebury, Nr Warrington, Cheshire
Jardine, Mr G.	Unitrition International, Olympia Mills, Barlby Road, Selby, N Yorks YO8 7AF
Jones, Dr E.	Dalgety Agriculture Ltd, 180 Aztec West, Almondsbury, Bristol BS12 4TH
Jones, Mr E.J.	Marius International Ltd, Salamander Quay, Quay West, Park Lane, Harefield, Uxbridge, Midx UB9 6NZ
Jones, Miss F.	Cherry Valley Farms Ltd, North Kelsey Moor, Lincoln LN7 6HH
Jones, Mr M.G.S.	ADAS Wolverhampton, Woodthorne, Wergs Road, Wolverhampton WV6 8TQ
Jones, Mr O.	Bio Agri Mix (Western) Ltd, 217 Douglas Woods Place S.E., Calgary, Alberta, Canada T2Z 1K6
Jones, Dr P.	Smithkline Beecham Anim Health, Hunters Chase, Walton Oaks, Tadworth, Surrey KT20 7NT
Kennedy, Mr G.	BASF plc, Earl Road, Cheadle Hulme, Cheshire, SK8 6QG
Keys, Mr J.	32 Holbrook Road, Stratford-upon-Avon, Warwickshire
Khan, Miss N.	University of Nottingham, School of Agriculture, Sutton Bonington, Loughborough, Leics LE12 5RD
Koch, Mr F.	Degussa AG, PO Box 1345, D-6450 Hanau 1
Kwakkel, Mr R.P.	Agricultural University, Dept Animal Nutrition, Haagsteeg 4, 6708 PM Wageningen, The Netherlands
Lamming, Prof G.E.	University of Nottingham, School of Agriculture, Sutton Bonington, Loughborough, Leics LE12 5RD
Lane, Mr P.	Parnutt Foods Ltd, Hadley Road, Woodbridge Industrial Estate, Sleaford, Lincs NG34 7EG
Law, Mr J.R.	Sheldon Jones Agriculture, Priory Mill, West Street, Wells, Somerset BA5 2HL
Lee, Dr P.	ADAS Reading, Coley Avenue, Reading RG1 6DE
Lima, Mr S.	Felleskjopet Rogaland Agder, PO Box 208, 4001 Stavanger, Norway

Lonsdale, Dr C.R.	Chapman & Frearson Ltd, Victoria Street, Grimsby, South Humberside DN31 1PX
Lowe, Mr J.A.	Gilbertson & Page, PO Box 321, Welwyn Garden City, Herts AL7 1LF
Lowe, Dr R.	Frank Wright Ltd, Blenheim House, Blenheim Road, Ashbourne, Derbyshire DE6 1HA
Lyons, Dr P.	Alltech Inc, Biotechnology Centre, 3031 Catnip Hill Pike, Nicholasville, Ky 40356, USA
Mackie, Mr I.L.	SCATS (Eastern Region), Robertsbridge Mill, Robertsbridge, East Sussex
MacMahon, Mr M.J.	Agribusiness Solutions, The Shribbery, Erdington Road, Aldridge, W Midlands WS9 8UH
Madsen, Dr J.	The Royal Veterinary & Agricultural University, Rolighedsvej 23, DK-1958 Frederiksberg C, Denmark
Mafo, Mr A.	BP Nutrition (UK) Ltd, Wincham, Northwich, Cheshire CW9 6DF Major, Mr N.C. BOCM Pauls Ltd, PO Box 39, 47 Key Street, Ipswich IP4 1BX
Malandra, Dr F.	Sildamin Spa, Sostegno di Spessa, 27010 Pavia, Italy
Marriage, Mr P.	W & H Marriage & Sons Ltd, Chelmer Mills, Chelmsford, Essex CM1 1PN
Marsden, Dr M.	J Bibby Agriculture Ltd, Head Office, Oxford Road, Adderbury, Banbury
Marsden, Dr S.	Dalgety Agriculture Ltd, 180 Aztec West, Almondsbury, Bristol BS12 4TH
Martyn, Mr S.	International Additives Ltd, Old Gorsey Lane, Wallasey, Merseyside
McIlmoyle, Dr W.A.	Agricultural Consultants, 2 Gregg Street, Lisburn BT27 5AN
McLean, Mr D.R.	W L Duffield & Sons Ltd, Saxlingham Thorpe Mills, Norwich NR15 1TY
Miller, Mr C.	Waterford Foods plc, Waterford, Ireland,
Miller, Dr E.L.	Deptartment of Clinical Veterinary Medicine, Nutrition Laboratory, 307 Huntingdon Road, Cambridge CB3 0JQ
Mills, Mr C.	University of Nottingham, School of Agriculture, Sutton Bonington, Loughborough, Leics LE12 5RD
Mitchell, Mr P.P.	Lopen Feed Mills, Lopen, South Petherton, Somerset
Moore, Mr D.R.	David Moore Flavours Ltd, 29 High Street. Harpenden, Herts AL5 2RU
Morgan, Mr J.	Lloyds Animal Feeds Ltd, Morton, Oswestry SY10 8BH
Morris, Mr W.	BOCM Pauls Ltd, PO Box 39, 47 Key Street, Ipswich
Morrow, Dr A.	BP Nutrition (N. Ireland), 36 Ship Street, Belfast BT15 1JL
Mounsey, Mr H.G.	HGM Publications, Abney House, Baslow, Derbyshire DE45 1RZ
Mounsey, Mr S.P.	HGM Publications, Abney House, Baslow, Bakewell, Derbyshire DE45 1RZ
Mudd, Dr A.J.	Cyanamid UK, Fareham Road, Gosport, Hants PO13 224329
Munford, Dr A.	University of Exeter, MSOR Department, Laver Building, Exeter
Murray, Mr F	Dairy Crest Ingredients, Philpot House, Rayleigh Essex

Newbold, Dr J.R.	BOCM Pauls Ltd, PO Box 30, 47 Key Street, Ipswich IP4 1BX
Newcombe, Mrs J.O.	University of Nottingham, School of Agriculture, Sutton Bonington, Loughborough, Leics LE12 5RD
Newman, Mr G.	Travellers Rest, Timberscombe, Minehead TA24 7UK
Nolan, Mr J.	International Additives Ltd, Old Gorsey Lane, Wallasey, Merseyside
O'Beirne, Mr P.	Cyanamid UK, Fareham Road, Gosport, Hants PO13 0AS
O'Brien, Mr J.	Veterinary Medicines Directorate, Woodham Lane, New Haw, Addlestone, Surrey KT15 3NB
O'Grady, Dr J.	IAWS Group plc, 151 Thomas Street, Dublin 8
Overend, Dr M.A.	Nutec Ltd, Eastern Avenue, Lichfield, Staffs
Owers, Dr M.	BOCM Pauls Ltd, PO Box 39, 47 Key Street, Ipswich, IP4 1BX
Packington, Mr A.	Colborn Dawes Nutrition, Heanorgate, Heanor, Derbys
Palazzo, Mr C.	Via Granaretto, Torrinpietra, Roma, Italy
Palmann, Mr K.	Smithkline Beecham Anim.Health, Hunters Chase, Walton Oaks, Tadworth, Surrey KT20 7NT
Papasolomontos, Dr S.	Dalgety Agriculture Ltd, 180 Aztec West, Almondsbury, Bristol BS12 4TH
Partridge, Dr G.G.	BP Nutrition (UK) Ltd, Wincham, Northwich, Cheshire CW9 6DF
Pass, Mr R.	United Distillers, 33 Ellersly Road, Edinburgh EH12 6JW
Pearce, Mr D.	Degussa Ltd, Paul Ungerer House, Stanley Green Trading Estate, Wilmslow, Cheshire SK9 6LA
Pearson, Mr A.	Hoechst Animal Health, Walton Manor, Walton, Milton Keynes MK7 7AJ
Perrott, Mr G.J.	British Sugar plc, PO Box 11, Oundle Road, Peterborough, PE2 9QX
Perry, Mr F.G.	BP Nutrition (UK) Ltd, Wincham, Northwich, Cheshire CW9 6DF
Petersen, Miss S.T.	University of Nottingham, School of Agriculture, Sutton Bonington, Loughborough, Leics LE12 5RD
Phillips, G M.R.	Greenway Farm, Charlton Kings, Cheltenham, Glos
Pickford, Mr J.R.	Bocking Hall, Bocking Church Street, Braintree, Essex CM7 5JY
Pike, Dr I.H.	I.A.F.M.M., Hoval House, Orchard Parade, Mutton Lane, Potters Bar, Herts EN6 3AR
Piva, Dr G.	Faculty of Agriculture, Via E Parmense, 84, 29100 Piacenza, Italy
Plowman, Mr G.B.	G W Plowman & Son Ltd, Selby House, High Street, Spalding, Lincs
Poornan, Mr P.	Lys Mill Ltd, Watlington, Oxon
Putnam, Mr M.	Roche Products Ltd, PO Box No 8, Welwyn Garden City, Herts AL7 3AY
Raine, Dr H.	J Bibby Agriculture Ltd, Head Office, Oxford Road, Adderbury, Banbury

Raper, Mr G.J.	Laboratories Pancosma (UK) Ltd, Crompton Road Ind. Estate, Ilkeston, Derbys DE7 4BG
Reckitt, Mr J.	Agil Ltd, Hercules 2, Calleva Park, Aldermaston, Berks RG7 4QW
Redshaw, Mr M.	University of Nottingham, School of Agriculture, Sutton Bonington, Loughborough, Leics LE12 5RD
Reeve, Dr A.	ICI Nutrition, Alexander House, Crown Gate, Runcorn WA7 2UP
Reeve, Mr J.G.	R S Feed Blocks, Orleigh Mill, Bideford, Devon Retter, Dr W.C. Heygate & Sons Ltd, Bugbrooke, Northampton
Richards, Mr P.	Novo Nordisk UK, 4 St Georges Yard, Castle Street, Farnham, Surrey GU9 7LW
Rigg, Mr G.J.	Elanco Animal Health, Dextra Court, Chapel Hill, Basingstoke
Rinta-Harri, Mr A.	Berner Ltd, PO Box 15, SF-00131 Helsinki, Finland
Roberts, Mr J.C.	Harper Adams Agricultural College, Edgmond, Newport, Shropshire TF10 8NB
Robinson, Mr D.K.	Favor Parker Ltd, The Hall, Stoke Ferry, Kings Lynn, Norfolk
Roele, Mr D.	Kemin Europa NV, Industriezone Wolfstee, 2200 Herentals, Belgium
Rosen, Dr G.	Field Investigation & Nutrition Services Ltd, 66 Bathgate Road, London SW19 5PH
Ross, Mr E.	BOCM Pauls Ltd, PO Box 39, 47 Key Street, Ipswich IP4 1BX
Rossillo, Mr J.	University of Nottingham, School of Agriculture, Sutton Bonington, Loughborough, Leics LE12 5RD
Round, Mr J.	J Bibby Agriculture Ltd, Head Office, Oxford Road, Adderbury, Banbury
Rulquin, Dr H.	INRA Saint Gilles, 35590 Saint Gilles, France
Russell, Miss S.	Farmlab Ltd, Whetstone Magna, Lutterworth Road, Leicester
Santoma, Dr G.	Correctores Y Servicios, Agropecuarios, C/ Mejia Lequercia 22-24, 08028 Barcelona, Spain
Scholman, Ir G.J.	Provimi B V, PO Box 5063, 3008 AB Rotterdam, The Netherlands
Shipton, Mr P.	Dardis & Dunns Coarse Feeds Ld, Ashbourne County Meath, Rep of Ireland
Shorrock, Dr C.	FSL Bells Ltd, Hartham Park, Corsham, Wilts
Shrimpton, Dr D.H.	International Milling, Turret House, 171 High Street, Rickmansworth, Herts Silvester, Ms L. W J Oldacre Ltd, Cleeve Hall, Bishops Cleeve, Glos GL52 4RP
Speight, Mr D.	Newmans, Rainton, Thirsk, N Yorkshire
Stainsby, Mr A.K.	Brandsby Agricultural Trading Association, Railway Street, Malton, N Yorkshire YO17 0NU
Statham, Mr R.	Pet's Choice, Greenbank Mill, Blackburn
Stebbens, Dr H.R.	Crina UK, 78 Coombe Road, New Malden, Surrey KT3 4QS
Stewart, Dr C.S.	Rowett Research Institute, Nutrition Division, Bucksburn, Aberdeen AB2 9SB
Stickney, Dr K.	Farm Nutrition Ltd, The Airfield, Topcroft, Bungay, Suffolk.

Stranks, Mr M.	76 Church Lane, Backwell, Bristol BS19 3JL
Sumner, Dr R.	Midland Shires Farmers Ltd, Defford Mill, Worcester, WR8 9DF
Swarbrick, Mr J.	Borculo Whey Products UK Ltd, Bryman Four Estate, River Lane, Saltney, Chester CH4 8RQ
Sylvester, Mr D.	Cyanamid UK, Fareham Road, Gosport, Hants PO13 0AS
Taylor, Dr A.	Roche Products Ltd, Heanorgate, Heanor, Derbys
Thompson, Mr D.H.	Rightfeeds Ltd, Castlegarde, Cappamore, Co Limerick, Ireland
Thompson, Dr F.	Rumenco, Stretton House, Derby Road, Burton-on- Trent, Staffs
Thompson, Mr J.G.	Feed Flavours Europe, Waterlip Mill, Cranmore, Shepton Mallet
Tibble, Mr S.J.	S C Associates Ltd, The Limes, Sowerby Road, Sowerby, Thirsk, N Yorks YO7 1HX
Tonks, Mr W.P.	Park Tonks Ltd, Abington House, Gt Abington, Cambridge CB1 6AS
Torvi, Miss J.	Stormollen, 5270 Vaksdal, Norway
Tripney, Mrs V.	Sun Valley Feed Mill, Tram Inn, Allensmore, Hereford, HR2 9AW
Twigge, Mr J.	BP Nutrition (UK) Ltd, Wincham, Northwich, Cheshire CW9 6DF
Tyler, Mr A.H.	Roche Products Ltd, Heanorgate, Heanor, Derbyshire
Uprichard, Mr J.	BP Nutrition (N. Ireland), 36 Ship Street, Belfast BT15 1JL
Vaiani, Dr S.	Vaiani s.a.s., Via L Hanara 3, 42100 Reggio Emelia, Italy
van der Ploeg, Mr H.	Stationsweg 4, 3603EE Maarssen, The Netherlands
Vander Elst, Mr P.	EMC Gelbium, Square de Meeus 1, B-1040 Brussels
Vromant, Mr F.	N V Radar, Dorpssiraat 4, B-9800 DEINZE, Belgium
Wakeman, Miss W.G.	BOCM Pauls Ltd, PO Box 39, 47 Key Street, Ipswich IP4 1BX
Ward, Mr M.	Britphos Ltd, Rawdon House, Yeadon, Leeds
Watkins, Dr B.A.	Purdue University, Department of Food Science, Smith Hall, West Layayette, IN 47909, USA
Webster, Mrs M.	Format International Ltd, Format House, Poole Road, Woking, GU22 1DY
Williams, Mr C.W.	Rhone Poulenc Chemicals Ltd, ABM Brewing Enzymes, Poleacre Lane, Woodley, Stockport SK6 1PQ
Williams, Mr D.J.	International Molasses, Shell Road, Royal Edward Dock, Avonmouth, Bristol BS11 9BW
Williams, Dr D.R.	BOCM Silcock, PO Box 4, Barlby Road, Selby, Yorks
Wilson, Mr B.J.	Cherry Valley Farms Ltd, North Kelsey Moor, Caistor, Lincoln LN7 6HH
Winwood, Mr J.	Rhone Poulenc Chemicals Ltd, ABM Brewing & Enzymes. Poleacre Lane, Woodley, Stockport SK6 1PQ
Wiseman, Dr J.	University of Nottingham, School of Agriculture, Sutton Bonington, Loughborough, Leics LE12 5RD
Woodward, Dr K.N.	Veterinary Medicines Directorate, Woodham Lane, New Haw, Addlestone, Surrey KT15 3NB

Woolford, Dr M.	Alltech UK, Units 16/17, Abenbury Way, Wrexham Industrial Estate, Wrexham, Clwyd
Yeo, Dr G.W.	Premier Nutrition Products Ltd, Unit 8, Walkmill Business Park, Cannock, Staffs WS11 3XE
Youdan, Dr J.	Nutrimix, Boundary Industrial Estate, Boundary Road, Lytham, Lancs FY8 5HU

INDEX